REDOX MECHANISMS IN INORGANIC CHEMISTRY

ELLIS HORWOOD SERIES IN INORGANIC CHEMISTRY

Series Editor: Dr JOHN BURGESS, University of Leicester
Consulting Editor: Ellis Horwood, MBE

Inorganic chemistry is a flourishing discipline in its own right and also plays a key role in many areas of organometallic, physical, biological, and industrial chemistry. This series is developed to reflect these various aspects of the subject from all levels of undergraduate teaching into the upper bracket of research.

Alcock, N.W.	**BONDING AND STRUCTURE: Structural Principles in Inorganic and Organic Chemistry**
Almond, M.J.	**SHORT-LIVED MOLECULES**
Barrett, J.	**UNDERSTANDING INORGANIC CHEMISTRY: The Underlying Physical Principles**
Beer, P.	**HOSTS, GUESTS AND INCLUSION CHEMISTRY**
Burger, K.	**BIOCOORDINATION CHEMISTRY: Coordination Equilibria in Biologically Active Systems**
Burgess, J.	**IONS IN SOLUTIONS: Basic Principles of Chemical Interactions**
Burgess, J.	**METAL IONS IN SOLUTION**
Burgess, J.	**INORGANIC SOLUTION CHEMISTRY**
Cardin, D.J., Lappert, M.F. & Raston, C.L.	**CHEMISTRY OF ORGANO-ZIRCONIUM AND -HAFNIUM COMPOUNDS**
Caulcutt, R.	**DATA ANALYSIS IN THE CHEMICAL INDUSTRY: Volume 1, Basic Techniques**
Constable, E.C.	**METALS AND LIGAND REACTIVITY**
Crichton, R.R.	**INORGANIC BIOCHEMISTRY OF IRON METABOLISM**
Filov, V.A.	**HARMFUL CHEMICAL SUBSTANCES: Volume 1: Elements in Groups I–IV of the Periodic Table and their Inorganic Compounds**
Harrison, P.G.	**TIN OXIDE HANDBOOK**
Hartley, F.R., Burgess, C. & Alcock, R.M.	**SOLUTION EQUILIBRIA**
Hay, R.W.	**BIO-INORGANIC CHEMISTRY**
Hay, R.W.	**REACTION MECHANISMS OF METAL COMPLEXES**
Housecroft, C.E.	**BORANES AND METALLOBORANES: Structure, Bonding and Reactivity**
Kendrick, M.J., May, M.T. & Robinson, K.D.	**METALS IN BIOLOGICAL SYSTEMS**
Lappert, M.F., Sanger, A.R., Srivastava, R.C. & Power, P.P.	**METAL AND METALLOID AMIDES**
Lappin, G.	**REDOX MECHANISMS IN INORGANIC CHEMISTRY**
Maddock, A.	**MÖSSBAUER SPECTROSCOPY**
Massey, A.G.	**MAIN GROUP CHEMISTRY**
Massey, A.G.	**TRANSITION METAL CHEMISTRY**
McGowan, J. & Mellors, A.	**MOLECULAR VOLUMES IN CHEMISTRY AND BIOLOGY: Applications Including Partitioning and Toxicity**
Parish, R.V.	**NMR, NQR, EPR, AND MÖSSBAUER SPECTROSCOPY IN INORGANIC CHEMISTRY**
Raithby, R.R.	**TRANSITION METAL CLUSTER CARBONYLS: Structure, Synthesis, Reactivity**
Roche, L.P.	**THE CHEMICAL ELEMENTS: Chemistry, Physical Properties and Uses in Science and Industry**
Romanowski, W.	**HIGHLY DISPERSED METALS**
Snaith, R. & Edwards, P.	**LITHIUM AND ITS COMPOUNDS: Structures and Applications**
Tsitsishvili, G.V., Andronikashvili, G. Kirov, G.N., and Filizova, L.D.	**NATURAL ZEOLITES**
Williams, P.A.	**OXIDE ZONE GEOCHEMISTRY**

REDOX MECHANISMS IN INORGANIC CHEMISTRY

GRAHAM LAPPIN
Department of Chemistry,
University of Notre Dame,
Indiana, USA

ELLIS HORWOOD
NEW YORK LONDON TORONTO SYDNEY TOKYO SINGAPORE

First published in **1994** by
Ellis Horwood Limited
Market Cross House, Cooper Street,
Chichester, West Sussex, PO19 1EB, England
A division of
Simon & Schuster International Group

© Ellis Horwood Limited **1994**

All rights reserved. No part of this publication may be reproduced, stored in a retrieval system, or transmitted, in any form, or by any means, electronic, mechanical, photocopying, recording or otherwise, without prior permission, in writing, from the publisher

Printed and bound in Great Britain
by Hartnolls, Bodmin

British Library Cataloguing in Publication Data

A catalogue record for this book is available from the British Library

ISBN 0-13-770751-7   Hbk
ISBN 0-13-770017-2   Pbk

Library of Congress Cataloging-in-Publication Data

Available from the publisher

1 2 3 4 5   97 96 95 94 93

Table of contents

Preface 7

1 Introduction 9
    1.1   Introduction 9
    1.2   Ligand field theory 10
    1.3   Substitution reactions 13
    1.4   Thermodynamic aspects 15
    1.5   Studies with thermodynamically unstable complexes 22
    1.6   Temperature and pressure dependence of reduction potentials 23
    1.7   Kinetic considerations 25
    1.8   Activation parameters 28
    1.9   Medium effects 31
    1.10 Isotope effects 36
Questions 38
References 40

2 The outer-sphere mechanism 42
    2.1   Introduction 42
    2.2   The electron transfer precursor 44
    2.3   The electron transfer step 51
    2.4   The Marcus linear free energy relationship 55
    2.5   Self-exchange rates 56
    2.6   Applications of Marcus Theory 64
    2.7   Photoinduced electron transfer 67
    2.8   Non-aqueous media 70
    2.9   Stereoselectivity in electron transfer 72
    2.10 Theoretical details 75
    2.11 The activation barrier $\Delta G^\ddagger$ 77
    2.12 Activation parameters and isotope effects 83
    2.13 The pre-exponential factors 84
    2.14 Electronic and structural effects 88
    2.15 Electron transfer involving large structural changes 95

6 Table of contents

Questions 108
References 110

3 The inner-sphere mechanism 120
   3.1 Introduction. 120
   3.2 Bridging ligands 121
   3.3 Double bridge formulation 127
   3.4 Dependence on oxidant. 129
   3.5 Aqua-ion reductants 131
   3.6 Anionic reductants 138
   3.7 Stereoselectivity in inner-sphere electron transfer reactions 140
   3.8 Theoretical aspects 141
   3.9 Electron transfer through organic structural units 147
   3.10 Intramolecular electron transfer 155
Questions 163
References 165

4 Intramolecular electron transfer 170
   4.1 Introduction. 170
   4.2 Optical electron transfer 171
   4.3 Chemically induced intramolecular electron transfer. 180
   4.4 Long-range electron transfer. 184
   4.5 Non-adiabatic electron transfer 193
   4.6 Reactions of metalloproteins 195
   4.7 Long-range electron transfer involving metalloproteins 205
Questions 210
References 211

5 Multiple electron transfer 215
   5.1 Introduction 215
   5.2 Reactions of thallium(III)/(I) 216
   5.3 Reactions of platinum(IV)/(II) 223
   5.4 Oxo-ion reagents. 228
   5.5 'Atom transfer' reactions 232
   5.6 The role of adduct formation in non-complementary reactions. 235
   5.7 Non-metallic reagents. 238
   5.8 Molecular oxygen and hydrogen peroxide 240
   5.9 Reactions of halogens, pseudohalogens and related species 248
   5.10 Reactions of inorganic radicals 259
   5.11 Reactions of halogenate ions 263
   5.12 Non-metallic reagents. 267
Questions 274
References 275

Index 282

Preface

This book grew out of conversations at a meeting of the Royal Society of Chemistry Inorganic Mechanisms Discussion Group in 1986. It was conceived as part of the Ellis Horwood Series in Inorganic Chemistry. While writing the text I have frequently been asked what the book is about. It is perhaps easier to begin the answer by describing what the book is not. It is not a book on kinetics and mechanisms. There are much better texts on that topic, those by Espenson and by Moore and Pearson are my favorites. Neither is it a book about electron transfer nor about the theory of electron transfer reactions. There has been a rash of excellent monographs on this topic recently. Rather, the focus of the book is an examination of trends in mechanism which are found in inorganic redox chemistry. Electron transfer processes are considered, but so also are atom transfer processes, and the limits where one class of reaction ends and another begins are described. The text is designed for senior undergraduates and beginning graduate students who want to know more about the chemistry of redox reactions. Some background in mechanistic chemistry is assumed and the book was intended to supplement *Inorganic Reaction Mechanisms* by R. W. Hay, also published by Ellis Horwood as part of the series in Inorganic Chemistry. The development of theory is minimized and is set in a context where the descriptive and experimental approaches are dominant, and I hope that this points out some of the limits of current thinking. Many examples of reactions are included in the text and references are primarily to the original literature, which I feel is important for a book of this nature. Some of the compilations of data are extensive — beyond that required to illustrate the material in the text — but I hope that they may be of some use to more experienced investigators.

  I would like to thank all those who have aided me in the preparation of the book, my graduate students and Mr. Robert M. L. Warren in particular, Dr. John Burgess for his advice and encouragement, Miss Lisa Briaris of Ellis Horwood for her patience, and the National Science Foundation for their generous support of the research which has sparked my interests.

to
Nancy, Andrew and Elizabeth

1

Introduction

1.1 INTRODUCTION

Reduction and oxidation processes in inorganic chemistry involve the transfer of charge between reactants. The reactions may appear quite simple and involve formal electron transfer (eqs (1.1)–(1.3)) or formal atom transfer (eqs (1.4) and (1.5)) or may be more complex, involving several reagents (eq. (1.6)).

$$[Co(phen)_3]^{3+} + [Ru(NH_3)_6]^{2+} \rightarrow [Co(phen)_3]^{2+} + [Ru(NH_3)_6]^{3+} \quad (1.1)$$

$$[(NC)_5Fe^{III}CNRu^{II}(NH_3)_5]^- \rightarrow [(NC)_5Fe^{II}CNRu^{III}(NH_3)_5]^- \quad (1.2)$$

$$[Tl_{aq}]^{3+} + 2\,[Fe(H_2O)_6]^{2+} \rightarrow [Tl_{aq}]^+ + 2\,[Fe(H_2O)_6]^{3+} \quad (1.3)$$

$$[IrCl_6]^{2-} + [Cr(H_2O)_6]^{2+} \rightarrow [IrCl_5(H_2O)]^{2-} + [Cr(H_2O)_5Cl]^{2+} \quad (1.4)$$

$$OCl^- + I^- \rightarrow OI^- + Cl^- \quad (1.5)$$

$$2\,[Co(NH_3)_5(OH_2)]^{3+} + SO_2 \xrightarrow{[H^+]} 2\,[Co(H_2O)_6]^{2+} + 10\,NH_4^+ + SO_4^{2-} \quad (1.6)$$

Detailed studies of these reactions reveal that even the most complex processes can be analyzed in a sequence of simple elementary steps which constitute the mechanism. The illustration of this sequence of steps for redox processes is the primary focus of this book. An attempt has been made to view inorganic redox chemistry from a descriptive perspective. There are many texts and review articles which emphasize different aspects of redox chemistry [1–4] and particularly the relationship of experiment to theory. However, it is also important to remember that inorganic chemistry harbors many exceptions to the general trends and mechanistic redox chemistry has its share. Accordingly, the book is arranged to consider the simplest processes first and add layers of complexity as the chapters progress. Theory is kept to the minimum possible to provide a context for the descriptive element and the reader is referred to detailed review articles for much of the derivative material.

Some knowledge of kinetics and mechanistic interpretation is assumed, and the book was designed to be read in conjunction with a more general text on inorganic reaction mechanisms [5]. Chapter 1 is a brief review of key points which are amplified in the succeeding chapters. Most information about the mechanism of a reaction comes from the experimentally derived rate law. Other clues to the mechanism come from comparisons of the rates of related reactions, investigations of the temperature dependence and ionic strength dependence of the reaction rates, from isotope effects, and from the detection of reaction intermediates and kinetically controlled reaction products. These extra-kinetic features have all proved to be of importance in the determination of redox mechanisms.

The mechanism is defined as the sequence of elementary steps which lead from reactants to products. Fortunately the variety of elementary steps which involve charge transfer is quite limited. The most useful classification involves the relationship between charge transfer and the ligand substitution properties of the reaction centers. When charge transfer takes place only in conjunction with ligand substitution into the inner coordination sphere of a metal ion or non-metal center, the reaction is defined as inner-sphere; otherwise the reaction is outer-sphere. The occurrence of these two mechanisms in reactions between metal ion complexes is discussed in Chapters 2 and 3. Despite the formal differences, they are in fact closely related and the relationship between them is explored in Chapter 4 together with a number of examples of mechanisms from biological systems which require the participation of metal ions. Chapter 5 deals with more complex reactions between metal ion complexes and reactions involving non-metallic reagents.

1.2 LIGAND FIELD THEORY

Much of this book features electron transfer reactions which involve transition metal ion complexes. As with most aspects of the chemistry of transition metal ions, ligand field effects play an important role in determining the redox mechanism and some brief introductory remarks on this topic seem appropriate [6]. When the set of d orbitals is placed in a non-spherically symmetric ligand field of well-defined geometry, the degeneracy of the five orbitals is lifted. For example in an octahedral field, the two e_g orbitals ($d_{x^2-y^2}$ and d_{z^2}) overlap directly with ligand orbitals and, along with the 4p and 4s orbitals are involved in the σ-bonding framework. The three t_{2g} orbitals (d_{xy}, d_{xz}, d_{yz}) are either non-bonding or π-bonding with respect to this framework, Fig. 1.1. The metal-centered d electrons are distributed between the π and σ^* orbitals depending on the strength of the binding interaction, measured as the ligand field splitting parameter for the octahedral field, Δ_o. For each electron in the π set of orbitals, there is an effective ligand field stabilization of $\frac{2}{5}\Delta_o$ while for each electron in the antibonding σ^* set, there is ligand field destabilization of $\frac{3}{5}\Delta_o$. Thus ligand field stabilization increases with increasing d-electron population from d^1 to d^3. At d^4, there is a choice depending on whether Δ_o is small, in which case the configuration is $\pi^3\sigma*^1$ (high spin), or large enough to overcome the energy required for spin pairing in which case the configuration is $\pi^4\sigma*^0$ (low spin). It is the balance between Δ_o and the spin-pairing energy which determines whether a high-spin or low-spin configuration is preferred.

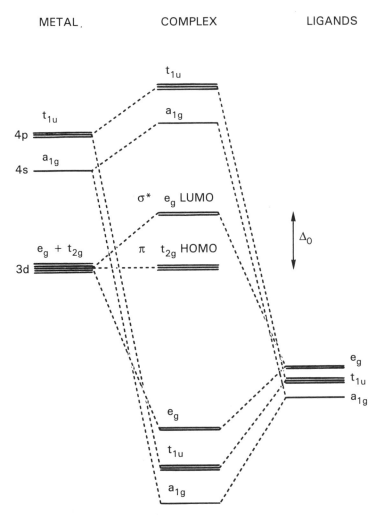

Fig. 1.1. Molecular orbital energy levels for a transition metal ion in an octahedral field showing the interaction with the σ-bonding and π-bonding ligand framework.

In Table 1.1, the ligand field stabilization energy (LFSE) is presented for different electronic configurations depending on whether Δ_o is smaller (high spin) or larger (low spin) than the energetic requirements for spin pairing. The magnitude of Δ_o is determined by the nature of the ligand and increases for the spectrochemical series of donor atoms in eq. (1.7).

$$I^- < Br^- < Cl^- < F^- \approx OH^- \approx HCO_2^- < ox^{2-} < H_2O < py \approx$$
$$NH_3 < en < bpy < phen < CN^- \qquad (1.7)$$

Table 1.1. Ligand field stabilization energy for octahedral complexes.

High spin			Low spin	
Configuration		LFSE	Configuration	LFSE
d^1	$\pi^1\sigma^{*0}$	$\frac{2}{5}\Delta_o$		
d^2	$\pi^2\sigma^{*0}$	$\frac{4}{5}\Delta_o$		
d^3	$\pi^3\sigma^{*0}$	$\frac{6}{5}\Delta_o$		
d^4	$\pi^3\sigma^{*1}$	$\frac{3}{5}\Delta_o$	$\pi^4\sigma^{*0}$	$\frac{8}{5}\Delta_o$
d^5	$\pi^3\sigma^{*2}$	$\frac{0}{5}\Delta_o$	$\pi^5\sigma^{*0}$	$\frac{10}{5}\Delta_o$
d^6	$\pi^4\sigma^{*2}$	$\frac{2}{5}\Delta_o$	$\pi^6\sigma^{*0}$	$\frac{12}{5}\Delta_o$
d^7	$\pi^5\sigma^{*2}$	$\frac{4}{5}\Delta_o$	$\pi^6\sigma^{*1}$	$\frac{9}{5}\Delta_o$
d^8	$\pi^6\sigma^{*2}$	$\frac{6}{5}\Delta_o$		
d^9	$\pi^6\sigma^{*3}$	$\frac{3}{5}\Delta_o$		
d^{10}	$\pi^6\sigma^{*4}$	$\frac{0}{5}\Delta_o$		

In addition, the value of Δ_o generally increases with increasing oxidation number of the metal ion. Metal ion complexes of the second and third row transition elements are invariably low spin since there is also an increase in Δ_o on going down a column in the periodic table. Similar ligand field arguments can be made for the distribution of electrons in the ligand fields of other geometries. Tetrahedral complexes which occur for some first-row transition metal ion complexes are generally high-spin because Δ_t is much smaller than Δ_o. While there is no distinction between high and low spin for d^8 octahedral complexes, in the presence of a strong square-planar ligand field, the degeneracy of both the e_g and t_{2g} orbital sets is removed, and a spin-paired square-planar arrangement results, Fig. 1.2. Special attention must also be paid to configurations where a degenerate set of orbitals is occupied unequally. The Jahn–Teller theorem indicates that the geometry of such complexes will distort to remove the degeneracy. Hence, especially in high-spin d^4, low-spin d^7 and d^9 configurations where there is uneven occupation of the σ^* orbitals, tetragonal distortion of the six-cooordinate geometry is encountered. In this work, where the coordination sphere of a metal ion complex is well characterized, it is represented as clearly as possible, for example $[Fe(H_2O)_6]^{3+}$. Where the complex is less well characterized, a less specific designation is provided, for example $[Eu_{aq}]^{2+}$ representing the solvated europium(II) ion in aqueous solution.

Ligand fields impose both energetic and symmetry conditions on reactions of transition metal ion complexes. The energetic constraints arise from the fact that in attaining a transition state for a reaction, the ligand field is altered and this can provide an energetic barrier in addition to the other requirements of the mechanism. Symmetry constraints arise in electron transfer reactions where there is a requirement for overlap between metal-centered donor and acceptor orbitals. Both can have a significant effect on the course of the reaction.

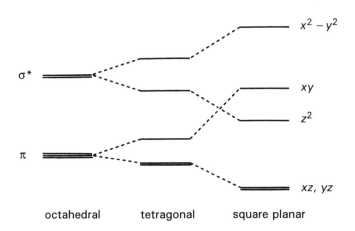

Fig. 1.2. Effect on the metal centered molecular orbit energy levels as a rsult of a descent in symmetry of the ligand field from octahedral to tetragonal and square planar.

1.3 SUBSTITUTION REACTIONS

The rate of ligand substitution at a metal or non-metal center has an important role in determining the nature of a redox reaction mechanism. Much of the first four chapters of this book is concerned with reactions involving metal ion complexes, and substitution reactions of these species will be considered chiefly in this chapter. A limited discussion of substitution at non-metal centers is reserved for Chapter 5. Typical rates of exchange of solvent water into the inner-coordination spheres of some representative metal ion complexes (eq. (1.8)) are presented in Table 1.2. These rates give a good indication of the

$$[Fe(H_2O)_6]^{2+} + H_2O* \quad [Fe(H_2O)_5H_2O*]^{2+} + H_2O \quad (1.8)$$

relative lability or inertness of a metal center with respect to substitution and provide a guide to differentiating between mechanisms of electron transfer in which ligand substitution plays a key role and mechanisms where substitution is precluded.

Table 1.2. Rates of exchange of solvent water into the inner-coordination spheres of metal ions at 25.0 °C

Complex	k^a (s^{-1})	$\Delta H^\ddagger$ (kJ mol^{-1})	$\Delta S^\ddagger$ (J K^{-1} mol^{-1})	$\Delta V^\ddagger$ (cm^3 mol^{-1})	Ref.
$[V(H_2O)_6]^{2+}$	8.7×10^1	62	−0.4	−4.1	7
$[Cr(H_2O)_6]^{2+}$	5×10^8				8
$[Mn(H_2O)_6]^{2+}$	2.1×10^7	33	6	−5.4	9
$[Fe(H_2O)_6]^{2+}$	4.4×10^6	41	21	+3.8	9
$[Co(H_2O)_6]^{2+}$	3.2×10^6	47	37	+6.1	9
$[Ni(H_2O)_6]^{2+}$	3.2×10^4	57	32	+7.2	10
$[Cu(H_2O)_6]^{2+}$	$\approx 5 \times 10^9$				8
$[Ru(H_2O)_6]^{2+}$	1.8×10^{-2}	88	16	−0.4	11
$[Ti(H_2O)_6]^{3+}$	1.8×10^5	43	1	−12.1	12
$[V(H_2O)_6]^{3+}$	5.0×10^2	49	−28	−8.9	13
$[Cr(H_2O)_6]^{3+}$	2.4×10^{-6}	109	12	−9.6	14
$[Fe(H_2O)_6]^{3+}$	1.6×10^2	64	12	−5.4	15
$[Ru(H_2O)_6]^{3+}$	3.5×10^{-6}	90	−48	−8.3	11
$[Rh(H_2O)_6]^{3+}$	2.2×10^{-9}	131	29	−4.2	16
$[Co(H_2O)_6]^{3+}$	$\approx 1 \times 10^{-1}$				17
$[Cr(H_2O)_5OH]^{2+}$	1.8×10^{-4}	110	55	+2.7	14
$[Fe(H_2O)_5OH]^{2+}$	1.2×10^5	42	5	+7	15
$[Ru(H_2O)_5OH]^{2+}$	5.9×10^{-4}	96	15	+0.9	11
$[(NH_3)_5Co(H_2O)]^{3+}$	5.7×10^{-6}	111	28	+1.2	18
$[(NH_3)_5Cr(H_2O)]^{3+}$	5.2×10^{-5}	97	0	−5.8	19
$[(NH_3)_5Rh(H_2O)]^{3+}$	8.7×10^{-6}	103	3	−4.1	19
$[(NH_3)_5Ru(H_2O)]^{3+}$	2.3×10^{-4}	92	−8	−4.0	20
$[(NH_3)_5Ir(H_2O)]^{3+}$	6.1×10^{-8}	118	11	−3.2	21
$[(CN)_5Co(H_2O)]^{2-}$	$\approx 1.0 \times 10^{-3}$				22
$[(CN)_5Fe(H_2O)]^{2-}$	$\approx 1.0 \times 10^{-2}$				23
$[(CN)_5Fe(H_2O)]^{3-}$	5.0×10^2				24

^{a}Second-order rate constants can be approximated by dividing the first-order rate constant by 55.5 ([H$_2$O]).

There is a strong correlation of the rates with ligand field stabilization energy since the substitution process necessarily involves a change in geometry around the metal ion center. Low-spin complexes exchange ligands very slowly, whereas the ions

$[Cr(H_2O)_6]^{2+}$ and $[Cu(H_2O)_6]^{2+}$ which are subject to tetragonal distortion with four shorter equatorial bonds and two longer axial bonds as a result of the Jahn–Teller effect, have high lability arising from the weaker axial coordination and the rapid vibrational interchange of the unique axis. The hydrolyzed ions such as $[Fe(H_2O)_5OH]^{2+}$ are significantly more labile than the hexa-aqua species and this has important mechanistic consequences in reactions where substitution at the metal center is involved.

1.4 THERMODYNAMIC ASPECTS

Thermodynamic aspects of reduction and oxidation are of fundamental importance. Thermodynamics determines the ultimate products of reactions and, indirectly, has a profound effect on the course of the reaction. Thermodynamic information is expressed in the form of a reduction potential such as that for $[Co(bpy)_3]^{3+/2+}$ (eq. (1.9)). The potential for this redox couple is given by the Nernst equation (eq. (1.10)), with a formal reduction potential, $E° = 0.31$ V (vs n.h.e.). Throughout this book, formal reduction potentials which refer to specific conditions of temperature and ionic strength are employed rather than standard potentials which refer to conditions of unit activity and 25.0°C.

$$[Co(bpy)_3]^{3+} + e^- \rightleftharpoons [Co(bpy)_3]^{2+} \tag{1.9}$$

$$E = E° - RT/nF \ln \frac{[[Co(bpy)_3]^{2+}]}{[[Co(bpy)_3]^{3+}]} \tag{1.10}$$

Measurement of reduction potentials is generally carried out either by potentiometric titration with a standard oxidant or reductant, or more conveniently by electrochemical methods, the most common of which is cyclic voltammetry. A cyclic voltammogram showing the reduction of $[Co(bpy)_3]^{3+}$ in 0.10 M NaNO$_3$ media is shown in Fig. 1.3. This is a reversible couple and the formal potential, 0.31 V, measured against the normal hydrogen electrode, can be obtained from the potential midway between the maxima in the oxidation and reduction waves. The separation between the peaks is approximately $60/n$ mV at 25.0°C. A listing of the reduction potentials for selected first-row transition metal ion complexes and aqua ions is given in Table 1.3.

Table 1.3. Reduction potentials for metal complexes and aqua ions at 25.0°C

Complex	$E°$ (V)		Complex	$E°$ (V)	
$[V(H_2O)_6]^{3+/2+}$	$\pi^2\sigma^{*0}/\pi^3\sigma^{*0}$	-0.20			
$[Cr(H_2O)_6]^{3+/2+}$	$\pi^3\sigma^{*0}/\pi^3\sigma^{*0}$	-0.41	$[Cr(bpy)_3]^{3+/2+}$	$\pi^6\sigma^{*0}/\pi^6\sigma^{*1}$	-0.23
$[Mn(H_2O)_6]^{3+/2+}$	$\pi^3\sigma^{*1}/\pi^3\sigma^{*2}$	1.77	$[Mn(bpy)_3]^{3+/2+}$	$\pi^6\sigma^{*0}/\pi^6\sigma^{*1}$	1.00

(continues)

Table 1.3. *(continued)*

Complex	$E°$ (V)		Complex	$E°$ (V)	
$[Fe(H_2O)_6]^{3+/2+}$	$\pi^3\sigma*^2/\pi^4\sigma*^2$	0.77	$[Fe(bpy)_3]^{3+/2+}$	$\pi^5\sigma*^0/\pi^6\sigma*^0$	1.00
$[Co(H_2O)_6]^{3+/2+}$	$\pi^6\sigma*^0/\pi^5\sigma*^2$	1.89	$[Co(bpy)_3]^{3+/2+}$	$\pi^6\sigma*^0/\pi^6\sigma*^1$	0.31
			$[Ni(bpy)_3]^{3+/2+}$	$\pi^6\sigma*^1/\pi^6\sigma*^2$	1.70

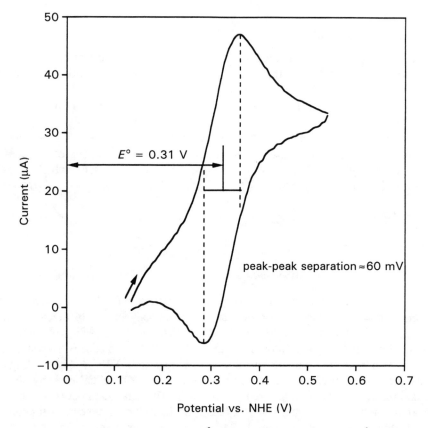

Fig. 1.3. Cyclic voltammogram of a 1×10^{-3} M aqueous solution of $[Co(bpy)_3]^{2+}$ in 0.1 M $NaNO_3$, scan rate 20 mV s^{-1}. The initial scan is the upper curve from left to right and shows a peak corresponding to oxidation of $[Co(bpy)_3]^{2+}$ to $[Co(bpy)_3]^{3+}$. When the scan is reversed, lower curve, $[Co(bpy)_3]^{3+}$ is reduced.

As the d-electron occupation is increased across the periodic table, there is an underlying reduction in stability of the higher oxidation state, the result of incomplete d-electron shielding on the ionization potential, and the consequent increase in reduction potential [25]. On this trend are superimposed ligand field effects. In the high-spin aqua complexes for example, the ions $[Fe(H_2O)_6]^{3+}$ and $[Mn(H_2O)_6]^{2+}$ are stabilized by the low ionization potential for d^5 ions while $[Cr(H_2O)_6]^{3+}$ has the $\pi^3\sigma*^0$ configuration with favorable ligand field stabilization energy. Among the low-spin tris-bipyridyl complexes, the $\pi^6\sigma*^0$ for $[Fe(bpy)_3]^{2+}$ and $[Co(bpy)_3]^{3+}$ are strongly stabilized. Higher oxidation states are favored among second- and third-row transition elements.

Complexation affects reduction potentials in a predictable fashion. Consider a ligand L which binds to both oxidized, A^{ox}, and reduced, A^{red}, forms of a metal ion according to eqs (1.11) and (1.12). The change in the reduction potential is given by eq. (1.13), amounting to approximately $60/n$ mV for each order of magnitude difference in the stability constants at 25.0°C.

$$A^{red} + L \rightleftharpoons A^{red}L \quad K_{red} = \frac{[A^{red}L]}{[A^{red}][L]} \tag{1.11}$$

$$A^{ox} + L \rightleftharpoons A^{ox}L \quad K_{ox} = \frac{[A^{ox}L]}{[A^{ox}][L]} \tag{1.12}$$

$$E^\circ_{A^{ox/red}L} - E^\circ_{A^{ox/red}} = \frac{RT}{nF} \ln \frac{K_{red}}{K_{ox}} \tag{1.13}$$

An example of this behavior is the complexation of $[Fe(CN)_5(OH_2)]^{2-/3-}$ by NH_3, where $K_{ox} = 6.8 \times 10^4$ M^{-1} and $K_{red} = 2.1 \times 10^4$ M^{-1} at 25.0°C and 0.10 M ionic strength. Stabilization of the higher oxidation state lowers the reduction potential from 0.370 V for $[Fe(CN)_5(OH_2)]^{2-}$ to 0.340 V for $[Fe(CN)_5(NH_3)]^{2-}$ [26]. More dramatic is the reduction of $[Ru(edta)(H_2O)]^-$ in the presence of N-methylpyrazinium cation, Mepz$^+$, where the binding constants are 15 M^{-1} and 5×10^{10} M^{-1} for K_{ox} in eq. (1.14) and K_{red} in eq. (1.15) [27]. The change in coordination is relatively sluggish and this results in an electrochemical 'square scheme' where the reduction potentials of both $[Ru(edta)(H_2O)]^-$, 0.00 V, and $[Ru(edta)(Mepz)]$, 0.55 V, can be measured.

$$[Ru(edta)(H_2O)]^- + N\bigcirc NMe^+ \rightleftharpoons [Ru(edta)N\bigcirc NMe] + H_2O \tag{1.14}$$

$$\updownarrow + e^- \qquad\qquad\qquad\qquad \updownarrow + e^-$$

$$[Ru(edta)(H_2O)]^{2-} + N\bigcirc NMe^+ \rightleftharpoons [Ru(edta)N\bigcirc NMe]^- + H_2O \tag{1.15}$$

One of the most important changes in complexation which occurs with metal ion complexes involves hydrolysis or changes in the degree of ligand protonation, which lead to a pH dependence of the reduction potential and can change the reactivity of the complex significantly. The aqua complexes of the transition metal ions are very prone to this behavior (eqs (1.17) and (1.18)). In this instance the $[Fe(H_2O)_6]^{3+}$ is more acidic than

$$[Fe(H_2O)_6]^{3+} + e^- \rightleftharpoons [Fe(H_2O)_6]^{2+} \tag{1.16}$$

$$[Fe(H_2O)_6]^{3+} \rightleftharpoons [Fe(H_2O)_5(OH)]^{2+} + H^+ \quad (1.17)$$

$$[Fe(H_2O)_6]^{2+} \rightleftharpoons [Fe(H_2O)_5(OH)]^+ + H^+ \quad (1.18)$$

$$[Fe(H_2O)_5(OH)]^{2+} + e^- \rightleftharpoons [Fe(H_2O)_5(OH)]^+ \quad (1.19)$$

$[Fe(H_2O)_6]^{2+}$, with acidity constants 2.0×10^{-3} M and 3.2×10^{-10} M respectively at 1.0 M ionic strength, so that the reduction potential for $[Fe(H_2O)_5(OH)]^{2+}$ is 0.36 V [28]. However, the pH dependence in this instance is complicated by the subsequent hydrolysis and oligomerization of $[Fe(H_2O)_5(OH)]^{2+}$ above pH 4.0. Simpler examples are known. The ruthenium(III) complex $[Ru(bpy)_2(py)(H_2O)]^{3+}$ exhibits $K_a^{ox} = 0.14$, while the reduced form shows evidence for hydrolysis only at much higher pH with $K_a^{red} = 6.31 \times 10^{-11}$. Consequently the reduction potential shows a pH dependence, as shown in Fig. 1.4, which may be described by eq. (1.23) at 25°C and 1.0 M ionic strength. The slope of the line is approximately 60 mV for each pH unit [29].

$$[Ru(bpy)_2(py)OH_2]^{3+} \rightleftharpoons [Ru(bpy)_2(py)OH]^{2+} + H^+ \quad pK_a^{ox} = 0.85 \quad (1.20)$$

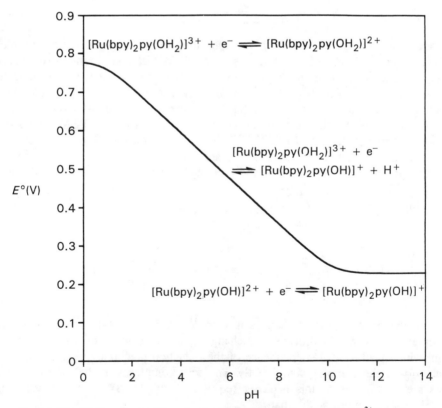

Fig. 1.4. The pH dependence of the reduction potential of $[Ru(bpy)_2(py)(H_2O)]^{3+}$. Note how the pK_a values can be identified by the breaks in slope

$$[\text{Ru(bpy)}_2(\text{py})\text{OH}_2]^{2+} \rightleftharpoons [\text{Ru(bpy)}_2(\text{py})\text{OH}]^+ + \text{H}^+ \quad pK_a^{red} = 10.20 \quad (1.21)$$

$$[\text{Ru(bpy)}_2(\text{py})\text{OH}_2]^{3+} + e^- \rightleftharpoons [\text{Ru(bpy)}_2(\text{py})\text{OH}_2]^{2+} \quad E = 0.78 \text{ V} \quad (1.22)$$

$$E = 0.78 - \frac{RT}{nF} \ln\left(\frac{[\text{H}^+] + K_a^{ox}}{[\text{H}^+] + K_a^{red}}\right) \quad (1.23)$$

Hydrolysis is not restricted to aqua ligands. Reduction of the 2-pyridinecarboxaldehyde complex (eq. (1.24)) is also pH-dependent over the range pH 1–6 [30].

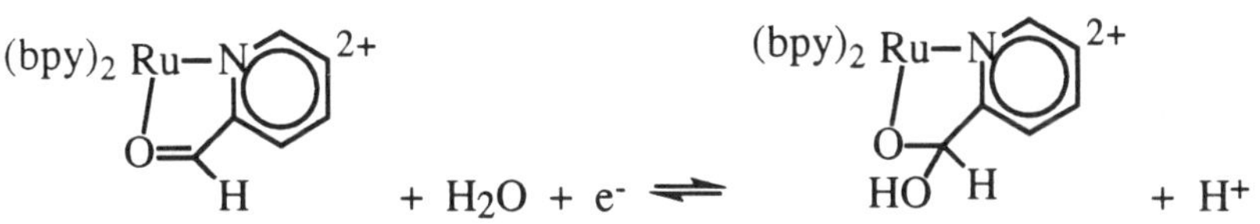

(1.24)

Higher oxidation states are more prone to hydrolysis than lower oxidation states. Oxo complexes are readily formed, and consequently the situation is more complex in redox processes which involve transfer of more than one unit of charge. Reduction of the ruthenium(IV) oxo complex, $[\text{Ru(bpy)}_2(\text{py})\text{O}]^{2+}$, takes place in two thermodynamically discernible one-electron processes which are pH-dependent (eqs (1.25)–(1.26)), and which are conveniently summarized in a Latimer diagram for 1 M [H$^+$] (Scheme 1.1).

$$[\text{Ru(bpy)}_2(\text{py})\text{O}]^{2+} + \text{H}^+ + e^- \rightleftharpoons [\text{Ru(bpy)}_2(\text{py})\text{OH}]^{2+} \quad E^\circ = 0.53 \text{ V}$$
(1.25)

$$[\text{Ru(bpy)}_2(\text{py})\text{OH}]^{2+} + \text{H}^+ + e^- \rightleftharpoons [\text{Ru(bpy)}_2(\text{py})\text{OH}_2]^{2+} \quad E^\circ = 0.42 \text{ V}$$
(1.26)

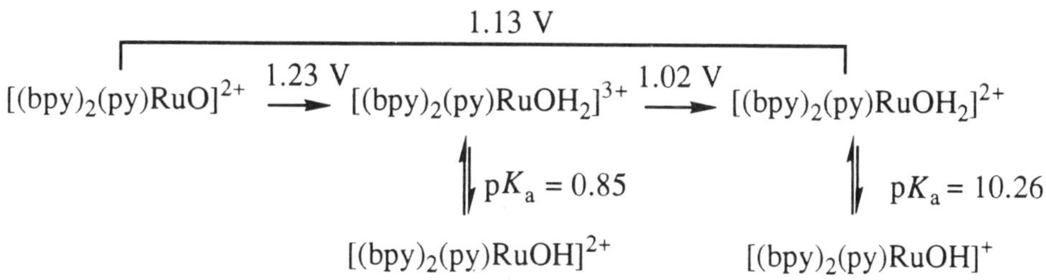

Scheme 1.1.

In this case the intermediate oxidation state is thermodynamically stable but in other instances one or more intermediate species may be thermodynamically unstable, and the observed thermodynamic potential is a multi-electron process. For example the formal nickel(IV) complex $[Ni(Me_2L)]^{2+}$ undergoes a two-electron reduction at low pH (eq. (1.27)) [31].

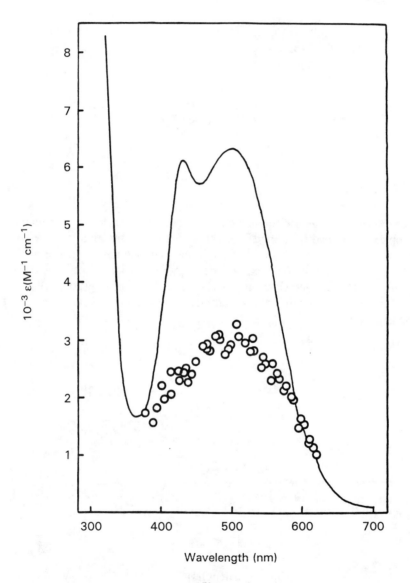

Fig. 1.5. Absorbance spectrum of $[Ni^{IV}(Me_2L)]^{2+}$ (solid line) and the transient $[Ni^{III}(Me_2L)H]^{2+}$ (open circles), 0.075 s after reaction with Me_2COH at pH 2.5, 10% Me_2CHOH, 21°C, from Baral, S.; Lappin, A. G. *J. Chem. Soc., Dalton Trans.* 1986, 2214 with permission.

$$[Ni(Me_2L)]^{2+} + 2H^+ + 2e^- \underset{}{\overset{0.94 \text{ V}}{\rightleftharpoons}} V[Ni(Me_2LH_2)]^{2+} \quad (1.27)$$

Information on the intermediate single-electron transfer processes can be obtained from kinetic measurements where the intermediate oxidation state is produced as a spectroscopically characterized reactive transient (Fig. 1.5) [32]. In general, rapid detection methods are required to investigate the chemistry of the intermediates since disproportionation processes are rapid. Pulse radiolysis and flash photolysis have proved to be particularly useful for the study of these species, and are briefly introduced in section 1.5.

The strong dependence on the nature of ligands of reduction potentials has prompted attempts to correlate ligand effects to allow the prediction of reduction potential data on complexes where direct measurement is not readily possible. Such a treatment depends on individual ligands having a uniform effect on reduction potential, independent of other ligands present and the electronic configuration of the metal ion. One of the more successful attempts [33] involves the assignment of a contribution, E_L, from each ligand in the inner coordination sphere of a six-coordinate metal complex. Values for E_L, the ligand electrochemical series, are presented in Table 1.4.

Table 1.4. Values for the ligand electrochemical series, E_L, for selected ligands

H_2O	0.04	F^-	−0.42	ox^{2-}	−0.17
py	0.25	Cl^-	−0.24	bpy	0.26
NH_3	0.07	Br^-	−0.22	phen	0.26
OH^-	−0.59	I^-	−0.24	en	0.06
CN^-	0.02	N_3^-	−0.30	terpy	0.25
O_2CH^-	−0.30	SCN^-	−0.06		

The correlations work well for electron transfer involving a π–π change where the electron is transferred to and from a non-bonding or weakly anti-bonding orbital and the corresponding structural changes are small. Sample correlations for ruthenium(III)/(II) and chromium (III)/(II) potentials in aqueous solution are given in (eqs (1.28) and (1.29)).

$$E_{Ru} = 1.14 \, [\Sigma E_L] - 0.35 \quad (1.28)$$

$$E_{Cr} = 0.58 \, [\Sigma E_L] - 1.12 \quad (1.29)$$

For example, the complex $[Ru(NH_3)_4(bpy)]^{3+}$ has $\Sigma E_L = 4 \times 0.07 + 2 \times 0.26$, from which the reduction potential is calculated to be 0.56 V compared with the observed value of 0.51 V. Although the values for the slopes and intercepts in eqs (1.28) and

22 Introduction   [Ch. 1

(1.29) are empirically derived, the former are largely determined by the relative ligand binding strengths of the two oxidation states and reflect the ligand field and charge, whereas the intercepts are affected by a number of factors including the ionization potential of the metal and solvation effects. Correlations for complexes of cobalt(III)/(II) and copper(II)/(I) where there are significant structural changes would be most useful, particularly since the reduction potentials in a large number of cobalt(III) complexes are inaccessible. However, to date, no such correlations are available.

1.5 STUDIES WITH THERMODYNAMICALLY UNSTABLE COMPLEXES

The generation and detection of thermodynamically unstable complexes depends ultimately on the controlled oxidation or reduction of a species in a thermodynamically stable state on a timescale which exceeds the rate at which disproportionation or some further reaction can take place. In some cases this may be accomplished chemically and there are examples in Chapter 5 to illustrate this point. More generally, however, large amounts of energy from the rapid application of radiation to solutions can be converted to oxidizing or reducing power. Examples using light energy for this purpose are discussed in section 2.7. In this section relevant details of the radiolysis of water are presented.

The irradiation of water with high-energy particles or γ-rays can be summarized by eq. (1.30) where the yields or G values are quoted for each 100 eV of energy [34].

$$4\ H_2O \longrightarrow 2.6\ e^-_{aq} + 2.6\ OH\cdot + 0.6\ H\cdot + 2.6\ H^+ + 0.4\ H_2 + 0.7\ H_2O_2 \quad (1.30)$$

A number of highly reactive radicals are produced with oxidizing species such as OH· and H· and reducing species such as e^-_{aq} in approximately equal amounts. These reactive species will rapidly recombine in the absence of other reagents. However, in the presence of N_2O, a useful scavenger for the solvated electron (eq. (1.31)),

$$e^-_{aq} + N_2O + H_2O \rightarrow N_2 + OH\cdot + OH^- \quad (1.31)$$

solutions can be produced which contain predominantly the oxidizing species OH·. Similarly, in the presence of the hydrogen carbonate ion or alcohols which act as scavengers for H· and OH·, solutions with the reducing radicals CO_2^- and ·CH_2OH are produced (eqs (1.32)–(1.33)).

$$HCO_2^- + OH\cdot\ \text{or}\ H\cdot\ \rightarrow CO_2^- + H_2O\ \text{or}\ H_2 \quad (1.32)$$

$$CH_3OH + OH\cdot\ \text{or}\ H\cdot\ \rightarrow\ \cdot CH_2OH + H_2O\ \text{or}\ H_2 \quad (1.33)$$

These reactions are rapid and, depending on the conditions, solutions which contain species which are either oxidizing or reducing can be produced within a few microseconds.

Reactions of the radicals with various metal ions have been studied extensively [35]. The rates vary over a significant range but are generally very rapid. For example the reaction of $[Ni([15]aneN_5)]^{2+}$ with $OH\cdot$ yields the nickel(III) complex, $[Ni([15]aneN_5)]^{3+}$, with a rate constant of 5×10^9 M^{-1} s^{-1} at 23 °C [36].

1.6 TEMPERATURE AND PRESSURE DEPENDENCE OF REDUCTION POTENTIALS

The temperature dependence of reduction potentials yields significant information on the enthalpy and entropy of the reaction. Values for the thermodynamic parameters are obtained by means of eq. (1.34), where the parameters are measured relative to those for the normal hydrogen electrode, and this must be taken into account when the absolute values are considered. Comparison of $\Delta H° = -129$ kJ mol^{-1} and $\Delta S° = -87$ J K^{-1} for reduction of $[Fe(phen)_3]^{3+}$ (eq. (1.35)) and $\Delta H° = -40$ kJ mol^{-1} and $\Delta S° = -11$ J K^{-1} mol^{-1} for reduction of $[Co(phen)_3]^{3+}$ (eq. (1.36)) indicates that the latter process is less exothermic and has a less unfavorable entropy change. The difference in the entropy term requires some comment since the complexes have similar structures and charges in both oxidation states and solvation changes are expected to be similar for the two metal ions. It is not clear from the thermodynamic data whether reduction of $[Fe(phen)_3]^{3+}$ or $[Co(phen)_3]^{3+}$ has the larger change since the contribution from the hydrogen half-reaction masks the process.

$$-n \Delta E° F = \Delta G° = \Delta H° - T\Delta S° \tag{1.34}$$

$$[Fe(phen)_3]^{3+} + \tfrac{1}{2} H_2 \underset{}{\overset{1.072 \text{ V}}{\rightleftharpoons}} [Fe(phen)_3]^{2+} + H^+ \tag{1.35}$$

$$[Co(phen)_3]^{3+} + \tfrac{1}{2} H_2 \underset{}{\overset{0.337 \text{ V}}{\rightleftharpoons}} [Co(phen)_3]^{2+} + H^+ \tag{1.36}$$

An absolute value of 113 J K^{-1} mol^{-1} is estimated for ΔS for the standard half-reaction (eq. (1.37)), and reaction entropies for a number of individual half-reaction couples have been evaluated, mainly from relatively simple measurements with non-isothermal electrochemical cells [37]. Reaction entropy, $\Delta S°_{rc}$, defined in eq. (1.38), has an advantage in interpretation over $\Delta S°$ since it refers to a specific half-reaction, and a number of values are presented in Table 1.5.

$$H^+ + e^- \rightleftharpoons \tfrac{1}{2} H_2 \tag{1.37}$$

$$\Delta S^\circ_{rc} = \Delta S^\circ - 113 \tag{1.38}$$

Table 1.5. Reduction potentials and reaction entropies for selected redox couples at 25.0 °C

Redox couple	μ (M)	E° (V)	ΔS°_{rc} (J K^{-1} mol^{-1})	Ref.
$[Co(OH_2)_6]^{3+/2+}$	—	1.93	≈ 190	37
$[Fe(OH_2)_6]^{3+/2+}$	0.2	0.74	180	37
$[Ru(OH_2)_6]^{3+/2+}$	0.3	0.23	151	37
$[V(OH_2)_6]^{3+/2+}$	0.2	− 0.23	155	37
$[Cr(OH_2)_6]^{3+/2+}$	1.0	− 0.42	205	37
$[Ru(NH_3)_6]^{3+/2+}$	0.2	0.06	75	37
$[Ru(en)_3]^{3+/2+}$	0.1	0.19	54	37
$[Co(en)_3]^{3+/2+}$	1.0	− 0.21	155	37
$[Co(sep)]^{3+/2+}$	0.1	− 0.29	79	37
$[Co(phen)_3]^{3+/2+}$	0.05	0.38	92	37
$[Co(bpy)_3]^{3+/2+}$	0.05	0.31	92	37
$[Ru(bpy)_3]^{3+/2+}$	0.1	1.29	0	37
$[Fe(bpy)_3]^{3+/2+}$	0.05	1.09	8	37
$[Ru(NH_3)_5 py]^{3+/2+}$	0.1	0.32	67	37
$[Co(edta)]^{-/2-}$	0.1	0.37	− 30	38
$[Cu(H_3G_4)]^{-/2-}$	1.0	0.63	− 64	39
$[Ni(H_3G_4)]^{-/2-}$	1.0	0.83	63	39
$[Fe(phen)_3]^{3+/2+}$	0.1	1.07	3	40
$[Co(phen)_3]^{3+/2+}$	0.1	0.38	98	40

The magnitude of ΔS°_{rc} gives information about changes in structure, charge and solvation on going from the oxidized to the reduced state. Reduction of $[M(H_2O)_6]^{3+}$ generally yields substantially larger values for ΔS°_{rc} than for $[M(NH_3)_6]^{3+}$, the result of greater changes in solvation. In a number of the complexes of cobalt(III), values for ΔS°_{rc} are higher than for isostructural complexes, other metal ions reflecting the larger structural changes associated with population of the antibonding orbital in the $\pi^6\sigma*^0$ to $\pi^6\sigma*^1$ change. The magnitude of ΔS°_{rc} is also markedly affected by the charge on the redox couple. For anionic complexes, reduction involves an increase in charge and consequently an increase in order as electrostriction of the solvent increases, whereas with cationic complexes the opposite is true. Large values are obtained when there is a significant structural change in the complex during the redox process. Attention is drawn to the values for eqs (1.39) and (1.40) which are approximately equal and opposite in sign. Reduction

$$[Ni(H_{-3}G_4)(OH_2)_2]^- + e^- \rightleftharpoons [Ni(H_{-3}G_4)]^{2-} + 2\,H_2O \qquad (1.39)$$

$$[Cu(H_{-3}G_4)]^- + e^- + 2\,H_2O \rightleftharpoons [Cu(H_{-3}G_4)(OH_2)_2]^{2-} \qquad (1.40)$$

of the tetragonal $[Ni(H_{-3}G_4)(OH_2)_2]^-$ to give the square-planar $[Ni(H_{-3}G_4)]^{2-}$ involves release of two water molecules whereas reduction of the square-planar $[Cu(H_{-3}G_4)]^-$ to give the tetragonal $[Cu(H_{-3}G_4)(OH_2)_2]^{2-}$ involves an uptake of two water molecules. Release of a single water of hydration gives an entropy contribution of +30 J K^{-1} mol^{-1}.

The effect of pressure on a reaction is measured by the reaction volume, $\Delta V°$ (eq. (1.41)). This parameter gives a useful measure of coordination and solvation changes which take place during charge transfer. For example the volume change for the electron transfer reaction in eq. (1.42) is readily calculated from the molar volumes of the reactants and products to be 22.8 cm^3 mol^{-1} [41]. Some perspective on reaction volume is obtained by considering the volume of solvent water, which is somewhat less than 18 cm^3 mol^{-1}.

$$\left(\frac{\partial \Delta G°}{\partial P}\right)_T = \Delta V° \qquad (1.41)$$

$$[Co(bpy)_3]^{3+} + [V(H_2O)_6]^{2+} \rightleftharpoons [Co(bpy)_3]^{2+} + [V(H_2O)_6]^{3+} \qquad (1.42)$$

1.7 KINETIC CONSIDERATIONS

The most important piece of information regarding the mechanism of a chemical reaction is the experimental rate law which describes how the rate of the reaction depends on the concentrations of the participating species. The rate of a chemical reaction is defined as the rate of disappearance of reactants with time or the rate of appearance of products with time. Due account must be taken of the stoichiometric coefficients of the reagents. For example, in the reduction of $[Ni(Me_2L)]^{2+}$ by $[Fe(H_2O)_6]^{2+}$ in acidic solution (eq. (1.43)) [42].

$$[Ni(Me_2L)]^{2+} + 2\,[Fe(H_2O)_6]^{2+} + 2\,H^+ \rightleftharpoons [Ni(Me_2LH_2)]^{2+} + 2\,[Fe(H_2O)_6]^{3+} \qquad (1.43)$$

the rate is defined conventionally as $-d[[Ni(Me_2L)]^{2+}]/dt = -\frac{1}{2}d[[Fe(H_2O)_6]^{2+}]/dt = d[[Ni(Me_2LH_2)]^{2+}]/dt = \frac{1}{2}d[[Fe(H_2O)_6]^{3+}]/dt$. In some instances the rate law is very simple and can be determined fully from just a few experiments. In other instances many hundreds of experiments may be required. This is not a book on kinetics and for the many tools which are available to the kineticist in order to determine reaction

Most mechanistic studies in solution are planned to minimize the mathematical computation required. In the reduction of $[Co(phen)_3]^{3+}$ by $[Ru(NH_3)_6]^{2+}$ (eq. (1.44)),

$$[Co(phen)_3]^{3+} + [Ru(NH_3)_6]^{2+} \rightleftharpoons [Co(phen)_3]^{2+} + [Ru(NH_3)_6]^{3+} \qquad (1.44)$$

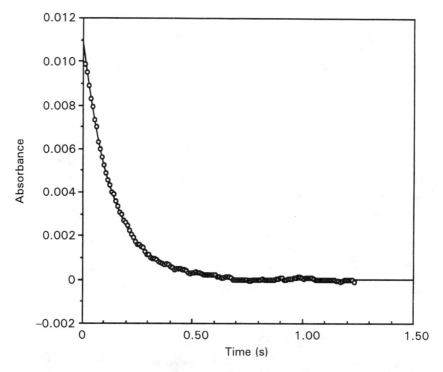

Fig. 1.6. Absorbance change at 455 nm as a function of time for the reaction of 5.02×10^{-4} M [Ru(NH3)6]2+ with 5×10^{-5} M [Co(phen)s]3+, at pH 8, 25°C, $\mu = 0.1$ M (NaCl). The solid line is a fit to eq. (1.46) with $k_{obsd} = 7.5$ s^{-1}

the rate can be measured conveniently by monitoring the absorbance, A, at 455 nm, where $[Co(phen)_3]^{3+}$ has an extinction coefficient of 99 M^{-1} cm^{-1}. In the general case, not only is the concentration of $[Co(phen)_3]]^{3+}$ changing during the course of the reaction but the concentration of $[Ru(NH_3)_6]^{2+}$ is also changing and the mathematical function describing the trace of A against time can be complex. Simplification results from using a large excess of $[[Ru(NH_3)_6]^{2+}]$ over $[[Co(phen)_3]^{3+}]$ such that over the time course of the reaction $[[Ru(NH_3)_6]^{2+}]$ remains effectively constant. Under these conditions, where $[Co(phen)_3]^{3+}$ is present at an initial concentration of 5×10^{-5} M, much less than the concentrantion of the reductant, 5.02×10^{-4} M, a plot of absorbance against time, t, is shown in Fig. 1.6. It is described by the expression eq. (1.45), where A_0 is the initial absorbance ($t = 0$), and A_∞ is the absorbance after an infinite time. The absorbance change is linearly related to the change in $[Co(phen)_3]^{3+}$ concentration, and dividing by the extinction coefficient yields eq. (1.46), which is the integrated form of a first-order differential equation (eq. (1.47)).

$$A_t = (A_0 - A_\infty) \exp(-k_{obsd}\, t) + A_\infty \qquad (1.45)$$

$$[[Co(phen)_3]^{3+}]_t = [[Co(phen)_3]^{3+}]_0 \exp(-k_{obsd}\, t) \qquad (1.46)$$

$$\text{Rate} = -d[[Co(phen)_3]^{3+}]/dt = k_{obsd}\, [[Co(phen)_3]^{3+}] \qquad (1.47)$$

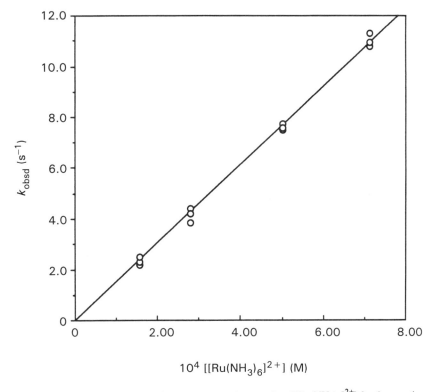

Fig. 1.7. Plot of pseudo-first-order rate constant k_{obsd} against $[[Ru(NH_3)_6]^{2+}]$ for the reaction of $[Ru(NH_3)_6]^{2+}$ with $[Co(phen)_3]^{3+}$, at pH 8, 25°C, $\mu = 0.1$ M (NaCl).

Although under the pseudo-first-order conditions the change in concentration of the oxidant, $[Ru(NH_3)_6]^{2+}$, is negligible, the rate is dependent on the concentration of this species. When $[[Ru(NH_3)_6]^{2+}]$ is varied, the value of the pseudo-first-order rate constant, k_{obsd}, varies as shown in Fig. 1.7. This linear dependence leads to the conclusion that the rate law shows a first-order dependence on both reactants (eq. (1.48)), second-order overall with a second-order rate constant $k = 1.5 \times 10^4$ M^{-1} s^{-1}. The bulk of the studies discussed in this book were examined under similar pseudo-first-order conditions. Other rates were determined directly under second-order conditions or pseudo-second-order conditions while in some instances the rate laws are much more complex.

$$-d[[Co(phen)_3]^{3+}]/dt = k\, [[Co(phen)_3]^{3+}]\, [[Ru(NH_3)_6]^{2+}] \qquad (1.48)$$

Interpretation of the rate law shown in eq. (1.48) is straightforward. There is one independent term which indicates a single pathway for the reaction and the reaction order in each reagent indicates that exactly one $[Co(phen)_3]^{3+}$ and one $[Ru(NH_3)_6]^{2+}$ are required to participate in the rate-limiting step for the reaction. The rate at which two reagents can diffuse together in solution is rapid: approximately 10^{10} M^{-1} s^{-1} for neutral species and approximately 10^9 M^{-1} s^{-1} for the similarly charged reactants in eq. (1.48). The observed rate constant, 1.5×10^4 M^{-1} s^{-1}, is significantly lower than this and indicates that there is a barrier to reaction which prevents conversion to products with every encounter. More complex rate laws demand more complex mechanisms and many examples are to be found in the text.

1.8 ACTIVATION PARAMETERS

Additional information about the mechanism of the reaction is deduced indirectly from activation parameters. These are commonly derived from the postulates of transition state theory. Consider the simple reaction between $[Co(phen)_3]^{3+}$ and $[Ru(NH_3)_6]^{2+}$ which is thermodynamically favored with $\Delta G = -30.0$ kJ mol^{-1}. The reaction coordinate diagram (Fig. 1.8) describes the progress of this reaction from reactants to products, indicated by the parameter, x, as a function of the free energy requirement, G, for the system. There is an energetic barrier for the reaction and the top of this barrier represents the transition state with a composition $\{[Co(phen)_3]^{3+},[Ru(NH_3)_6]^{2+}\}^{\ddagger}$ and a lifetime of the order of a molecular vibration, 10^{-13} s. Not all of the reactant pairs will have sufficient energy to overcome this energy barrier. Energy distribution in the assembly of the reactants is determined by the Boltzmann function and the free energy of activation, $\Delta G^{\ddagger}$, may be defined in terms of eq. (1.49), the difference in free energy between the isolated reactants and the transition state. In turn, $\Delta G^{\ddagger}$ can be divided into contributions from the enthalpy of activation, $\Delta H^{\ddagger}$, and the entropy of activation, $\Delta S^{\ddagger}$ (eq. (1.50)).

$$k = \frac{RT}{Nh} \exp\left(\frac{-\Delta G^{\ddagger}}{RT}\right) \qquad (1.49)$$

$$k = \frac{RT}{Nh} \exp\left(\frac{-\Delta H^{\ddagger}}{RT}\right) \exp\left(\frac{-\Delta S^{\ddagger}}{R}\right) \qquad (1.50)$$

These activation parameters are generally evaluated from the temperature dependence of the reaction rate from a plot of $\ln(k/T)$ against T^{-1}. The enthalpy of activation

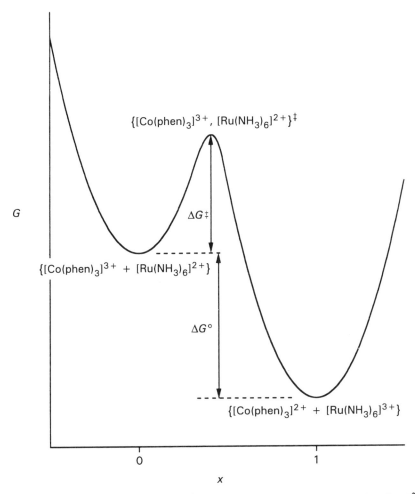

Fig. 1.8. Reaction coordinate diagram showing the progress of the reaction of $[Ru(NH_3)_6]^{2+}$ with $[Co(phen)_3]^{3+}$ from reactants to products as a function of the gree energy requirement, G, of the system.

$\Delta H^{\ddagger}$ is related to the activation energy, E_a, evaluated from the Arrhenius treatment as shown in eq. (1.51), and $\Delta S^{\ddagger}$ is related to the pre-exponential factor, A, as shown by eq. (1.52).

$$\Delta H^{\ddagger} = E_a - RT \qquad \text{(kJ mol}^{-1}) \qquad (1.51)$$

$$\Delta S^{\ddagger} = 19.15\,(\log A - 13.23) \qquad \text{(J K}^{-1}\text{ mol}^{-1}) \qquad (1.52)$$

The reaction volume ΔV° is defined in eq. (1.41). Applying this expression to $\Delta G^{\ddagger}$, the volume of activation $\Delta V^{\ddagger}$ can be defined (eq. (1.53)), and this quantity has found considerable use.

$$\left(\frac{\partial \ln k}{\partial P}\right)_T = \frac{\partial (\Delta G^{\ddagger}/RT)}{\partial P} = -\frac{\Delta V^{\ddagger}}{RT} \tag{1.53}$$

Activation volumes are generally evaluated from a plot of ln k against P. Assuming that the activation volume is independent of pressure, this corresponds to the integrated form of eq. (1.53), given by eq. (1.54). However, it frequently happens that $\Delta V^{\ddagger}$ shows a dependence on pressure and the compressibility of activation, $\Delta \beta^{\ddagger}$, is defined in eq. (1.55).

$$\ln k = \ln k^{\circ} - \frac{\Delta V^{\ddagger}}{RT} P \tag{1.54}$$

$$\Delta \beta^{\ddagger} = -\frac{d \Delta V^{\ddagger}}{dP} \tag{1.55}$$

Interpretation of the activation parameters is often difficult. In substitution reactions, it might be expected that a reaction in which a covalent bond is broken in the rate-limiting step would have a large positive value for $\Delta H^{\ddagger}$, and a large positive value for $\Delta S^{\ddagger}$ whereas a reaction in which a covalent bond is formed in the rate limiting step would have a small value for $\Delta H^{\ddagger}$, and a large negative value for $\Delta S^{\ddagger}$. If bond breaking occurs, the volume of the activated complex is larger than that of the reactants so that $\Delta V^{\ddagger}$ should be positive, whereas if bond formation occurs, $\Delta V^{\ddagger}$ should be negative. These expectations allow interpretation of the data for solvent exchange in Table 1.2 in terms of bond cleavage (dissociative) or bond formation (associative) mechanisms. However, with the exception of $\Delta V^{\ddagger}$, interpretation is difficult since changes in solvation between the reactants and the transition state can play a dominant role. Thus although there is an increase in entropy when a bond is broken, there is also increased solvation of the free ligand, which can dominate the process. Similarly when a bond is formed, the release of solvent can cause $\Delta S^{\ddagger}$ to be positive.

Complications from solvation phenomena are highlighted when there are changes in the overall charge product between the reactants and the transition state, and this is exactly the situation in reactions which involve charge transfer. An increase in the charge product on going from reactants to the transition state should require increased electrostriction of the solvent and a negative value for $\Delta S^{\ddagger}$, while a decrease in charge product should involve solvent release and a positive $\Delta S^{\ddagger}$. Such solvent effects will mask the contributions inherent from the charge transfer process itself. Even $\Delta V^{\ddagger}$ data, which have proved so useful in the interpretation of substitution rate data, are less effective in the interpretation of redox processes. Calculation of activation parameters has met with some success for the simplest process, outer-sphere electron transfer, and this is discussed in section 2.12, but much work remains to be done in this area. Fortunately the experimental data has been useful in providing empirical ranges of parameters which characterize different reaction types. Activation parameters are useful for distinguishing between reactions which may be controlled by substitution or electron transfer. Likewise, atom transfer can be distinguished from electron transfer, as is apparent in Chapter 5.

The use of reaction coordinate diagrams to describe the course of a chemical reaction tends to be less well advanced in inorganic mechanistic chemistry than in organic chemistry. The diagrams are not used extensively in this text; nevertheless they can provide some important information and it is worthwhile devoting some attention to their interpretation. The Hammond postulate [45] relates the position of the transition state to the driving force for the reaction. Thus a reaction which is very exoenergetic will have a transition state which resembles the reactants in both energy and geometry (Fig. 1.9), while a reaction which is endoenergetic will have a transition state which is more product-like. Changes which affect the thermodynamic driving force can therefore be expected to change the position of the transition state.

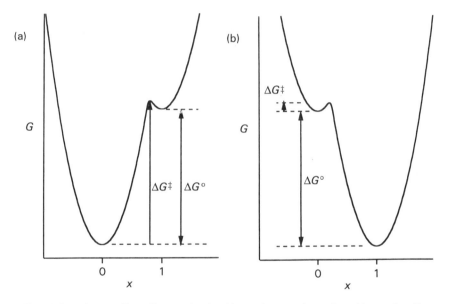

Fig. 1.9. Reaction coordinate diagram showing (a) an endoenergetic reaction with a product-like transition state, and (b) an exoenergetic reaction with a reactant-like transition state.

1.9 MEDIUM EFFECTS

Reaction rates involving charged species are dependent on the ionic composition of the reaction medium [46]. The most general effect is due to the change in activity of the reacting species with changes in the ionic composition of the reaction medium. In carrying out studies where the concentrations of reagents are varied, it is important to maintain a medium in which the ionic composition is sensibly constant. Thus changes in the concentrations of reactants must be compensated by changes in the concentrations of an inert supporting electrolyte such as $NaClO_4$, $LiClO_4$,

$LiO_3SC_6H_4CH_3$, NaO_3SCF_3, $Bu_4^nNBF_4$, or simply NaCl. The activity coefficient for an ion of charge z_A is given by the Debye–Hückel expression (eq. (1.56)), where ionic strength, μ, is defined in eq. (1.57). In this expression, $[A_i]$ is the concentration of the ith component of the solution and z_A is its charge.

$$\log \gamma_A = \frac{-0.509\, z_A^2\, \mu^{1/2}}{1 + \mu^{1/2}} \tag{1.56}$$

$$\mu = \frac{1}{2} \sum_i [A_i]\, z_A^2 \tag{1.57}$$

Ionic strength is the most widely used measure of composition but other quantities, including constant counter-ion concentration, have also been employed. For a reaction between two charged species with a second-order rate law (eq. (1.58)), the Davies modification of the Debye–Hückel expression for the dependence of the reaction rate on ionic strength (eq. (1.59)) is generally employed. Typical values for the parameters are $a = 0.59$, $b = 0.33$, and r is the distance of closest approach of the ions in ångströms, conveniently taken as 3 Å.

$$\text{Rate} = k\,[A][B] \tag{1.58}$$

$$\ln k = \ln k_0 + \frac{2 z_A z_B a \sqrt{\mu}}{1 + br\sqrt{\mu}} \tag{1.59}$$

The application of the Davies equation in the description of the ionic strength dependencies of reactions (1.60)–(1.62) is shown in Fig. 1.10 [47–49]. The fits to the data are quite satisfactory although some modifications are required to deal with higher ionic strengths (eq. (1.63)) where the parameter c is empirically determined.

$$[Co(ox)_3]^{3-} + [Fe(H_2O)_6]^{2+} \rightarrow \tag{1.60}$$

$$[Ru(bpy)(NH_3)_4]^{3+} + [Ru(phen)(NH_3)_4]^{2+} \rightarrow \tag{1.61}$$

$$[Co(NH_3)_5Cl]^{2+} + [Cu_{aq}]^+ \rightarrow \tag{1.62}$$

$$\ln k = \ln k_0 + \frac{2 z_A z_B a \sqrt{\mu}}{1 + br\sqrt{\mu}} + c\mu \tag{1.63}$$

In the reactions of many highly charged complexes the application of the Davies equation is severely limited because of specific ion effects on the reactions where the background electrolyte is insufficiently inert. The parameter c may be considered a measure of this latter interaction [50, 51]. However, the specific ion effects take a number of forms. Ion-multiplet formation is important in reactions between similarly charged ions. Thus in the reaction between $[Fe(CN)_6]^{4-}$ and $[IrCl_6]^{2-}$ (eq. (1.64)), the rate constant shows a

$$[Fe(CN)_6]^{4-} + [IrCl_6]^{2-} \rightarrow [Fe(CN)_6]^{3-} + [IrCl_6]^{3-} \tag{1.64}$$

dependence on the cation in the order $Cs^+ > Rb^+ > NH_4^+ > K^+ > Na^+ > Li^+$ [52]. Reduction of the neutral $[Cu(H_{-3}Aib_3a)]$ by $[Fe(CN)_6]^{4-}$ shows the opposite trend [53] and the effects are ascribed to ion association with the reductant. Whether the rate acceleration reflects a reduction in charge or some electronic effect has not, however, been resolved. In reactions of cationic species, specific effects of anions are detected and these are ascribed to ion pair formation and the reduction of electrostatic repulsions between the reactants [54, 55], or to electronic effects [56].

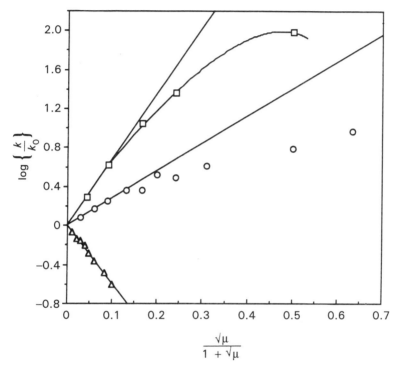

Fig. 1.10. Plots of $\log(k/k_o)$ for the reactions between $[Fe(H_2O)_6]2+$ and $[Co(ox)_3]^{3-}$ (triangles, Ref. [47]); $[Ru(bpy)(NH_3)_4]^{3+}$ and $[Ru(phen)(NH_3)_4]^{2+}$ (squares, Ref. [48]); and $[Cu_{aq}]^+$ and $[Co(NH_3)_5Cl]^{2+}$ (circles, Ref. [49]). Linear fits of eq. (1.59) with (br) set to 1.0 are shown for the data at lower ionic strength. The curved fit for the reaction between $[Ru(bpy)(NH_3)_4]^{3+}$ and $[Ru(phen)(NH_3)_4]^{2+}$ was obtained from eq. (1.63) with $(br) = 1.41$ and $c = -0.11$.

The hydrolysis of the aqua-ion reagents discussed in section 1.3 results in highly specific medium effects on changing $[H^+]$. Reduction of $[Fe(H_2O)_6]^{3+}$ by $[Cr(H_2O)_6]^{2+}$ follows the rate law in eq. (1.65) in acidic solution where $[H^+]$ is varied from 0.07 M to 1.0 M [57].

$$-\frac{d[[Fe(H_2O)_6]^{3+}]}{dt} = \left\{k_1 + \frac{k'}{[H^+]}\right\} [[Fe(H_2O)_6]^{3+}] [[Cr(H_2O)_6]^{2+}] \qquad (1.65)$$

The rate law indicates the presence of two pathways, one with a transition state of composition $\{[Fe(H_2O)_6]^{3+},[Cr(H_2O)_6]^{2+}\}^{\ddagger}$ and the other with a transition state of composition $\{H_{-1},[Fe(H_2O)_6]^{3+},[Cr(H_2O)_6]^{2+}\}^{\ddagger}$, where H_{-1} indicates that a proton has been lost from the formulation of the reactants under the experimental conditions. There is a proton ambiguity in that both $[Fe(H_2O)_6]^{3+}$ and $[Cr(H_2O)_6]^{2+}$ can lose H^+ in a hydrolysis reaction with K_a values 1.69×10^{-3} and 2×10^{-9} respectively. At 1.0 M ionic strength with $NaClO_4$ as supporting electrolyte, $k_1 = 2.3 \times 10^3$ M^{-1} s^{-1} and $k' = 5.4 \times 10^3$ s^{-1}. In the proposed mechanism (eqs (1.66)–(1.68)), the derived rate law is eq. (1.69) where $[H^+] > K_a$, and the subscript T refers to total concentration of the species, and the k' pathway is ascribed to reaction of $[Fe(H_2O)_5OH]^{2+}$ with $k_2 = 3.3 \times 10^6$ M^{-1} s^{-1}. If hydrolysis of $[Cr(H_2O)_6]^{2+}$ is considered as the source of the $[H^+]^{-1}$ dependence, the rate constant for the reaction of $[Fe(H_2O)_6]^{3+}$ with $[Cr(H_2O)_5OH]^+$ is 3×10^{12} M^{-1} s^{-1}, faster than the diffusion limit, and so this explanation can be discarded.

$$[Fe(H_2O)_6]^{3+} \underset{}{\overset{K_a}{\rightleftharpoons}} [Fe(H_2O)_5OH]^{2+} + H^+ \tag{1.66}$$

$$[Fe(H_2O)_6]^{3+} + [Cr(H_2O)_6]^{2+} \overset{k_1}{\rightarrow} [Fe(H_2O)_6]^{2+} + [Cr(H_2O)_6]^{3+} \tag{1.67}$$

$$[Fe(H_2O)5OH]^{2+} + [Cr(H_2O)_6]^{2+} \overset{k_2}{\rightarrow} [Fe(H_2O)_6]^{2+} + [Cr(H_2O)_5OH]^{2+} \tag{1.68}$$

$$-\frac{d[[Fe(H_2O)_6]^{3+}]_T}{dt} = \left\{\frac{k_1[H^+] + k_2 K_a}{[H^+] + K_a}\right\} [[Cr(H_2O)_6]^{2+}][[Fe(H_2O)_6]^{3+}]_T \tag{1.69}$$

When the reaction is examined with $LiClO_4$ as supporting electrolyte, the k_1 term is much reduced (570 M^{-1} s^{-1}) and indeed is zero within the experimental uncertainty of the experiment whereas k_2 is largely unaffected with a value of 7.3×10^3 M^{-1} s^{-1} [58]. The explanation is that there is a medium effect as a result of the imperfect substitution of the alkali metal cations for H^+ over the range of experimental conditions. There are large differences in activity coefficients ($\gamma_{\pm}$) for H^+ compared with Na^+, and these differences are smaller for H^+ compared with Li^+. For example in 1.0 M solution at 25.0°C, $\gamma_{\pm} = 0.823$ ($HClO_4$), $\gamma_{\pm} = 0.629$ ($NaClO_4$), and $\gamma_{\pm} = 0.887$ ($LiClO_4$) [59]. The use of Li^+ over Na^+ as the background electrolyte is to be preferred in experiments in acidic media, but even so, activity coefficients in mixed H^+/Li^+ media vary considerably with $[H^+]$ even at constant ionic strength. Artifacts such as these are commonly observed and are particularly important if the reaction shows a strong dependence on $[H^+]^{-1}$ and extrapolation is to a small intercept [60–62].

Besides these important cation effects, the reaction between $[Fe(H_2O)_6]^{3+}$ and $[Cr(H_2O)_6]^{2+}$ also shows a dependence on anion [57]. In the presence of Cl^-, an additional term (eq. (1.70)) is detected in the rate law indicating a transition state of composition $\{[Fe(H_2O)_6]^{3+},[Cr(H_2O)_6]^{2+},Cl^-\}^{\ddagger}$.

$$k'' [[Fe(H_2O)_6]^{3+}] [[Cr(H_2O)_6]^{2+}] [Cl^-] \quad (1.70)$$

However, the value of k'' depends on whether chloride ion is equilibrated with $[Fe(H_2O)_6]^{3+}$ before reaction, $k'' = 5.8 \times 10^7$ M^{-2} s^{-1} or not, $k'' = 2 \times 10^4$ M^{-2} s^{-1}, so that two chloride ion catalyzed pathways can be distinguished. When $[Fe(H_2O)_6]^{3+}$ is equilibrated with Cl^-, the reactant is $[Fe(H_2O)_5Cl]^{2+}$, which has a stability constant of 2.9 M^{-1}; however, the rate of substitution of Cl^- into the inner coordination sphere of $[Fe(H_2O)_6]^{3+}$ is slower than electron transfer and when equilibration is not permitted, the reactant is an outer-sphere complex, $\{[Fe(H_2O)_6]^{3+},Cl^-\}$. Again, there are many other examples of this type of behavior.

The dependence on [H$^+$] of the rate in eq. (1.65) can be ascribed to the stoichiometric change in the reactant from $[Fe(H_2O)_6]^{3+}$ to $[Fe(H_2O)_5OH]^{2+}$, as a result of the equilibrium in eq. (1.66). The change in the equilibrium is independent of the source of the hydrogen ion since the equilibrium is fully established on a timescale much shorter than the subsequent electron transfer. The rate dependence refers specifically to [H$^+$]$^{-1}$ and since there is an inverse relationship, equivalent to a rate acceleration by [OH$^-$], this is referred to as specific base catalysis of the reaction. However, acidity dependencies can arise from different sources. In the oxidation of SO_2 by H_2O_2 in aqueous solution, the experimental rate law shows a complex dependence on [H$^+$] (eq. (1.72)) [63]. In part, this dependence is due to the stoichiometric change in eq. (1.72), where HSO_3^- is identified as the active form of the reductant (eq. (1.73)), giving rise to a general base term where [HSO$_3^-$]$_T$ refers to the total

$$-\frac{d[HSO_3^-]_T}{dt} = \left\{\frac{k_1(k_2[H^+] + k_3[HA])}{k_{-1} + k_2[H^+] + k_3[HA]}\right\} \left\{\frac{K_a}{K_a = [H^+]}\right\} [H_2O_2] [HSO_3^-]_T \quad (1.71)$$

$$[H_2SO_3] \underset{}{\overset{K_a}{\rightleftharpoons}} [HSO_3^-] [H^+] \quad (1.72)$$

$$[HSO_3^-] = \frac{K_a}{K_a + [H^+]} [HSO_3^-]_T \quad (1.73)$$

concentration [HSO$_3^-$] + [H$_2$SO$_3$], under pH conditions where the latter species predominates in solution. The rate terms indicate that there are two pathways. One pathway, $k_1k_2[H^+]$, has a dependence on the specific acid, H$^+$, with an activated complex of composition,$\{H^+,H_2O_2,HSO_3^-\}^{\ddagger}$, while the other, $k_1k_3[HA]$, has a dependence on a general acid, HA, with an activated complex, $\{HA,H_2O_2,HSO_3^-\}^{\ddagger}$. When a general acid catalysis of this sort is observed in the rate law, it cannot be explained by a stoichiometric change in the position of an equilibrium, but instead indicates the active participation of HA in the rate-limiting step for the reaction.

36 Introduction [Ch. 1]

The mechanism must involve rate-limiting H^+ transfer, and is shown in eqs (1.74)–(1.76).

$$H_2O_2 + HSO_3^- \underset{k_{-1}}{\overset{k_1}{\rightleftharpoons}} HOOSO_2^- + H_2O \qquad (1.74)$$

$$HOOSO_2^- + H^+ \overset{k_2}{\rightarrow} H^+ + HSO_4^- \qquad (1.75)$$

$$HOOSO_2^- + HA \overset{k_3}{\rightarrow} HA + HSO_4^- \qquad (1.76)$$

To avoid a termolecular process, an adduct, $HOOSO_2^-$, is proposed and decomposition involves H^+ transfer. Normally, H^+ transfer is very rapid [64], but in this instance, one of the reactants in the process is a steady-state species available in only very low concentrations.

Medium effects are not restricted to ionic strength and specific ion interactions. Changes in solvent also have a very important effect on both the thermodynamics and the kinetics of reactions and these are discussed primarily in Chapter 2.

1.10 ISOTOPE EFFECTS

As noted in section 1.1, some redox reactions are accompanied by changes in coordination and others involve a formal atom transfer. Thus while bond cleavage and bond formation need not be a requirement for simple electron transfer processes, in some instances such processes may be involved. Primary isotope effects provide an important mechanistic probe of this involvement.

Consider a reaction in which an O—H bond is broken. The reaction profile is shown in Fig. 1.11. In this case, the ordinate, x, is a measure of the stretching of the O—H bond which is considered to be fully cleaved in the transition state. When the corresponding reactant containing an O—D bond is used, a primary isotope effect arises because the zero-point energies of the two bonds differ. In a simplified analysis, if the bonds are approximated by harmonic oscillators, the ratio of the stretching frequencies, v_H and v_D, is given by eq. (1.77), where μ_H and μ_D are the reduced masses of the systems containing hydrogen and deuterium respectively [64].

$$\frac{v_H}{v_D} = \left(\frac{\mu_D}{\mu_H}\right)^{1/2} \qquad (1.77)$$

For hydrogen and deuterium, this ratio, v_H/v_D, will approach $2^{1/2}$, but for heavier atoms, isotope effects are much smaller and consequently are more difficult to detect. Thus isotope effects are of much less importance in the study of Cl or even O atom transfer.

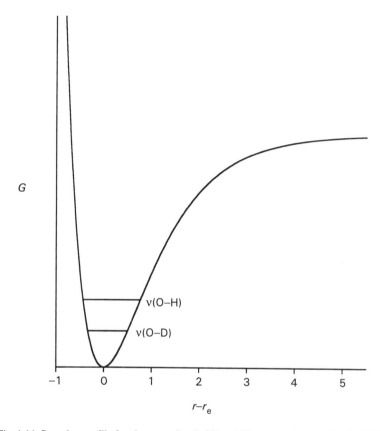

Fig. 1.11. Reaction profile for cleavage of an O–H bond. The zero point energies for O–H and O–D which are proportional to $(1/\mu_H)^{1/2}$ and $(1/\mu_D)^{1/2}$ respectively are shown.

The effect on the activation energy can be calculated from the expression for the zero-point energy, eq. (1.78), where h is Planck's constant and c is the speed of light. For an O—H bond with $\nu_H = 3300$ cm^{-1} or 1.00×10^{14} s^{-1}, $E_H = 2.00 \times 10^4$ J mol^{-1}, and with $\nu_D = 2^{1/2}\, \nu_H$, $E_D = 1.41 \times 10^4$ J mol^{-1}. The value for k_H/k_D calculated from eq. (1.79) at 298 K is 10.8.

$$E_H - E_D = -\tfrac{1}{2} hc\, (\nu_H - \nu_D) \tag{1.78}$$

$$\frac{k_H}{k_D} \approx \exp\left(-\frac{E_H - E_D}{RT}\right) \tag{1.79}$$

This value should represent a maximum in the isotope effect for cleavage of an O—H bond. When incomplete bond cleavage is involved, the isotope effect will be reduced. Similarly if partial bond formation is also a component of the activation process, smaller isotope effects can be expected. In practice much larger effects can be detected since this semi-classical approach does not take into account the ability of the atom to tunnel below the energy barrier.

Nuclear tunneling is a quantum mechanical effect and is a consequence of the wave nature of particles [65]. The ability of a particle to tunnel through a barrier depends on whether the wavelength of the particle, λ, given by eq. (1.80), where m is the mass of the particle and E_k is its kinetic energy, is comparable to the width of the activation barrier.

$$\lambda = \frac{h}{2mE_k} \qquad (1.80)$$

The tunneling correction required is inversely proportional to the the square root of the mass of the particle and will therefore be much larger for H than D. Tunneling will be less important for heavier atoms but will be an important mechanism for electronic transmission.

QUESTIONS

1.1 Correlate the solvent exchange rate data in Table 1.2 with electronic configuration.

1.2 The ligand electrochemical series (Table 1.4) shows some parallels with the spectrochemical series (eq. (1.7)), but does not correlate exactly. Amplify this statement and provide a rationalization for the observation.
Estimate reduction potentials for the following complexes in aqueous solution:

$[Ru(bpy)_2(en)]^{3+}$, $[Ru(bpy)(en)_2]^{3+}$, $[Cr(NH_3)_5I]^{2+}$, $[Cr(terpy)(ox)I]$.

1.3 Use the reduction potential and acidity constant data in section 5.8 to determine the apparent reduction potential for reduction of O_2 by one, two and four electrons at pH 7.0.

1.4 The following data refers to the reduction of the formal nickel(IV) complex, $[Ni(Me_2L)]^{2+}$, to $[Ni(Me_2LH_2)]^{2+}$ which occurs in two single-electron steps. Further protonation of any of the complexes cannot be detected above pH 0.
Calculate the pH at which the one-electron reductions of nickel(IV) and nickel(III) have the same potential.

$[Ni(Me_2L)]^{2+} + e^- \rightleftharpoons [Ni(Me_2L)]^+ \qquad E° = 0.65$ V

$[Ni(Me_2L)]^+ + e^- \rightleftharpoons [Ni(Me_2L)] \qquad E° = 0.42$ V

$$pK_{a1} = 5.8$$

$[Ni(Me_2LH_2)]^{2+} \rightleftharpoons [Ni(Me_2LH)]^+ + H^+$

$$pK_{a2} = 7.7$$

$[Ni(Me_2LH)]^+ \rightleftharpoons [Ni(Me_2L)] + H^+$

$$pK_{a3} = 4.1$$
$$[Ni(Me_2LH)]^{2+} \rightleftharpoons [Ni(Me_2L)]^+ + H^+$$

(Lappin, A. G.; Laranjeira, M. C. M. *J. Chem. Soc., Dalton Trans.* 1982, 1861–1865.)

1.5 The binding constants for the binding of picolinate, pic⁻, with $[Fe(H_2O)_6]^{3+}$ are $K_1 = 1.1 \times 10^6$ M⁻¹, $K_2 = 7.2 \times 10^6$ M⁻¹ and $K_3 = 2.3 \times 10^4$ M⁻¹, while those for $[Fe(H_2O)_6]^{2+}$ are $K_1 = 7.9 \times 10^4$ M⁻¹, $K_2 = 1.3 \times 10^4$ M⁻¹ and $K_3 = 2.0 \times 10^3$ M⁻¹. Given that the reduction potential for $[Fe(H_2O)_6]^{3+}$ is 0.77 V, calculate reduction potentials for $[Fe(H_2O)_4(pic)]^{2+}$, $[Fe(H_2O)_2(pic)_2]^+$ and $[Fe(pic)_3]$.

1.6 Second-order rate constants for the reaction of $[CoW_{12}O_{40}]^{5-}$ with $[Fe(CN)_6]^{4-}$ at 25 °C and 0.10 M [H⁺] are presented in the table as a function of [K⁺] and µ.

[K⁺] (M)	µ (M)	k (M⁻¹ s⁻¹)
0.058	0.165	1.4×10^4
0.108	0.215	2.4×10^4
0.208	0.315	4.8×10^4
0.358	0.465	8.3×10^4
0.508	0.615	1.16×10^5

Treat this data according to eq. (1.59), and discuss the mechanistic implications of any deviations from this idealized behavior.
(Das-Sharma, M; Gangopadhyay, S.; Ali, M.; Banergee, P. *J. Chem. Res. (S)* 1993, 122–123.)

1.7 From the following data, calculate $\Delta H^\ddagger$, $\Delta S^\ddagger$, and $\Delta V^\ddagger$ for the reaction:

$$[Co(sep)]^{3+} + [*Co(sep)]^{2+} \xrightarrow{k} [Co(sep)]^{2+} + [*Co(sep)]^{3+}$$

Second-order rate constants at 0.20 M ionic strength

T (°C)	P (MPa)	k (kg mol⁻¹ s⁻¹)	T (°C)	P (MPa)	k (kg mol⁻¹ s⁻¹)
11.8	0.1	2.39	17.8	0.1	3.11
20.0	0.1	3.26	25.0	0.1	5.08
34.0	0.1	8.71	25.0	27.0	5.37
25.0	47.2	5.86	25.0	52.7	5.73
25.0	77.2	6.26	25.0	104.0	6.51
25.0	129.8	7.11	25.0	152.7	7.50
25.0	178.1	8.20	25.0	203.2	8.52

(Doine, H.; Swaddle, T. W. *Inorg. Chem.* 1991, 30, 1858–1862.)

REFERENCES

[1] Cannon, R. D. *Electron Transfer Reactions,* Butterworth, 1980.
[2] Zuckerman, J. J. (Ed.) *Inorganic Reactions and Methods* Vol 15, VCH, 1986.
[3] Taube, H. *Electron Transfer Reactions of Complex Ions in Solution,* Academic Press, 1970.
[4] Lippard, S. J. *Progress in Inorganic Chemistry* Vol 30, Wiley Interscience, 1983.
[5] Hay, R. W. *Inorganic Mechanisms,* Ellis Horwood — PTR Prentice Hall, 1992.
[6] Cotton, F. A.; Wilkinson, G. *Advanced Inorganic Chemistry,* 5th Edition, John Wiley, New York, 1988.
[7] Ducommun, Y.; Zbinden, D.; Merbach, A. E. *Helv. Chim. Acta* 1982, **65,** 1385–1396.
[8] Bechtler, A.; Breitschwerdt, K. G.; Tamm, K. *J. Chem. Phys.* 1970, **52,** 2975–2982.
[9] Ducommun, Y.; Newman, K. E.; Merbach, A. E. *Inorg. Chem.* 1980, **19,** 3696–3703.
[10] Bechtold, D. B.; Liu, G.; Dodgen, H. W.; Hunt, J. P. *J. Phys. Chem.* 1978, **82,** 333–337.
[11] Rapaport, I.; Helm, L.; Merbach, A. E.; Bernhard, P.; Ludi, A. *Inorg. Chem.* 1988, **27,** 873–879.
[12] Hugi, A. D.; Helm, L.; Merbach, A. E. *Inorg. Chem.* 1987, **26,** 1763–1768.
[13] Hugi, A. D.; Helm, L.; Merbach, A. E. *Helv. Chim. Acta* 1985, **68,** 508–521.
[14] Xu, F.-C.; Krouse, H. R.; Swaddle, T. W. *Inorg. Chem.* 1985, **24,** 267–270.
[15] Grant, M.; Jordan, R. B. *Inorg. Chem.* 1981, **20,** 55–60.
[16] Laurenczy, G.; Rapaport, I.; Zbinden, D.; Merbach, A. E. *Magn. Reson. Chem.* 1991, **29,** S45–S51.
[17] Conocchioli, T. J.; Nancollas, G. H.; Sutin, N. *Inorg. Chem.* 1965, **5,** 1–5.
[18] Hunt, H. R.; Taube, H. *J. Am. Chem. Soc.* 1958, **80,** 2642–2646.
[19] Swaddle, T. W.; Stranks, D. R. *J. Am. Chem. Soc.* 1972, **94,** 8357–8360.
[20] Doine, H.; Ishihara, K.; Krouse, H. R.; Swaddle, T. W. *Inorg. Chem.* 1987, **26,** 3240–3242.
[21] Tong, S. B.; Swaddle, T. W. *Inorg. Chem.* 1974, **13,** 1538–1539.
[22] Basolo, F.; Pearson, R. G. *Mechanisms of Inorganic Reactions,* 2nd Edition, Wiley, 1967.
[23] Espenson, J. H.; Woleneuk, S. G. *Inorg. Chem.* 1972, **11,** 2034–2341.
[24] Toma, H. E.; Malin, J. M.; Giesbrecht, E. *Inorg. Chem.* 1973, **12,** 2080–2083.
[25] Johnson, D. A. *Some Thermodynamic Aspects of Inorganic Chemistry,* 2nd Edition, Cambridge University Press, 1981.
[26] Toma, H. E.; Batista, A. A.; Gray, H. B. *J. Am. Chem. Soc.* 1982, **104,** 7509–7515.
[27] Araki, K. Shu, C.-F.; Anson, F. C. *Inorg. Chem.* 1991, **30,** 3043–3047.
[28] Martell, A. E.; Smith, R. M. *Critical Stability Constants* Vol 4, Plenum, 1976.
[29] Moyer, B. A.; Meyer, T. J. *Inorg. Chem.* 1981, **20,** 436–444.
[30] Blaho, J. K.; Goldsby, K. A. *J. Am. Chem. Soc.* 1990, **112,** 6132–6133.
[31] Mohanty, J. G.; Chakravorty, A. *Inorg. Chem.* 1976, **15,** 2912–2916.
[32] Baral, S.; Lappin, A. G. *J. Chem. Soc., Dalton Trans.* 1985, 2213–2215.
[33] Lever, A. B. P. *Inorg. Chem.* 1990, **29,** 1271–1285.

[34] Buxton, G. V.; Sellers, R. M. *Coord. Chem. Rev.* 1977, **22**, 195–274.
[35] Buxton, G. V.; Sellers, R. M. *Compilation of Rate Constants for Reactions of Metal Ions in Unusual Valency States,* NSRDS, 1978.
[36] Fabbrizzi, L.; Cohen, H.; Meyerstein, D. *J. Chem. Soc., Dalton Trans.* 1983, 2125–2126.
[37] Yee, E. L.; Cave, R. L.; Guyer, K. L.; Tyma, P. D.; Weaver, M. J. *J. Am. Chem. Soc.* 1979, **101**, 1131–1137.
[38] Ogino, H.; Ogino, K. *Inorg. Chem.* 1983, **22**, 2208–2211.
[39] Youngblood, M. P.; Margerum, D. W. *Inorg. Chem.* 1980, **19**, 3068–3072.
[40] Taniguchi, V. T.; Sailasuta-Scott, N.; Anson, F. C.; Gray, H. B. *Pure Appl. Chem.* 1980, **52**, 2275–2281.
[41] Bansch, B.; Martinez, P.; van Eldik, R. *J. Phys. Chem.* 1992, **96**, 234–238.
[42] Macartney, D. H.; McAuley, A. *Inorg. Chem.* 1983, **22**, 2062–2066.
[43] Espenson, J. H., *Chemical Kinetics and Reaction Mechanisms,* McGraw-Hill, 1981.
[44] Moore, J. W.; Pearson, R. G. *Kinetics and Mechanism,* Wiley-Interscience, 1981.
[45] Hammond, G. S. *J. Am. Chem. Soc.* 1955, **77**, 334–338.
[46] Pethybridge, A. D.; Prue, J. E. *Progress Inorg. Chem.* 1972, **17**, 327–390.
[47] Barrett, J.; Baxendale, J. H. *Trans. Faraday Soc.* 1956, **52**, 210–217.
[48] Brown, G. M.; Sutin, N. *J. Am. Chem. Soc.* 1979, **101**, 883–892.
[49] Parker, O. J.; Espenson, J. H. *J. Am. Chem. Soc.* 1969, **91**, 1968–1974.
[50] Rosseinsky, D. R. *Comments Inorg. Chem.* 1984, **3**, 153–170.
[51] Rosseinsky, D. R.; Stead, K.; Coston, T. P. J.; Glidle, A. *J. Chem. Soc., Chem. Commun.* 1986, 70–71.
[52] Bruhn, H.; Nigam, S.; Holzwarth, J. F. *Farad. Discuss. Chem. Soc.* 1982, **74**, 129–140.
[53] Anast, J. M.; Margerum, D. W. *Inorg. Chem.* 1982, **21**, 3494–3501.
[54] Przystas, T. J.; Sutin, N. *J. Am. Chem. Soc.* 1973, **95**, 5545–5555.
[55] Warren, R. M. L.; Lappin, A. G.; Mehta, B. D.; Neumann, H. M. *Inorg. Chem.* 1990, **29**, 4185–4189.
[56] Endicott, J. F.; Ramasami, T. *J. Am. Chem. Soc.* 1982, **104**, 5252–5254.
[57] Dulz, G.; Sutin, N. *J. Am. Chem. Soc.* 1964, **86**, 829–832.
[58] Carlyle, D. W.; Espenson, J. H. *J. Am. Chem. Soc.* 1969, **91**, 599–606.
[59] Robinson, R. A.; Stokes, R. H. *Electrolyte Solutions,* Butterworth, 1955.
[60] Birk, J. P.; Espenson, J. H. *Inorg. Chem.* 1968, **7**, 991–998.
[61] Pennington, D. E.; Haim, A. *Inorg. Chem.* 1967, **6**, 2138–2146.
[62] Davies, G. *Inorg. Chem.* 1971, **10**, 1155–1159.
[63] McArdle, J. V.; Hoffmann, M. R. *J. Phys. Chem.* 1983, **87**, 5425–5429.
[64] Bell, R. P. *The Proton in Chemistry,* Chapman and Hall, 1973.
[65] Bell, R. P. *The Tunnel Effect in Chemistry,* Chapman and Hall, 1980.

2

The outer-sphere mechanism

2.1 INTRODUCTION

The second-order rate constant for reduction of $[Co(phen)_3]^{3+}$ by $[Ru(NH_3)_6]^{2+}$ is 1.5×10^4 M^{-1} s^{-1} at 1.0 M ionic strength and 25.0°C [1], more rapid than the normal rates of substitution of ligands into the inner-coordination spheres of either reagent. Consequently, it is deduced that electron transfer takes place with the inner-coordination spheres intact, though not necessarily undisturbed. This statement provides a working definition of the outer-sphere mechanism. No bond making or bond breaking is required, and the simplicity of the mechanism, where a single electron is transferred from one metal center to another, makes this process highly amenable to the application of theory. Indeed, the rapid development of the field is due in no small part to the strong interplay between the predictions of theory and the results of experiment [2–5].

The reaction can be broken down into a sequence of elementary steps. For effective interaction, the two reactants must be in close proximity and the first step is the diffusion together of the oxidant and reductant to form an assembly which is a precursor to electron transfer (eq. (2.1)). This is followed by the electron transfer itself (eq. (2.2)), and the subsequent, generally rapid, dissociation of the successor complex (eq. (2.3)). In this instance, as in most of the cases which will be examined, the electron transfer step is rate limiting, but very fast electron transfer reactions, approaching the diffusion limit, have also been noted.

$$[Co(phen)_3]^{3+} + [Ru(NH_3)_6]^{2+} \underset{k_{-o}}{\overset{k_o}{\rightleftharpoons}} \{[Co(phen)_3]^{3+}, [Ru(NH_3)_6]^{2+}\} \quad (2.1)$$

$$\{[Co(phen)_3]^{3+}, [Ru(NH_3)_6]^{2+}\} \overset{k_{et}}{\rightarrow} \{[Co(phen)_3]^{2+}, [Ru(NH_3)_6]^{3+}\} \quad (2.2)$$

$$\{[Co(phen)_3]^{2+}, [Ru(NH_3)_6]^{3+}\} \underset{k_o'}{\overset{k_o'}{\rightleftharpoons}} [Co(phen)_3]^{2+} + [Ru(NH_3)_6]^{3+} \quad (2.3)$$

Application of the steady-state approximation to this mechanism gives eq. (2.4), which, far from the diffusion limit ($k_{et} \ll k_{-o}$) reduces to eq. (2.5) where $K_o (= k_o/k_{-o})$ is an equilibrium constant describing the formation of the precursor assembly. Such simple, second-order rate expressions are characteristic of outer-sphere electron transfer processes and as a result, most mechanistic information is obtained indirectly and not from the rate law.

$$\text{Rate} = \frac{k_o k_{et}}{k_{-o} + k_{et}} [[Co(phen)_3]^{3+}][[Ru(NH_3)_6]^{2+}] \quad (2.4)$$

$$\text{Rate} = K_o k_{et} [[Co(phen)_3]^{3+}][[Ru(NH_3)_6]^{2+}] \quad (2.5)$$

Rate constants and activation parameters for a selection of outer-sphere reactions involving metal ion complexes are presented in Table 2.1. The rates included in the Table vary from 4×10^9 M^{-1} s^{-1}, near to the diffusion limit, to 1×10^{-3} M^{-1} s^{-1}, a substantial range. Elucidation of the factors which govern this range is the subject of this chapter. A further observation is that the activation parameters in most instances show a modest enthalpy of activation and a strongly negative value for the entropy of activation. In a few instances where $\Delta H^{\ddagger}$ is larger, $\Delta S^{\ddagger}$ appears to be correspondingly less negative. Detailed interpretation of the enthalpy and entropy values is difficult because they are composite quantities, reflecting both the precursor formation and electron transfer steps. The problems of charge neutralization in the former process are considerable and reduce the ease of interpretation of activation parameters. The intuitively simpler $\Delta V^{\ddagger}$ is also dominated by charge neutralization and solvation changes.

Table 2.1. Rate constants for selected outer-sphere electron transfer reactions

Oxidant	Reductant	μ (M)	k (M^{-1} s^{-1})	$\Delta H^{\ddagger}$ (kJ mol^{-1})	$\Delta S^{\ddagger}$ (J K^{-1} mol^{-1})	$\Delta V^{\ddagger}$ (cm^3 mol^{-1})	Ref.
[Co(bpy)$_3$]$^{3+}$	[Cr(phen)$_3$]$^{2+}$	0.15	2.0×10^8				6
[Ru(bpy)$_3$]$^{3+}$	[Ru(NH$_3$)$_6$]$^{2+}$	1.0	3.7×10^9				7
[Co(phen)$_3$]$^{3+}$	[Co(terpy)$_2$]$^{2+}$	0.5	4.2×10^2	27.6	−100		8
[Co(phen)$_3$]$^{3+}$	[Ru(NH$_3$)$_6$]$^{2+}$	1.0	1.5×10^4	18	−105		1
[Co(bpy)$_3$]$^{3+}$	[Co(terpy)$_2$]$^{2+}$	0.01	3.0×10^1	21	−155	−9	9
[Co(phen)$_3$]$^{3+}$	[Ru(NH$_3$)$_5$py]$^{2+}$	0.5	1.9×10^3	21	−100		10
[Co(NH$_3$)$_6$]$^{3+}$	[Ru(NH$_3$)$_6$]$^{2+}$	0.2	1.1×10^{-2}	56	−1		11,12
[Fe(H$_2$O)$_6$]$^{3+}$	[Ru(NH$_3$)$_5$py]$^{2+}$	1.0	7.8×10^4	20	−84		13
[Ru(NH$_3$)$_5$py]$^{3+}$	[V(H$_2$O)$_6$]$^{2+}$	1.0	3.0×10^5	0	−138		1
[Co(phen)$_3$]$^{3+}$	[V(H$_2$O)$_6$]$^{2+}$	1.0	4.0×10^3	16	−121		1
[Co(terpy)$_2$]$^{3+}$	[V(H$_2$O)$_6$]$^{2+}$	1.0	3.8×10^3	7.9	−150	−1.8	14
[Co(NH$_3$)$_6$]$^{3+}$	[Cr(H$_2$O)$_6$]$^{2+}$	1.0	1.0×10^{-3}				15
[Ni(bpy)$_3$]$^{3+}$	[Fe(H$_2$O)$_6$]$^{2+}$	1.0	6.7×10^6	7	−92		16
[Co(en)$_3$]$^{3+}$	[V(pic)$_3$]$^{-}$	0.1	3.1×10^3	52	−4		17

(continues)

Table 2.1. (continued)

Oxidant	Reductant	μ (M)	k ($M^{-1} s^{-1}$)	$\Delta H^{\ddagger}$ (kJ mol^{-1})	$\Delta S^{\ddagger}$ (J K^{-1} mol^{-1})	$\Delta V^{\ddagger}$ (cm^3 mol^{-1})	Ref.
[Fe(bpy)$_3$]$^{3+}$	[Co(edta)]$^{2-}$	0.5	3.3×10^4	29	−63		18
[Co(edta)]$^{-}$	[Fe(pdta)]$^{2-}$	0.5	1.3×10^1	30	−128		19
[IrCl$_6$]$^{2-}$	[Ru(CN)$_6$]$^{4-}$	0.1	6.6×10^4	19	−88		20
[Co(ox)$_3$]$^{3-}$	[Ru(NH$_3$)$_6$]$^{2+}$	0.2	1.8×10^{-1}	45	−108		21
[Fe(CN)$_6$]$^{3-}$	[Co(phen)$_3$]$^{2+}$	0.1	6.0×10^6				22

2.2 THE ELECTRON TRANSFER PRECURSOR

The rate and activation parameters for outer-sphere reactions are complex quantities which represent contributions from both precursor formation and electron transfer steps. The reactions in Table 2.1 include examples with both favorable and unfavorable electrostatic interactions between the reactants, and this is reflected mainly in the precursor association term. In order to understand the electron transfer processes more clearly, it is of some importance to separate K_o from k_{et}, and a great deal of

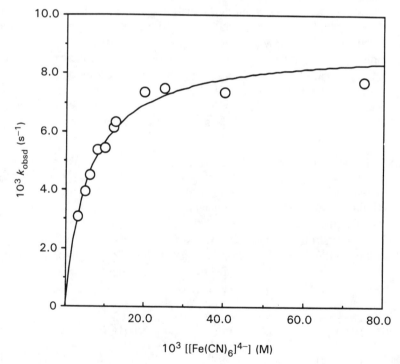

Fig. 2.1. Plot of pseudo-first-order rate constant k_{obsd} against [[Fe(CN)$_6$]$^{4-}$] for the reaction of [Fe(CN)$_6$]$^{4-}$ with [Co(NH$_3$0$_5$py)]$^{3+}$ at 25°C, μ = 1.0 M, from Ref. [32].

effort has gone into designing experiments to measure directly the rate of intramolecular electron transfer within the precursor assembly.

In suitable instances, where electron transfer is slow and the electrostatic and other interactions between the reactants are strongly favorable, it is possible to detect stoichiometric formation of ion pairs through the deviations from second-order behavior which this causes in the rate law. The reduction of $[Co(NH_3)_5py]^{3+}$ by $[Fe(CN)_6]^{4-}$ has been studied [23] under pseudo-first-order conditions with an excess of reaction, and a plot of the pseudo-first-order rate constant against $[[Fe(CN)_6]^{4-}]$ is shown in Fig. 2.1.

The dependence is described by a rate law of the form in eq. (2.6), where the subscript T refers to the total concentration of the oxidant, $K_o = 2400$ M^{-1} and $k_{et} = 1.5 \times 10^{-2}$ s^{-1}. The derived mechanism, eqs (2.8)–(2.9), is consistent with the rate law; the dominant form of the oxidant in solution changes from $[Co(NH_3)_5py]^{3+}$ to the rapidly formed ion pair $\{[Co(NH_3)_5py]^{3+}, [Fe(CN)_6]^{4-}\}$ as the concentration of $[Fe(CN)_6]^{4-}$ is increased.

$$-d[[Co(NH_3)_5py]^{3+}]_T/dt = k_{obsd}[[Co(NH_3)_5py]^{3+}]_T \tag{2.6}$$

$$k_{obsd} = \frac{K_o k_{et} [[Fe(CN)_6]^{4-}]}{1 + K_o[[Fe(CN)_6]^{4-}]} \tag{2.7}$$

$$[Co(NH_3)_5py]^{3+} + [Fe(CN)_6]^{4-} \overset{K_o}{\rightleftharpoons} \{[Co(NH_3)_5py]^{3+}, [Fe(CN)_6]^{4-}\} \tag{2.8}$$

$$\{[Co(NH_3)_5py]^{3+}, [Fe(CN)_6]^{4-}\} \overset{k_{et}}{\rightarrow} \{[Co(NH_3)_5py]^{2+}, [Fe(CN)_6]^{3-}\} \tag{2.9}$$

In eqs (2.8) and (2.9), the ion pair is considered to have the characteristics of the electron transfer precursor so that electron transfer occurs in this assembly. It must be pointed out that an alternative, kinetically indistinguishable mechanism is also possible. If the arrangement of the reactants in the ion pair is not conducive to electron transfer, then electron transfer will take place by a competing interaction (eq. (2.10)). The resulting rate law has the same form as eq. (2.7), but the limiting rate is no longer the rate of electron transfer within the precursor assembly and k_{obsd} is given by eq. (2.11). This is the so-called 'dead-end' mechanism where the observed adduct is not an intermediate in the reaction.

$$[Co(NH_3)_5py]^{3+} + [Fe(CN)_6]^{4-} \overset{k}{\rightarrow} [Co(NH_3)_5py]^{2+} + [Fe(CN)_6]^{3-} \tag{2.10}$$

$$k_{obsd} = \frac{k[[Fe(CN)_6]^{4-}]}{1 + K_o[[Fe(CN)_6]^{4-}]} \tag{2.11}$$

46 The outer-sphere mechanism [Ch. 2]

In fact the interactions between the ions form an assembly of ion pairs where the intimate interactions differ. Not all structures will lead to electron transfer. If the majority of the structures do lead to electron transfer then the ion pair resembles the precursor and is an intermediate in the course of the reaction, but if they do not then the ion pair is a 'dead-end' complex as shown in Scheme 2.1. Unless there is other evidence to the contrary, the simpler mechanism (eqs (2.8)–(2.9)), is generally assumed to hold.

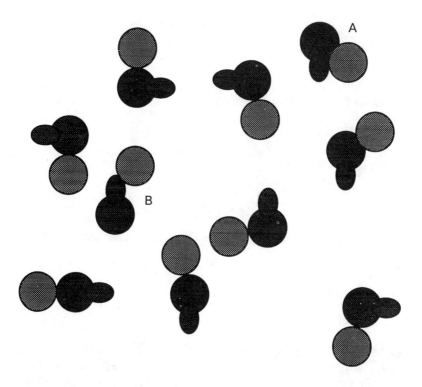

Scheme 2.1 An assembly of ion pairs such as might be found for the reaction between $[Ru(NH_3)_5py]^{3+}$ (darker shapes) with $[Fe(CN)_6]^{4-}$ (lighter shapes). If contact with the pyridine in $[Ru(NH_3)_5py]^{3+}$ is required for reaction, then only A and B represent reactant pairs and the ion pair assembly is best considered as representing a 'dead-end' complex. On the other hand if contact with the NH_3 is required, then only structure B is unreactive and the assembly can be considered a reaction intermediate.

Data for a number of reactions in which limiting first-order behavior is observed are presented in Table 2.2. In some cases, spectrophotometric evidence for the intermediates has also been obtained. This can be of some value in understanding the energetics of electron transfer, and further details and discussion of this topic can be found in section 4.2.

Table 2.2. Ion association and electron transfer rate constants at 25.0°C

Oxidant	Reductant	μ (M)	K_o (M^{-1})	k_{et} (s^{-1})	Ref.
$[Co(NH_3)_5py]^{3+}$	$[Fe(CN)_6]^{4-}$	0.1	2400	1.5×10^{-2}	23
$[Co(NH_3)_5(OH_2)]^{3+}$	$[Fe(CN)_6]^{4-}$	0.1	1500	1.9×10^{-1}	24
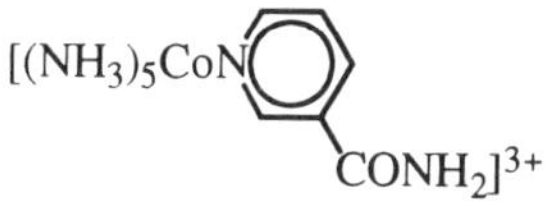 $[(NH_3)_5CoN\text{-}C_6H_4\text{-}CONH_2]^{3+}$	$[Fe(CN)_6]^{4-}$	0.1	2500	5.0×10^{-2}	25
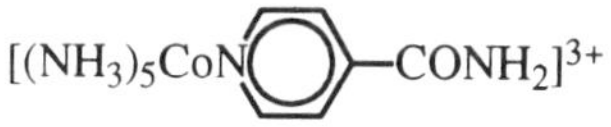 $[(NH_3)_5CoN\text{-}C_6H_4\text{-}CONH_2]^{3+}$	$[Fe(CN)_6]^{4-}$	0.1	5300	3.6×10^{-2}	25
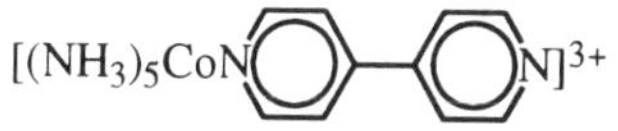 $[(NH_3)_5CoN\text{-}C_6H_4\text{-}C_5H_4N]^{3+}$	$[Fe(CN)_6]^{4-}$	0.1	2300	2.4×10^{-2}	25
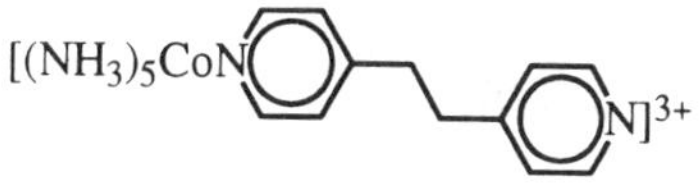 $[(NH_3)_5CoN\text{-}C_6H_4\text{-}CH_2\text{-}C_5H_4N]^{3+}$	$[Fe(CN)_6]^{4-}$	0.1	2800	1.0×10^{-2}	25
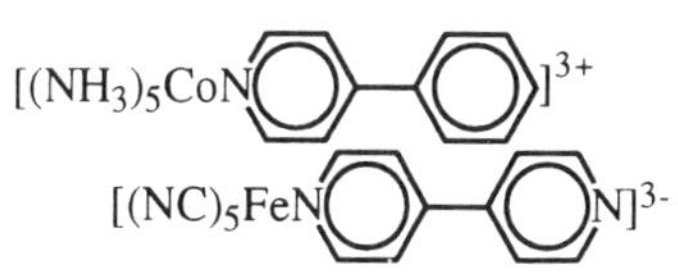 $[(NH_3)_5CoN\text{-}C_6H_4\text{-}C_6H_4N]^{3+}$ / $[(NC)_5FeN\text{-}C_6H_4\text{-}C_6H_4N]^{3-}$		0.077	1100	3.1×10^{-3}	26
$[Co(NH_3)_5O_2CCH_3]^{2+}$	$[Fe(CN)_6]^{4-}$	0.1	300	3.7×10^{-4}	27

(continues)

48 The outer-sphere mechanism [Ch. 2

Table 2.2. (continued)

Oxidant	Reductant	μ (M)	K_o (M^{-1})	k_{et} (s^{-1})	Ref.
	[Co(ox)$_3$]$^{3-}$	0.017	65	2.4×10^{-1}	28
[(NH$_3$)$_3$Co(OH)(OH)Co(NH$_3$)$_3$]$^{3+}$ (with pyridine-carboxylate bridge)					
		0.1	74	3.7×10^{-3}	29
	[Fe(dipic)$_2$]$^{2-}$				

In some cases it has also been possible to separate the thermodynamic parameters for the ion association from the activation parameters for electron transfer (Table 2.3), and these prove to be revealing. With one exception, the activation entropy for the intramolecular electron transfer process is positive. The large negative values which are observed in bimolecular reactions appear to be the result of the ion association. Formation of ion pairs between oppositely charged ions results in charge neutralization and the release of electrostricted solvent with an accompanying positive $\Delta S°$. For ion association between similarly charged ions, an increase in solvation can be expected and $\Delta S°$ is negative. The enthalpy values which correspond to these changes are expected to be small and positive for a decrease in solvation and negative for an increase in solvation [30]. However, specific interactions such as hydrogen bonding and stacking between the ions must also be taken into account. There is considerable disagreement about experimental values of $\Delta V°$ which are difficult to determine. Electrostricted solvent has a smaller molar volume than bulk solvent and so $\Delta V°$ should be negative for association between similarly charged ions and positive for oppositely charged ions [31]. Relatively small values for this quantity are interpreted to mean that there is little solvent release and hence little charge neutralization in the formation of the ion-pairs. Fortunately values for $\Delta V^{\ddagger}$ are in better agreement and represent a decrease in electrostriction with the transfer of charge since the successor complexes have a lower charge product that the precursor. These activation parameters are considered in section 2.12 but clearly, the problems of charge neutralization are considerable and reduce the effectiveness of the interpretation.

Table 2.3. Thermodynamic parameters for formation of precursor ion pairs at 25.0 °C

Reaction Pair		μ (M)	K_o (M^{-1})	$\Delta H°$ (kJ mol^{-1})	$\Delta S°$ (J K^{-1} mol^{-1})	$\Delta V°$ (cm^3 mol^{-1})	Ref.
[Co(NH$_3$)$_5$(OH$_2$)]$^{3+}$	[Fe(CN)$_6$]$^{4-}$	0.5	480			-15	32–34
[Co(NH$_3$)$_5$(OH$_2$)]$^{3+}$	[Fe(CN)$_6$]$^{4-}$	0.5	194			3.5	35
[Co(NH$_3$)$_5$ py]$^{3+}$	[Fe(CN)$_6$]$^{4-}$	1.0	168			23.4	32,36
[Co(NH$_3$)$_5$(OSMe$_2$)]$^{3+}$	[Fe(CN)$_6$]$^{4-}$	1.0	34	-8	-240	-11	32
[Co(NH$_3$)$_5$N$_3$]$^{2+}$	[Fe(CN)$_6$]$^{4-}$	1.0	49	2	-204	-16.5	37

(continues)

Table 2.3. (continued)

Reaction Pair		μ (M)	K_o (M^{-1})	$\Delta H°$ (kJ mol^{-1})	$\Delta S°$ (J K^{-1} mol^{-1})	$\Delta V°$ (cm^3 mol^{-1})	Ref.
[Co(NH$_3$)$_5$Cl]$^{2+}$	[Fe(CN)$_6$]$^{4-}$	1.0	38	28	−120	−3	37
[Co(phen)$_3$]$^{2+}$	[Co(ox)$_3$]$^{3-}$	0.017	65	−20	−33		28
[(NH$_3$)$_3$Co(OH)(OH)(O-py-O)Co(NH$_3$)$_3$]$^{3+}$	[Fe(dipic)$_2$]$^{2-}$	0.1	74	−38	−92		29

Table 2.4. Activation parameters for electron transfer within precursor ion pairs at 25.0°C

Reaction Pair		μ (M)	k_{et} (s^{-1})	$\Delta H^{\ddagger}$ (kJ mol^{-1})	$\Delta S^{\ddagger}$ (J K^{-1} mol^{-1})	$\Delta V^{\ddagger}$ (cm^3 mol^{-1})	Ref.
[Co(NH$_3$)$_5$(OH$_2$)]$^{3+}$	[Fe(CN)$_6$]$^{4-}$	0.5	1.2×10^{-1}	102	79	26.5	32–34
[Co(NH$_3$)$_5$(OH$_2$)]$^{3+}$	[Fe(CN)$_6$]$^{4-}$	0.5	9.3×10^{-2}	115	122	37.6	35
[Co(NH$_3$)$_5$py]$^{3+}$	[Fe(CN)$_6$]$^{4-}$	1.0	8.9×10^{-3}	118	113	29.8	32
[Co(NH$_3$)$_5$py]$^{3+}$	[Fe(CN)$_6$]$^{4-}$	1.0				23.9	36
[Co(NH$_3$)$_5$(OSMe$_2$)]$^{3+}$	[Fe(CN)$_6$]$^{4-}$	1.0	2.0×10^{-1}	84	25	34.4	32
[Co(NH$_3$)$_5$N$_3$]$^{2+}$	[Fe(CN)$_6$]$^{4-}$	1.0	6.2×10^{-4}	104	44	18.8	37
[Co(NH$_3$)$_5$Cl]$^{2+}$	[Fe(CN)$_6$]$^{4-}$	1.0	2.7×10^{-2}	85	11	25.9	37
[Co(phen)$_3$]$^{2+}$	[Co(ox)$_3$]$^{3-}$	0.017	2.4×10^{-1}	32	−148		28
[(NH$_3$)$_3$Co(OH)(OH)(O-py-O)Co(NH$_3$)$_3$]$^{3+}$	[Fe(dipic)$_2$]$^{2-}$	0.1	3.7×10^{-3}	103	54		29

Instances where precursor complexes are detected are rare and it is difficult to draw many conclusions from such a limited data set. However, the problem of separating precursor formation from electron transfer has been approached elegantly from other perspectives. It is possible to estimate ion association constants. The interaction between ions in solution can be represented [38] by the expression eq. (2.12), where r is the distance between the metal centers and $\omega(r)$ is the work required to bring the reactants from infinity to the distance r. In recognition of the fact that a single assembly defined by r is unlikely to represent the precursor for a dynamic process such as electron transfer, the widely accepted [5] expression for calculation of K_o is given in eq. (2.13), where δr is the spread of distances over which the reaction takes place, generally estimated to be around 0.8Å.

$$K_o(r) = N \int 4\pi r^2 \exp(-\omega(r)/k_B T)\, dr \tag{2.12}$$

$$K_o = (4\pi N r^2 \delta r) \exp(-\omega(r)/RT) \tag{2.13}$$

For a purely electrostatic interaction, the association constant K_o will be a function of the charges on the reactants, z_A and z_B, and the distance, r, between the metal centers. The quantity $\omega(r)$ is estimated by eq. (2.14), where e is the electronic charge, D_s is the static dielectric constant for the medium and $\beta = (8\pi N e^2/D_s kT)^{1/2}$.

$$\omega(r) = z_A z_B e^2 / D_s r (1 + \beta r) \tag{2.14}$$

Application of this expression presupposes some knowledge of the detailed structure of the precursor, in particular the distance, r, between the metal centers. In fact this is poorly known. Likewise, dipolar interactions, hydrogen bonding and hydrophobic stacking are not taken into account by this simple model. In Table 2.5, association constants obtained for interactions between complex ions are compared with those calculated with the use of eq. (2.13). The calculated values were obtained by assuming that the complexes are hard spheres and that the distance between the metal centers is the sum of the radii, a. Calculated values are for the most part an order of magnitude smaller than those experimentally observed. Substantially better agreement is obtained by allowing the spheres to interpenetrate and there is some experimental evidence from NMR studies of ion pairs in solution [28, 39, 40] to suggest that this may be a valid approach. For example the experimentally derived Co–Co distance in the ion pair $\{[Co(ox)_3]^{3-},[Co(phen)_3]^{2+}\}$ is around 6 Å, substantially closer than predicted by the hard-sphere model. However, it must be noted that eq. (2.13) is now almost universally accepted. In defense of its use it may be noted that there is no evidence that the strong ion pairs noted in experimental studies represent structures conducive to facile electron transfer.

Table 2.5. Ion-association constants at 25 °C for interactions between complex ions

Oxidant	Reductant	μ (M)	K_o (expt)	K_o (eq. (2.13))	Ref.
$[Co(ox)_3]^{3-}$	$[Co(phen)_3]^{2+}$	0.017	650	10	28
$[Co(en)_3]^{3+}$	$[Co(edta)]^-$	0.05	32	2	41
$[Co(NH_3)_5(py)]^{3+}$	$[Fe(CN)_6]^{4-}$	0.1	2400	86	23
$[Co(NH_3)_5(OH_2)]^{3+}$	$[Fe(CN)_6]^{4-}$	0.1	1500	99	24
		1.0	480	6	32
$[Co(NH_3)_5(py)]^{3+}$	$[Fe(CN)_6]^{4-}$	1.0	168	6	32

(continues)

Table 2.5. (continued)

Oxidant	Reductant	μ (M)	K_o (expt)	K_o (eq. (2.13))	Ref.
$[Co(NH_3)_5(DMSO)]^{3+}$	$[Fe(CN)_6]^{4-}$	1.0	34	6	32
$[Co(NH_3)_5(N_3)]^{2+}$	$[Fe(CN)_6]^{4-}$	1.0	49	3	37
$[Co(NH_3)_5(OAc)]^{2+}$	$[Fe(CN)_6]^{4-}$	0.1	300	12	27
$[Co(phen)_3]^{3+}$	$[Fe(CN)_6]^{4-}$	0.1	≤20	25	22

It is generally assumed that in the precursor, the two reacting complexes come into intimate contact thereby partly defining (at least for spherically symmetric complexes) the geometry required for electron transfer. Although this description appears to work very well in many instances, it is clearly very limited. If the complexes are not spherically symmetric, some idea of shape can be introduced by approximating the species as a spheroid with axes of length d_x, d_y, and d_z, such that a is then given by eq. (2.15). Values of a for some common complexes are presented in Table 2.6.

$$a = \tfrac{1}{2}(d_x d_y d_z)^{1/3} \tag{2.15}$$

Table 2.6. Apparent radii for use in calculating precursor complex stability

Complex	a (Å)	Complex	a (Å)
$[M(H_2O)_6]^{3+/2+}$	3.2	$[M(NH_3)_2(bpy)_2]^{3+/2+}$	5.6
$[M(NH_3)_6]^{3+/2+}$	3.3	$[M(bpy)_3]^{3+/2+}$	6.8
$[M(en)_3]^{3+/2+}$	3.8	$[M(phen)_3]^{3+/2+}$	7.2
$[M(NH_3)_5py]^{3+/2+}$	4.2	$[M(CN)_6]^{3-/4-}$	4.1
$[M(NH_3)_4(bpy)]^{3+/2+}$	4.4	$[M(cp)_2]^{+/0}$	3.8

Although metal–ligand distances differ for the different oxidation states and for ions from different rows in the periodic table, the radii are determined predominantly by the magnitude of the ligands and these approximate values serve as useful starting points for most calculations.

2.3 THE ELECTRON TRANSFER STEP

Returning to the reaction between $[Co(phen)_3]^{3+}$ and $[Ru(NH_3)_6]^{2+}$, the value for K_o is calculated to be approximately 0.25 M^{-1} at 1.0 M ionic strength, small because of the electrostatic repulsions between the ions. Although the ion association constant is small, formation of the ion pair is very rapid, generally close to the limit at which the ions can diffuse through solution since the energetic barrier in this step of the reaction is low. Energetic requirements for the formation of the precursor assembly are generally less demanding than those for the rate-limiting electron transfer, and attention will now focus on the dynamics of the electron transfer within the precursor.

As the reactants evolve to products in the electron transfer step within the precursor complex, some rearrangement of the inner and outer coordination spheres takes place. For example, the Co—N bond length in [Co(phen)$_3$]$^{3+}$ is 1.91 Å [42] whereas in the product [Co(phen)$_3$]$^{2+}$ it is significantly longer (2.11 Å) [43]. In addition there are less dramatic changes in the bond lengths and angles within the phen ligands and, more importantly, changes in the orientation and distribution of solvent water and anions in the outer coordination sphere of the complex. Similar considerations apply to the transformation of [Ru(NH$_3$)$_6$]$^{2+}$ to [Ru(NH$_3$)$_6$]$^{3+}$ where the Ru—N bond length decreases from 2.14 Å to 2.11 Å [44, 45]. Nuclear relaxation is slow relative to the movement of the electron, which means that an important contribution to the barrier for electron transfer will result from the changes in geometry of the complexes. The restrictions on electron transfer are very similar to the Frank–Condon restrictions in electronic spectroscopy.

In the case of electron transfer within a precursor assembly, this restriction is best understood by consideration of Fig. 2.2, where two parabolas represent cross-sections through potential energy surfaces representing the motions of the precursor, P, and successor, S, complexes along a coordinate which includes the structural rearrangement involved in oxidation and reduction. This coordinate is denoted as x and it varies from 0 to 1 as the electron is transferred. The two surfaces intersect at a point, T, which represents the transition state for thermally activated electron transfer. At this point, the geometries of the complexes in the reacting assembly will be intermediate between those of the oxidized and those of the reduced forms. The equations of the two parabolas are defined by coefficients, λ, which are determined by the structural reorganization, eq. (2.16), described above. Thus λ_P involves contraction of the Ru^{II}—N bonds and extension of the Co^{III}—N bonds with associated solvent reorganization, while λ_S involves extension of the Ru^{III}—N bonds, contraction of the Co^{II}—N bonds and the solvent reorganization associated with these changes. If the geometry changes are not substantial then $\lambda_P \approx \lambda_S$ and at the point of crossing, $T, x^\ddagger = \frac{1}{2}(1 + \Delta G°/\lambda)$ and hence the contribution to the activation energy from structural reorganization is given by eq. (2.17).

$$G_P = \lambda_P x^2 \qquad G_S = \lambda_S (1-x)^2 + \Delta G° \tag{2.16}$$

$$\Delta G^\ddagger = \frac{1}{4} \lambda (1 + \Delta G°/\lambda)^2 \tag{2.17}$$

Thus the energy required to reach the point of crossing within this static precursor complex depends on the quantity λ, the energy for structural reorganization, and $\Delta G°$, the free energy change for the overall reaction.

This dependence of $\Delta G^\ddagger$ on $\Delta G°$ can be readily appreciated in Fig. 2.3. For a constant reorganizational barrier, λ, the driving force, $\Delta G°$, is changed from very endoenergetic to very exoenergetic [46]. When $\Delta G°$ is large and positive, representing a thermodynamically unfavorable process, the activation barrier is also large and positive and, according to the Hammond Postulate, the activated complex will resemble the successor in geometry and solvation. As the driving force becomes more favorable the activation energy decreases. This is the 'normal region' for reaction energetics. Note that when $\Delta G° = 0$, then $\Delta G^\ddagger = \lambda/4$ and the geometry of the activated complex will be midway between that of the precursor and that of the successor.

When $\Delta G° = -\lambda$, the electron transfer process is activationless with $\Delta G^{\ddagger} = 0$, and at higher driving force with $\Delta G° > -\lambda$, $\Delta G^{\ddagger}$ again increases, this time in the 'inverted region' for reaction energetics. The implication of this is that in the normal region the rate should increase with increasing driving force, go through a maximum where the activation barrier is minimized, and then invert its dependence, decreasing with increasing driving force. Most of the reactions examined in this book are in the normal region but some evidence for the inverted region will be presented. The Hammond Postulate cannot apply to the inverted region and the energetics in this region must be considered quite differently.

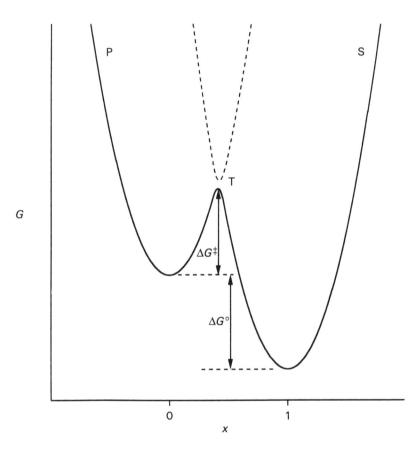

Fig. 2.2. Reaction coordinate diagram for electron transfer within a precursor assembly. Progress of the reaction from the precursor P to the successor S through the transition state T involves overcoming the Frank–Condon barrier, the inner-coordinate sphere and outer-coordination sphere changes measured by the reaction coordinate, x, which moves from 0 to 1. At the transition state, the reactants move from the precursor surface to the successor surface with unit probability (adiabatic process), but the coupling between the two surfaces is weak so that $\Delta G^{\ddagger}$ is determined solely by the Frank–Condon barrier and $\Delta G°$.

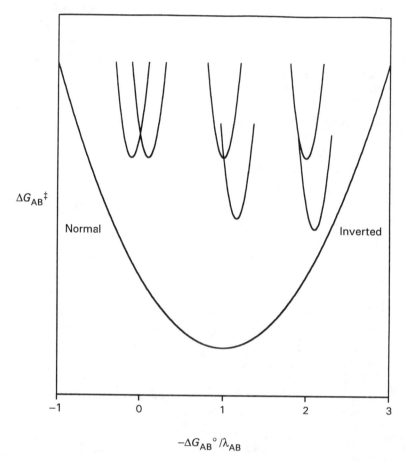

Fig. 2.3. Reaction coordinate diagram illustrating the dependence of $\Delta G^{\ddagger}_{AB}$ on ΔG°_{AB}. The three inserts show reaction coordinate diagrams for the values of $-\Delta G^{\circ}_{AB}/\lambda_{AB}$ indicated. In the normal region, when $-\Delta G^{\circ}_{AB}/\lambda_{AB} < 1$, the Marcus expresion is applicable and $\Delta G^{\ddagger}_{AB}$ decreases with increasing driving force, with a corresponding increase in rate. When $-\Delta G^{\circ}_{AB}/\lambda_{AB} = 1$, the electron transfer is activationless and the rate will reach a maximum value. When $-\Delta G^{\circ}_{AB}/\lambda_{AB} > 1$, $\Delta G^{\ddagger}_{AB}$ increases with increasing driving force and a decrease in rate is expected. This is the inverted region where the free energy relationship is inverted from its normal form. Note that the transition state geometry is not intermediate between those of the precursor and successor and the reaction is not adiabatic.

Returning to Fig. 2.2 in the normal region for electron transfer, the rate of the electron transfer can be calculated classically from the rate of the structural rearrangement, υ_{eff}, an effective frequency which determines the rate of transmission along the ordinate x, the height of the barrier, $\Delta G^{\ddagger}$, and the probability that, having reached the point T, the system will pass from the precursor curve to the successor curve, κ_{el} (eq. (2.18)). Once the Frank–Condon factors have been satisfied, the probability that electron transfer takes place depends on the interaction between the donor orbitals which contain the electron, and the acceptor orbitals to which it will be transferred. If the electronic coupling as a result of the orbital overlap between the two reactant complexes is very small, the probability of crossover from the precursor surface to the successor surface will be small ($\kappa_{el} \ll 1$) and the rate will be determined by both κ_{el} and the Frank–Condon factors involved in reaching the point of intersection. Under these conditions, the reaction is said to be non-adiabatic. However, there will be a point at which the coupling is sufficiently large so that when the reactants reach point T, the probability of going to products $\kappa_{el} \approx 1$. The energetics of the electron transfer are then determined by Frank–Condon factors alone, and the reaction is said to be adiabatic. Computation of the rate under such circumstances can be accomplished solely with a knowledge of atomic motions and the overall free energy change for the reaction. There is a third possibility. If the electronic coupling is very large, the reaction is adiabatic but the Frank–Condon factors are no longer the appropriate description for the barrier to electron transfer, and calculation of the rate constant is much more difficult.

$$k_{et} = \upsilon_{eff}\, \kappa_{el}\, \exp(-\Delta G^{\ddagger}/RT) \tag{2.18}$$

2.4 THE MARCUS LINEAR FREE ENERGY RELATIONSHIP

Although eq. (2.18) has a relatively simple form, computation of electron transfer rates relies on the evaluation of a number of complex factors associated with the motions of the atoms in the reactant complexes. Detailed discussion of these factors is postponed until section 2.10. Marcus [2] noted that if the reaction rates are governed by Frank–Condon factors alone then many of the unknown quantities can be cancelled by comparing the rates of a cross-reaction between two different reagents A^{ox} and B^{red} (eq. (2.19)), with the rates of the individual self-exchange processes (eqs (2.20)–(2.21)). These latter processes have $\Delta G^{\circ} = 0$.

$$[A^{ox}] + [B^{red}] \xrightarrow{k_{AB}} [A^{red}] + [B^{ox}] \tag{2.19}$$

$$[A^{ox}] + [A^{red}] \xrightarrow{k_{AA}} [A^{red}] + [A^{ox}] \tag{2.20}$$

$$[B^{ox}] + [B^{red}] \xrightarrow{k_{BB}} [B^{red}] + [B^{ox}] \tag{2.21}$$

Activation free energies can be evaluated from eq. (2.22), and if it is assumed that $\lambda_{AB} = \tfrac{1}{2}\,(\lambda_{AA} + \lambda_{BB})$ the resulting linear free energy relationship is given by eq. (2.24) which can be expressed in its well known rate constant form as eq. (2.26) where Z is a collision frequency for neutral molecules in solution, estimated as approximately 10^{11} M^{-1} s^{-1}. The strict assumptions of adiabatic electron transfer and weak orbital overlap can be relaxed to the condition that the activated state attained by a reaction in an outer-sphere reaction is independent of its reaction partner [47]. In other words, the Frank–Condon factors and electronic couplings associated with a particular complex in the cross-reaction are identical with those for the same complex in the self-exchange reaction.

$$\Delta G^{\ddagger}_{AB} = \tfrac{1}{4} \lambda_{AB} (1 + \Delta G^{\circ}_{AB}/\lambda_{AB})^2 \tag{2.22}$$

$$\Delta G^{\ddagger}_{AA} = \tfrac{1}{4} \lambda_{AA} \qquad \Delta G^{\ddagger}_{BB} = \tfrac{1}{4} \lambda_{BB} \tag{2.23}$$

$$\Delta G_{AB} = \tfrac{1}{2} (\Delta G^{\ddagger}_{AA} + \Delta G^{\ddagger}_{BB}) + \tfrac{1}{2}\Delta G^{\circ}_{AB}(1 + \alpha) \tag{2.24}$$

$$\alpha = \frac{\Delta G^{\circ}_{AB}}{4 (\Delta G^{\ddagger}_{AA} + \Delta G^{\ddagger}_{BB})} \tag{2.25}$$

$$k_{AB} = (k_{AA} \, k_{BB} \, K_{AB} \, f_{AB})^{1/2} \tag{2.26}$$

$$\log f = (\log K_{AB})^2 / 4 \log(k_{AA} k_{BB}/Z^2) \tag{2.27}$$

The Marcus expression eq. (2.24), can be modified to incorporate the electrostatic interactions involved in precursor complex formation eq. (2.28), where $\Delta G_{AA}* = \Delta G_{AA}^{\ddagger} - \omega_{AA}$, $\Delta G_{BB}* = \Delta G_{BB}^{\ddagger} - \omega_{BB}$, and $\Delta G_{AB}* = \Delta G_{AB}^{\ddagger} - \omega_{AB}$.

$$\Delta G_{AB}* = \tfrac{1}{2}(\Delta G_{AA}* + \Delta G_{BB}*) + \tfrac{1}{2}(\Delta G^{\circ}{}_{AB} + \omega_{AB} + \omega_{BA})(1 + \alpha) \tag{2.28}$$

$$\alpha = (\Delta G^{\circ}{}_{AB} + \omega_{AB} + \omega_{BA})/4(\Delta G_{AA}* + \Delta G_{BB}*) \tag{2.29}$$

This allows the relationship to be used under conditions where changes in the work terms required to bring the reactants into the precursor assembly do not cancel (similarly charged reactants) or cannot be ignored [1].

Activation parameters, $\Delta H^{\ddagger}$, $\Delta S^{\ddagger}$ have been widely determined for outer-sphere electon transfer reactions. Relationships between the activation parameters for electron transfer are also predicted by Marcus Theory, eqs (2.30) and (2.31), where α is given in eq. (2.29) [48].

$$\Delta H_{AB}* = \tfrac{1}{2}(\Delta H_{AA}* + \Delta H_{BB}*)(1 - 4\alpha^2) + \tfrac{1}{2}\Delta H^{\circ}{}_{AB}(1 + 2\alpha) \tag{2.30}$$

$$\Delta S_{AB}* = \tfrac{1}{2}(\Delta S_{AA}* + \Delta S_{BB}*)(1 - 4\alpha^2) + \tfrac{1}{2}\Delta S^{\circ}{}_{AB}(1 + 2\alpha) \tag{2.31}$$

The simple correlations of Marcus Theory work relatively well. They have been extensively tested and the ability to test depends on the ready availability of self-exchange rate data.

2.5 SELF-EXCHANGE RATES

The determination of the rates of electron self-exchange processes has been an important element of the test of theory. The rate, k_{AA}, associated with the dynamic equilibrium (eq. (2.20)), is a fundamental quantity for the reactant in question and is independent of free energy change. It involves no apparent reaction and this presents some difficulties for measurement. The most generally applicable means for measuring the rate is to distinguish the species in one oxidation state and to monitor the approach to the true equilibrium so that in practice there is at least a small entropic driving force corresponding to the statistical factor $RT \ln 2 = 1.8$ kJ mol^{-1} at 25.0 °C. The distinguishing feature should not markedly affect the reactivity, and several methods have been widely accepted.

(a) Isotopic labeling. This is the oldest of the methods and suffers only from the disadvantage that the rate is subject to isotope effects, generally small for isotopes of the heavier elements. The most sensitive studies have been carried out with radioactive tracers and suffer additional limitations on the rate imposed by the need to separate the oxidation states to measure the approach to equilibrium. For example in the well-studied exchange between $[Fe(H_2O)_6]^{3+}$ and $[^{55}Fe(H_2O)_6]^{2+}$, separation is accomplished by the addition of bpy to complex the iron(II) as $[Fe(bpy)_3]^{2+}$ followed by precipitation of iron(III) as $[Fe(OH)_3]$ [44].

$$[Fe(H_2O)_6]^{3+} + [^{55}Fe(H_2O)_6]^{2+} \xrightarrow{k_{ex}} [Fe(H_2O)_6]^{2+} + [^{55}Fe(H_2O)_6]^{3+} \quad (2.32)$$

The rate of exchange is given by eq. (2.33) which may be manipulated to give k_{ex} by the McKay relationship [49], (eq. (2.34)), where F is the fraction of exchange which has occurred at time t (eq. (2.35)).

$$\frac{-d[[^{55}Fe(H_2O)_6]^{2+}]}{dt} = k_{ex}[[^{55}Fe(H_2O)_6]^{2+}][[Fe(H_2O)_6]^{3+}]$$
$$- k_{ex}[[^{55}Fe(H_2O)_6]^{3+}][[Fe(H_2O)_6]^{2+}] \quad (2.33)$$

$$\ln(1 - F) = -k_{ex} \frac{([[Fe(H_2O)_6]^{3+}] + [[^{55}Fe(H_2O)_6]^{2+}])}{([[Fe(H_2O)_6]^{3+}][[^{55}Fe(H_2O)_6]^{2+}])} t \quad (2.34)$$

$$F = \frac{[[^{55}Fe(H_2O)_6]^{2+}]_t - [[^{55}Fe(H_2O)_6]^{2+}]_0}{[[^{55}Fe(H_2O)_6]^{2+}]_\infty - [[^{55}Fe(H_2O)_6]^{2+}]_0} \quad (2.35)$$

A further complication in this reaction and in others involving metal aqua ions is the presence of terms in the rate law with an $[H^+]^{-1}$ dependence (eq. (2.36)), where K_h is the hydrolysis constant for the more readily deprotonated higher oxidation state. The rate constant k'_{AA} is interpreted as reaction of the hydrolyzed form of the oxidant (eq. (2.37)).

$$\frac{d[[^{55}Fe(H_2O)_6]^{3+}]}{dt} = \left\{k_{AA} + k'_{AA} \frac{K_h}{[H^+]^{-1}}\right\} [[Fe(H_2O)_6]^{2+}] [[Fe(H_2O)_6]^{3+}]$$
$$(2.36)$$

$$[Fe(H_2O)_5OH]^{2+} + [^{55}Fe(H_2O)_6]^{2+} \rightarrow [Fe(H_2O)_5OH]^+ + [^{55}Fe(H_2O)_6]^{3+} \quad (2.37)$$

More recently, stable isotopes have been employed. The reaction between $[Ru(ND_3)_6]^{3+}$ and $[Ru(NH_3)_6]^{2+}$ has been monitored by the changes in the near infra-red N—H(D) overtones at 1550 nm [50], and exchange between $[Ni([14]aneN_4)(H_2O)_2]^{3+}$ and $[^{61}Ni([14]aneN_4)]^{2+}$ has been monitored by rapid quench EPR [51]. Monitoring changes in EPR line broadening due to the ^{61}Ni hyperfine interaction has also proved useful [52].

(b) Optically active complexes. This method was first suggested by Dwyer [53] and has been used in a number of investigations. The distinguishing feature is that the complex in one oxidation state is initially optically active. The method suffers in that the symmetry of the complex must be sufficiently low for optical activity and relies on the absence of stereoselectivity. This latter restriction has been shown to involve corrections of 10–20% at most [54, 55]. One example of the use of the technique is in the reaction between optically inert $[\Lambda\text{-Co(phen)}_3]^{3+}$ and labile, racemic $[Co(phen)_3]^{2+}$ [55]. Loss of optical activity ϕ can occur by two pathways, the intrinsic inversion process (eq. (2.38)), and a cobalt(II) catalyzed pathway. Provided the rate of racemization of the $[Co(phen)_3]^{2+}$ is rapid compared with the rate of electron transfer, then the rate of this latter pathway is $k_{\Lambda\Delta}$. The corresponding pathway between the two isomers of similar configuration $k_{\Lambda\Lambda}$ leads to no net loss of optical activity. The quantity $k_{\Lambda\Delta}$ is equal to the self-exchange rate only if $k_{\Lambda\Delta} = k_{\Lambda\Lambda}$; this is rarely true but in practice the differences in the two rate constants are small.

$$[\Lambda\text{--Co(phen)}_3]^{3+} \underset{}{\overset{k_{inv}^{III}}{\rightleftharpoons}} [\Delta\text{--Co(phen)}_3]^{3+} \tag{2.38}$$

$$[\Lambda\text{--Co(phen)}_3]^{3+} + [\Delta\text{--Co(phen)}_3]^{2+} \underset{}{\overset{k_{\Lambda\Delta}}{\rightleftharpoons}} [\Lambda\text{--Co(phen)}_3]^{2+} + [\Delta\text{--Co(phen)}_3]^{3+} \tag{2.39}$$

$$[\Lambda\text{--Co(phen)}_3]^{2+} \underset{}{\overset{k_{inv}^{II}}{\rightleftharpoons}} [\Delta\text{--Co(phen)}_3]^{2+} \tag{2.40}$$

$$-d\phi/dt = \left\{2k_{inv}^{III} + k_{\Lambda\Delta}[Co(phen)_3]^{2+}\right\}\phi \tag{2.41}$$

The method has proved extremely useful in the determination of the self-exchange rate for $[Co(sep)]^{3+/2+}$ where the two accessible oxidation states are neither labile nor undergo racemization [56]. In this case, the reaction can be monitored by the addition of $[\Delta\text{--Co(sep)}]^{3+}$ to $[\Lambda\text{--Co(sep)}]^{2+}$ and the rate constant $k_{\Lambda\Delta}$ is twice the self-exchange rate. Again, it has been shown that stereoselectivity in these reactions introduces a minor correction.

$$[\Delta\text{--Co(sep)}]^{3+} + [\Lambda\text{--Co(sep)}]^{2+} \rightarrow [\Delta\text{--Co(sep)}]^{2+} + [\Lambda\text{--Co(sep)}]^{3+} \tag{2.42}$$

$$-d\phi/dt = 2k_{\Lambda\Delta}[Co(sep)]^{2+}\phi \tag{2.43}$$

(c) Structural substitution. A common method for the estimation of self-exchange rates is the study of the reaction rates between structurally similar though chemically modified complexes to obtain pseudo-self-exchange rates. It relies on the substituent having a minimal effect on the reduction potential, $E°$, and reactivity of the complex. The method has the distinct advantage that the reaction can generally be monitored by standard kinetic methods. Examples of this are the reduction of $[Ru(NH_3)_5 (nic)]^{2+}$ by $[Ru(NH_3)_5 (isonic)]^+$ [13] and $[Ru(NH_3)_4(bpy)]^{3+}$ by $[Ru(NH_3)_4(phen)]^{2+}$ [57] where in both cases the difference in reduction potentials is less than 5 mV, and the reaction between $[Ni([9]aneN_3)_2]^{3+}$ with $[Ni(2\text{-Me-}[9]aneN_3)_2]^{2+}$ where the difference in reduction potentials is 5 mV [58].

(d) One of the most powerful methods for the determination of self-exchange rates takes advantage of the fact that the electron transfer takes place between the reactants in different electronic states. In particular when exchange is between a diamagnetic oxidation state and a paramagnetic oxidation state, the exchange will broaden the NMR spectrum of the diamagnetic complex [59, 60] or the EPR spectrum of the paramagnetic complex [52]. The effect of the addition of $[Cu^{II}(H_{-2}Aib_3)]^-$ on the ^1H NMR spectrum of the diamagnetic complex $[Cu^{III}(H_{-2}Aib_3)]$ is shown in Fig. 2.4. In the slow exchange limit, the self-exchange rate is given by eq. (2.44) where $1/T_2$ is the observed transverse relaxation rate and the subscripts D and DP refer to the solutions containing the diamagnetic complex and the mixture of the diamagnetic complex with its paramagnetic reaction partner. The transverse relaxation rate is related to $\Delta\upsilon$, the NMR peak width at half-height giving eq. (2.45).

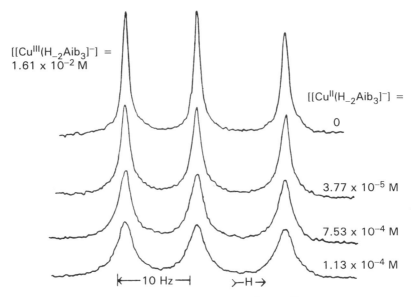

Fig. 2.4. 100 MHz NMR spectra of a solution of 1.61×10^{-2} M $[Cu^{III}(H_{-2}Aib_3)]$ at 25°C with various concentrations of $[Cu^{II}(H_{-2}Aib_3)]^-$, from Koval, C. A.; Margerum, D. W. *Inorg. Chem.* 1981, **20**, 2315 with permision. Line broadening, corresponding to a shotterning of T_2, occurs as the concentration of the paramagnetic ion increases.

Table 2.7. Self-exchange rates and reduction potentials for outer-sphere electron transfer reagents

Reaction	μ (M)	$E°$ (V)	k_{AA} (M^{-1} s^{-1})	$\Delta H^{\ddagger}$ (kJ mol^{-1})	$\Delta S^{\ddagger}$ (J K^{-1} mol^{-1})	$\Delta V^{\ddagger}$ (cm^3 mol^{-1})	Ref.
[V(OH$_2$)$_6$]$^{3+/2+}$	2.0	−0.26	1×10^{-2}	55	−105		61
[Cr(OH$_2$)$_6$]$^{3+/2+}$	1.0	−0.41	$<1.9 \times 10^{-5}$				62
[Mn(sar)]$^{3+/2+}$	0.1	0.52	39	38	−92		63
[MnO$_4$]$^{-/2-}$	1.1[a]	0.56	2.4×10^{-3}			−21	64,65
[Fe(H$_2$O)$_6$]$^{3+/2+}$	0.1	0.74	1.1	46.4	−88	−11[b]	44,66
[Fe(sar)]$^{3+/2+}$	0.1	0.09	7×10^5	0	−134		63
Fe([9]aneS$_3$)$_2$]$^{3+/2+}$	2.0[c]	≈1.4	3.2×10^6				67
[Fe(phen)$_3$]$^{3+/2+}$	0.4[d]	1.05	1.3×10^7	1.6	−102		68
[Fe(CN)$_6$]$^{3-/4-}$	0.1	0.42	1.9×10^4			22	22,69,70
[Co(OH$_2$)$_6$]$^{3+/2+}$	0.5	1.84	5	44	−92		71
[Co(en)$_3$]$^{3+/2+}$	0.98	−0.18	7.7×10^{-5}	59	−134		72
[Co(en)$_3$]$^{3+/2+}$	0.50[e]			58	−130	−20	73
[Co(NH$_3$)$_6$]$^{3+/2+}$	2.5[f]	0.06	8×10^{-6}				
	1.0	0.06	2×10^{-7}				75
[Co(sep)]$^{3+/2+}$	0.2	−0.30	5.1	40	−96		56
[Co(sep)]$^{3+/2+}$	0.2			41	−93	−6.4	76
[Co(sar)]$^{3+/2+}$	0.2			59	−42		54
[Co(diamsar)]$^{3+/2+}$	0.2	−0.44	2.1				54
[Co(diamsarH$_2$)]$^{5+/4+}$	0.2	−0.32	0.5				54
[Co(azamesar)]$^{3+/2+}$	0.2	0.03	0.024				56
[Co(amsartacn)]$^{3+/2+}$	0.2	−0.36	2.9				77
[Co(azacapten)]$^{3+/2+}$	0.2	−0.30	0.09				78
	0.2	0.01	2.2×10^4	29	−75		79
[Co(Me$_2$S6sar)]$^{3+/2+}$	0.1	0.27	2.8×10^4				80
[Co([9]aneN$_3$)$_2$]$^{3+/2+}$	0.1	−0.41	0.19				81
[Co([9]aneS$_3$)$_2$]$^{3+/2+}$	0.2	0.42	1.6×10^5	35	−29		76,82
[Co([9]aneS$_3$)$_2$]$^{3+/2+}$	0.1	0.42	1.3×10^5	25	−64	−4.8	55
[Co(phen)$_3$]$^{3+/2+}$	0.1	0.36	12	21	−156		

(continues)

Table 2.7. (continued)

Reaction	μ (M)	$E°$ (V)	k_{AA} (M^{-1} s^{-1})	$\Delta H^{\ddagger}$ (kJ mol^{-1})	$\Delta S^{\ddagger}$ (J K^{-1} mol^{-1})	$\Delta V^{\ddagger}$ (cm^3 mol^{-1})	Ref.
[Co(4,7-Me$_2$phen)$_3$]$^{3+/2+}$	0.1	0.16	1.2×10^2				55
[Co(5,6-Me$_2$phen)$_3$]$^{3+/2+}$	0.1	0.23	3.0×10^2				55
[Co(bpy)$_3$]$^{3+/2+}$	0.1	0.31	5.7	31	−127		28
[Co(edta)]$^{-/2-}$	0.2^g	0.37	1.39×10^{-4}	83	−88		83
	0.1	0.37	2×10^{-7}				84
[Co(edta)]$^-$/[Co(Hedta)]$^-$	0.5^h					−3.2	85
[CoW$_{12}$O$_{40}$]$^{5-/6-}$	0.21	1.0	0.72	18	−184		86
[Ni([9]aneN$_3$)$_2$]$^{3+/2+}$	1.0	0.94	2.7×10^3				58
[Ni(sar)]$^{3+/2+}$	0.1	0.86	4×10^3	21	−100		63
[Ni(Me$_2$L)]$^{2+/+}$	0.1	0.65	2.4×10^4				87
[Ni(H$_{-2}$Aib$_3$)]$^{0/-}$	0.1^i	0.84	4.5×10^2				52
[Cu(H$_{-2}$Aib$_3$)]$^{0/-}$	0.1	0.66	5.5×10^4	29	−54		59
[Cu(dmp)$_2$]$^{2+/+}$	0.002	0.59	2.0×10^5	24	−63		88
[Ru(OH$_2$)$_6$]$^{3+/2+}$	5.0	0.22	20	46	−66		89
[Ru(en)$_3$]$^{3+/2+}$	0.75	0.15	3.1×10^4	25	−71		90
[Ru(NH$_3$)$_6$]$^{3+/2+}$	0.013	0.07	8.2×10^2	43	−46		50
[Ru(NH$_3$)$_6$]$^{3+/2+}$	0.125^j	0.07	6.6×10^3				91
[Ru(NH$_3$)$_6$]$^{3+/2+}$	0.1	0.07	4.3×10^3				92
[Ru(sar)]$^{3+/2+}$	0.1	0.29	6×10^5				63
[Ru([9]aneN$_2$)$_3$]$^{3+/2+}$	0.1^k	0.37	5×10^4				93
[Ru(NH$_3$)$_5$py]$^{3+/2+}$	1.0	0.30	4.7×10^5	12	−92		13
[Ru(NH$_3$)$_5$py]$^{3+/2+}$	0.1	0.27	1.1×10^5				13
[Ru(NH$_3$)$_5$(isonic)]$^{3+/2+}$	1.0	0.38	2.7×10^4				13
[Ru(NH$_3$)$_4$bpy]$^{3+/2+}$	0.1	0.52	7.7×10^5	13	−88		57
[Ru(NH$_3$)$_2$(bpy)$_2$]$^{3+/2+}$	0.1		8.4×10^7				57
[Ru(bpy)$_3$]$^{3+/2+}$	0.1	1.26	4.2×10^8	32	−28		94

(continues)

Table 2.7. (continued)

Reaction	μ (M)	$E°$ (V)	k_{AA} (M^{-1} s^{-1})	$\Delta H^\ddagger$ (kJ mol^{-1})	$\Delta S^\ddagger$ (J K^{-1} mol^{-1})	$\Delta V^\ddagger$ (cm^3 mol^{-1})	Ref.
$[Ru(CN)_6]^{3-/4-}$	0.1	0.92	8.3×10^3	40	−36		20
$[Rh_2(OAc)_4(D_2O)_2]^{+/0}$	0.1	1.25	2.1×10^5				95
$[IrCl_6]^{2-/3-}$	0.10	0.89	2.3×10^5				96
$[Ce_{aq}]^{4+/3+}$	6.0[l]	1.7	$\leq 6 \times 10^{-2}$				97

[a]4°C. [b]0.5M, 2°C. [c]1°C. [d]3°C. [e]65°C. [f]40°C. [g]100°C. [h]85°C. [i]24°C. [j]4°C. [k]23°C. [l]0.1°C.

$$\left(\frac{1}{T_2}\right)_{DP} - \left(\frac{1}{T_2}\right)_{D} = k_{AA}\,[[Cu^{II}H_{-2}Aib_3)]^-] \tag{2.44}$$

$$\pi\,(\Delta\upsilon_{DP} - \Delta\upsilon_D) = k_{AA}\,[[Cu^{II}H_{-2}Aib_3)]^-] \tag{2.45}$$

A list of self-exchange rates determined by these methods together with $E°$ values is presented in Table 2.7.

The extensive collection of experimental data in Table 2.7 requires some comment. Inclusion in the table does not necessarily imply that the reaction is outer-sphere in nature, however, most of the reactions are thought to represent outer-sphere processes and they form the basis for discussion for the remainder of this chapter. The data are arranged in accord with the ligand environment. Rates and activation parameters show a wide variation and depend on both the ligand environment and the nature of the metal center. While these are all reactions in aqueous solution, experimental conditions also vary. Ionic strength generally has a substantial effect on the rates since they concern reactions between similarly charged complexes. Specific ion effects are frequently reported. For example the self-exchange rate for $[Co(phen)_3]^{3+/2+}$ changes from 4.9 M^{-1} s^{-1} in 0.1 M [Cl$^-$] to 12.0 M^{-1} s^{-1} in 0.1 M [NO$_3^-$] while the rate for $[Fe(CN)_6]^{3-/4-}$ varies by an order of magnitude on changing the supporting cation from K$^+$ to Me$_4$N$^+$. The metal aqua ions require particular attention since the rate laws consist of two terms, one independent of [H$^+$] and the other dependent on [H$^+$]$^{-1}$ (eq. (2.36)). It is the rate for the former term which is reported in Table 2.7. Although some ambiguity exists, the term dependent on [H$^+$]$^{-1}$ is generally interpreted in terms of reaction of the more readily hydrolyzed oxidant (eq. (2.46)), which is a pseudo-self-exchange rate.

$$[Fe(H_2O)_5OH]^{2+} + [Fe(H_2O)_6]^{2+} \rightleftharpoons [Fe(H_2O)_5OH]^{2+} + [Fe(H_2O)_6]^{2+} \tag{2.46}$$

Data for these reactions are included in Table 2.8. Values for $\Delta H^\ddagger$ and $\Delta S^\ddagger$ are similar to those for outer-sphere reactions but the value for $\Delta V^\ddagger$ for the reaction between $[Fe(H_2O)_5OH]^{2+}$ and $[Fe(H_2O)_6]^{2+}$ is considerably more positive than the value for the $[Fe(H_2O)_6]^{3+/2+}$ exchange, leading to the proposal that this represents an inner-sphere reaction. Further discussion of reactions of this type may be found in section 3.2.

Table 2.8. Rate and activation parameters for the [H$^+$]$^{-1}$ catalyzed pathway for self-exchange in metal aqua complexes

Reaction	μ (M)	k_{AA} (M^{-1} s^{-1})	$\Delta H^\ddagger$ (kJ mol^{-1})	$\Delta S^\ddagger$ (J K^{-1} mol^{-1})	$\Delta V^\ddagger$ (cm^3 mol^{-1})	Ref.
$[Fe(H_2O)_5OH]^{2+}/[Fe(H_2O)_6]^{2+}$	0.1	1360	28	−92	0.8	44,66
$[Co(H_2O)_5OH]^{2+}/[Co(H_2O)_6]^{2+}$	0.5	680	36	−96		71
$[Cr(H_2O)_5OH]^{2+}/[Cr(H_2O)_6]^{2+}$	1.0	0.7				62

2.6 APPLICATIONS OF MARCUS THEORY

In Fig. 2.5 is shown a plot of $\log(k_{AB})_{exp}$ against $\log(k_{AB})_{calc}$ for the data in Table 2.1 for which self-exchange rate data are available. With the simple electrostatic work function correction, agreement between theory and experiment is generally excellent, surprisingly so since factors such as hydrogen bonding and stacking interactions specific to certain ion pairs are ignored. This adherence to the Marcus expression is established as a criterion for outer-sphere processes since it can be argued that the rates of inner-sphere reactions will be much more dependent on the nature of the reaction partner. Further, it can be argued that inner-sphere processes will be faster than the outer-sphere reaction. If they are not, then the outer-sphere reaction will undoubtedly occur. However, there are some exceptions to this good agreement, particularly in reactions involving metal aqua complexes where the rates calculated by Marcus Theory are consistently larger than those obtained experimentally, particularly at large driving forces [1, 98].

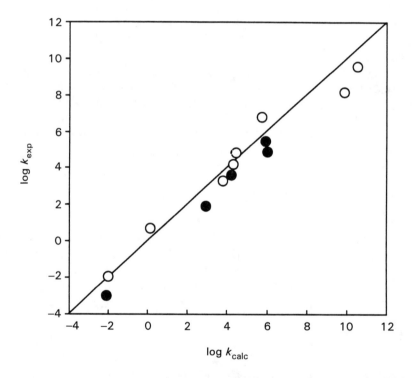

Fig. 2.5. Plot of $\log(k_{AB})_{exp}$ against $\log(k_{AB})_{calc}$ from Marcus Theory for reactions of the series of metal–ion complexes included in Table 2.1. Closed circles represent reactions involving aqua ions which, as a group, tend to fall below the line.

The Marcus expression is also of use for the computation of self-exchange rates for complexes where no direct determination is possible. This method relies on the measurement of rates for cross-reactions with well-characterized outer-sphere reactants where alternative mechanisms can be excluded. Particularly useful are [Ru(NH$_3$)$_6$]$^{3+/2+}$ and [Co(sep)]$^{3+/2+}$, but a number of others have also been employed. A partial list of such determinations corrected for electrostatic effects is presented in Table 2.9. Also included in this table are values for the self-exchange rates of the aqua ions calculated from cross-reaction data with the use of eq. (2.26). These values should be compared with the corresponding entries in Table 2.7.

Table 2.9. Outer-sphere self-exchange rates for reagents determined indirectly by Marcus Theory at 25 °C

Reductant	μ (M)	$E°$ (V)	k_{AA} (M^{-1} s^{-1})	Ref.
[V(pic)$_3$]$^{0/-}$	0.10	−0.41	1× 10^6	17
[Cr(H$_2$O)$_6$]$^{3+/2+}$	0.10	−0.41	≈ 1 × 10^{-10}	54
[Cr(bpy)$_3$]$^{3+/2+}$	0.10	−0.25	1 × 10^9	99
[Fe(H$_2$O)$_6$]$^{3+/2+}$	0.10	0.74	≈ 1 × 10^{-3}	63
[Fe(dipic)$_3$]$^{-/2-}$	0.10	0.36	2 × 10^7	100
[Fe(pic)$_3$]$^{0/-}$	0.10	0.48	5 × 10^5	101
[Fe(edta)]$^{2-/-}$	0.10	0.12	3 × 10^4	18
[Fe(CN)$_5$py]$^{2-/3-}$	0.05	0.48	7 × 10^5	102
[Co(H$_2$O)$_6$]$^{3+/2+}$	0.5	1.92	1 × 10^{-12}	103
[Co(terpy)$_2$]$^{3+/2+}$	0.10	0.31	4 × 10^2	104
[Co(Me$_4$en)$_2$]$^{3+/2+}$	1.0	0.28	1 × 10^{-7}	105
[Co(chxn)$_3$]$^{3+/2+}$	1.0	−0.16	8 × 10^{-5}	106
[Co(ox)$_3$]$^{3-/2-}$	0.10	0.57	3 × 10^{-7}	107
[Co(dipic)$_3$]$^{-/2-}$	0.10	0.75	1 × 10^{-5}	108
[Co(pic)$_3$]$^{0/-}$	0.10	0.43	2 × 10^{-3}	109
[Ni(bpy)$_3$]$^{3+/2+}$	1.00	1.72	1.5 × 10^3	16
[Ni([9]aneN$_3$)$_2$]$^{3+/2+}$	0.10	0.95	6 × 10^3 110	110
[Cu$_{aq}$]$^{2+/+}$	0.5	0.15	5 × 10^{-7}	111
[Ru(pic)$_3$]$^{0/-}$	0.10	0.40	1.4 × 10^8	109
[Ag$_{aq}$]$^{2+/+}$	0.10	1.98	2 × 10^{-9}	112
[Eu$_{aq}$]$^{3+/2+}$	0.10	−0.38	≈ 1 × 10^{-5}	54
[U$_{aq}$]$^{4+/3+}$	0.10	−0.63	≈ 6 × 10^{-5}	54
[(UO$_2$)$_{aq}$]$^{2+/+}$	0.10	0.04	≈ 10	113

The Marcus relationship has also proved most useful in mechanistic determinations in reactions where the reduction potentials, $E°$, and self-exchange rates of participating reagents are not known. It was noted [114] that the ratio of the rate constants for outer-sphere reactions of [Cr(H$_2$O)$_6$]$^{2+}$ and [V(H$_2$O)$_6$]$^{2+}$, $k_{[Cr(H_2O)_6]^{2+}}/k_{[V(H_2O)_6]^{2+}}$, was constant with a value of approximately 0.020, and that a mechanism could be assigned on this basis. The relationship may be recast [115] in a more useful form (eq. (2.47)) derived empirically from the outer-sphere reactions of cobalt(III) oxidants of the type [Co(NH$_3$)$_5$L]$^{2+}$ where L is an organic ligand. Reduction potentials cannot be determined for these complexes since the reduced forms are labile and rapidly dissociate.

$$\log k_{[V(H_2O)_6]^{2+}} = 1.1 \log k_{[Cr(H_2O)_6]^{2+}} + 1.85 \tag{2.47}$$

A word of caution, however, is required in that a similar relationship, eq. (2.48), is found when both reductants undergo a similar inner-sphere mechanism [116]. The distinguishing feature of the outer-sphere mechanism is the coefficient of the logarithm, which should be close to unity.

$$\log k_{[V(H_2O)_6]^{2+}} = 0.4 \log k_{[Cr(H_2O)_6]^{2+}} + 0.22 \qquad (2.48)$$

Clearly the fact that both ions are capable of reaction by both outer-sphere and inner-sphere mechanisms presents a complication and a better outer-sphere standard is $[Ru(NH_3)_6]^{2+}$ (eq. (2.49) and (2.50)). The Marcus relationship can be modified to predict the relationship between the rates of reactions of two reductants (eq. (2.51)).

$$[Co(NH_3)_5X]^{2+} + [M(OH_2)_6]^{2+} \xrightarrow{k_{AB}} \qquad (2.49)$$

$$[Co(NH_3)_5X]^{2+} + [Ru(NH_3)_6]^{2+} \xrightarrow{k_{AC}} \qquad (2.50)$$

$$\log k_{AC} = \log k_{AB} + \log \{(k_{CC}/k_{BB})(K_{BC})(f_{AC}/f_{AB})\} \qquad (2.51)$$

To a first approximation, the term f_{AC}/f_{AB} can be set equal to unity, particularly if the reduction potentials of the two reductant reagents are similar, and hence the second term in eq. (2.51) will be a constant. Where the reduction potentials and self-exchange rates of the two reductants differ markedly, or where they differ markedly in size and charge, the second term will vary. Thus the choice of a reference reductant is of considerable importance. Relationships of this sort have been obtained for a number of reductants (eqs (2.52)–(2.54)). In Fig. 2.6 are shown data for the reductions of a series of binuclear cobalt(III) complexes by $[Cr(OH_2)_6]^{2+}$ and $[V(OH_2)_6]^{2+}$ [117]. This analysis has been used extensively to demonstrate that reactions of $[V(OH_2)_6]^{2+}$ proceed by outer-sphere mechanisms. The relationships can also be used to estimate outer-sphere contributions in reactions which proceed predominantly by an inner-sphere mechanism. For example, in the $[Cr(H_2O)_6]^{2+}$ reduction of $[Co(NH_3)_5O_2CCF_3]^{2+}$ which takes place predominantly by an inner-sphere mechanism, the outer-sphere contribution to the rate is predicted to be 3.8% of the total by comparison with the rate for $[Ru(NH_3)_6]^{2+}$ [118].

$$\log k_{[Ru(NH_3)_6]^{2+}} = 1.05 \log k_{[V(H_2O)_6]^{2+}} + 0.48 \qquad (2.52)$$

$$\log k_{[Ru(NH_3)_6]^{2+}} = 1.05 \log k_{[Eu_{aq}]^{2+}} + 0.96 \qquad (2.53)$$

$$\log k_{[Ru(NH_3)_6]^{2+}} = 1.05 \log k_{[Cr(H_2O)_6]^{2+}} + 2.3 \qquad (2.54)$$

More recently, rate ratios have been used in the critical evaluation of rate data for the reduction of $[Co(NH_3)_6]^{3+}$ for which self-exchange rate data were suspect [75]. For a variety of outer-sphere reactions, the rate is approximately 10 times the rates for the corresponding reactions of [Co(en)$_3$]$^{3+}$ establishing that the self-exchange rate is approximately 2×10^{-7} M^{-1} s^{-1} at 25.0°C, four orders of magnitude faster than the rate determined by Stranks [119] but more in line with more recent determinations [74].

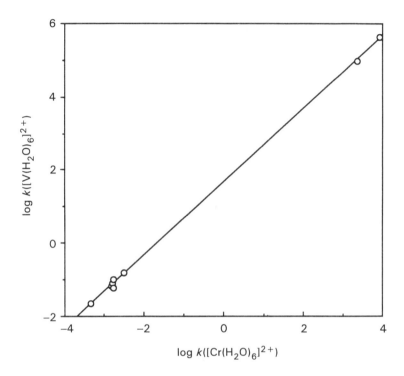

Fig. 2.6. Plot of log k([V[V(H$_2$O)$_6$]$^{2+}$) against k([Cr(H$_2$O)$_6$]$^{2+}$) for reduction of a number of binuclear cobalt(III) complexes from the data in Ref. [117]. Note that it is preferable to use a substitution inert ion such as [Ru(NH$_3$)$_6$]$^{2+}$ for these rate comparisons to avoid ambiguities with a reference reaction which may be inner-sphere.

2.7 PHOTOINDUCED ELECTRON TRANSFER

Irradiation of solutions of [Ru(bpy)$_3$]$^{2+}$ with 530 nm light produces the triplet metal–ligand charge transfer excited state, [*Ru(bpy)$_3$]$^{2+}$, which is long lived with a lifetime, τ_o, in aqueous solution on the order of 6.6×10^{-7} s [120]. The addition of photochemical energy to [Ru(bpy)$_3$]$^{2+}$ changes the thermodynamic driving force for reactions, including electron transfer, and gives the excited state a unique chemistry of its own. In this case, the geometry of the excited state is similar to that of the ground state and decay of the excited state is sufficiently slow to allow the new chemical reactivity to be probed. For example, the ground-state molecule is a poor reductant but on excitation to [*Ru(bpy)$_3$]$^{2+}$, this situation is dramatically changed [121, 122].

The energy of the excited state above the ground state can be calculated from the average of the energy for the singlet to triplet absorption (549 nm) and emission (610 nm) to be 2.12 eV. Consequently the reduction potential for eq. (2.55) is −0.84 V [123], obtained by the sum of the driving forces for eqs (2.56) and (2.57). As a result of the fact that the ground and excited states have similar geometry, the self-exchange rate for [Ru(bpy)$_3$]$^{3+}$/[*Ru(bpy)$_3$]$^{2+}$ is very fast, on the order of 1×10^9 M^{-1} s^{-1} [123]. The excited state is a more powerful reductant than [Cr(H$_2$O)$_6$]$^{2+}$ and extensive studies of the oxidative quenching of [*Ru(bpy)$_3$]$^{2+}$ have been carried out. However, considerable care is required in assigning mechanisms.

$$[Ru(bpy)_3]^{3+} + e^- \rightleftharpoons [*Ru(bpy)_3]^{2+} \tag{2.55}$$

$$[Ru(bpy)_3]^{3+} + e^- \rightleftharpoons [Ru(bpy)_3]^{2+} \tag{2.56}$$

$$[Ru(bpy)_3]^{2+} \rightleftharpoons [*Ru(bpy)_3]^{2+} \tag{2.57}$$

The excited state is not only a good reductant, but the excess energy may be lost by other quenching mechanisms. [*Ru(bpy)$_3$]$^{2+}$ is also a moderately powerful oxidant and so reductive quenching may also compete [124]. In addition, quenching by energy transfer mechanisms can be important where the quenching species has excited states of appropriate energy [125]. The reduction potentials are presented schematically in Scheme 2.2.

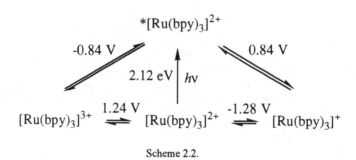

Scheme 2.2.

Most emphasis in this section is placed on oxidative quenching of [*Ru(bpy)$_3$]$^{2+}$ and a general mechanism for this process is presented in Scheme 2.3 with [Co(sep)]$^{3+}$ as a sample oxidant [126]. As with more conventional electron transfer reactions, the first step in the mechanism is the diffusion together of the reactants to form an encounter complex, k_1. In this complex, two things may happen. There may be energy transfer, k_3, ultimately regenerating the ground-state reactants, or electron transfer, k_2, may occur to give the products $\{[Ru(bpy)_3]^{3+}, [Co(sep)]^{2+}\}$, still trapped in their solvent cage. It must be realized that these products trapped in the solvent cage have a strong thermodynamic driving force to undergo back reaction, k_4, to regenerate the reactants. However, depending on the relative rates of k_4 compared with the rate at which the products can diffuse apart and escape from the cage, some free electron transfer products will result. In turn, these will react with solvent, with other components of the solution, or with each other. One common additive in solution is H_2edta^{2-} which undergoes a facile reaction with $[Ru(bpy)_3]^{3+}$ to give $[Ru(bpy)_3]^{2+}$ and H_2edta^-, thereby generating a solution of the reductant, in this case $[Co(sep)]^{2+}$.

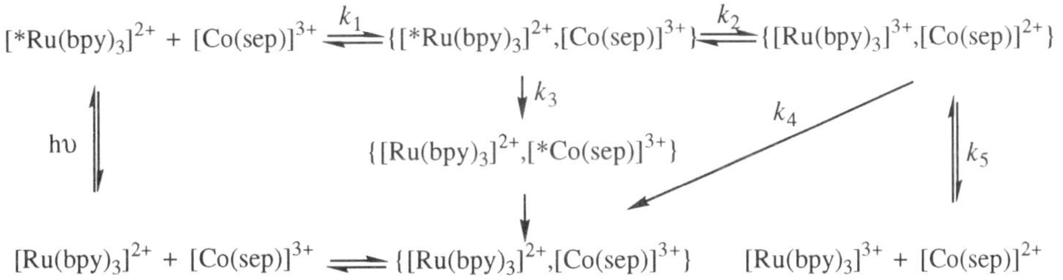

$$[*Ru(bpy)_3]^{2+} + [Co(sep)]^{3+} \underset{}{\overset{k_1}{\rightleftharpoons}} \{[*Ru(bpy)_3]^{2+},[Co(sep)]^{3+}\} \underset{}{\overset{k_2}{\rightleftharpoons}} \{[Ru(bpy)_3]^{3+},[Co(sep)]^{2+}\}$$

$$h\nu \downarrow \qquad \downarrow k_3 \qquad \qquad k_4 \qquad \qquad \updownarrow k_5$$

$$\qquad \qquad \{[Ru(bpy)_3]^{2+},[*Co(sep)]^{3+}\}$$

$$[Ru(bpy)_3]^{2+} + [Co(sep)]^{3+} \rightleftharpoons \{[Ru(bpy)_3]^{2+},[Co(sep)]^{3+}\} \qquad [Ru(bpy)_3]^{3+} + [Co(sep)]^{2+}$$

Scheme 2.3.

Quenching rates are close to the diffusion limit but can be corrected by eq. (2.58) where k_1 is the diffusion rate of approximately 2×10^9 M^{-1} s^{-1} for a +3 ion reacting with $[*Ru(bpy)_3]^{2+}$ at 0.2 M ionic strength, and $k_{qc} = K_1 k_2 + K_1 k_3$ is the corrected quenching rate constant. With the $[Co(sep)]^{3+}$, the rate constants for quenching exceed the rates expected on the basis of Marcus Theory [123] for electron transfer so that it may be concluded that a significant energy quenching contribution is present.

$$\frac{1}{k_q} = \frac{1}{k_1} + \frac{1}{k_{qc}} \qquad (2.58)$$

To separate the energy and electron transfer components, the quantum yield of the electron transfer products must be determined and this is given by eq. (2.59) where f_q is a correction factor for excited states unquenched by $[Co(sep)]^{3+}$ and $\Phi_{cage} = k_5/(k_4 + k_5)$ is the fraction of electron transfer products that escape from the cage in which they were formed. It is estimated that $k_{-5}/k_5 = 0.5$ M^{-1} [127]. A number of electron transfer rate constants are presented in Table 2.10.

$$\Phi_{Ru(III),Co(II)} = f_q \frac{k_2}{k_2 + k_3} \Phi_{cage} \qquad (2.59)$$

Table 2.10. Rate constants for reduction and oxidation by [*Ru(bpy)$_3$]$^{2+}$ at 25.0°C

Oxidant	µ (M)	k (M^{-1} s^{-1})	Ref.
[Fe(H$_2$O)$_6$]$^{3+}$	0.5	2.7 × 10^9	128
[Ru(NH$_3$)$_6$]$^{3+}$	0.5	2.7 × 10^9	129
[Co(bpy)$_3$]$^{3+}$	1.0	2.27 × 10^9	130
[Co(phen)$_3$]$^{3+}$	1.0	2.18 × 10^9	130
[Cr(bpy)$_3$]$^{3+}$	0.5	2.5 × 10^9	131
[Co(sep)]$^{3+}$	0.15	3.4 × 10^7	132
[Co(sar)]$^{3+}$	0.15	3 × 10^6	132
Reductive quenching			
[Ru(NH$_3$)$_6$]$^{2+}$	0.5	2.4 × 10^9	124
[Fe(CN)$_6$]$^{4-}$	0.5	3.5 × 10^9	124
[Eu$_{aq}$]$^{2+}$	0.5	2.8 × 10^7	124

While both oxidative and reductive electron transfer mechanisms are found to be the exclusive quenching mechanisms with a number of important reagents, energy quenching mechanisms are prevalent with some of the more sluggish oxidants. In particular the percentages of the total quenching process which proceed by electron transfer for [Co(sep)]$^{3+}$, [Co(en)$_3$]$^{2+}$ and [Co(NH$_3$)$_6$]$^{3+}$ are 23%, 11%, 45% respectively [132, 133].

Other excited state reagents which have been examined extensively are the ^{2}E d-d excited state, [*Cr(bpy)$_3$]$^{3+}$ and its derivatives [99, 134] and [*Cu(dmp)$_2$]$^+$. [*Cr(bpy)$_3$]$^{3+}$ is a useful oxidant with a reduction potential of 1.44 V (eq. (2.60)) [135], and has proved to be valuable for studies of [Cr(bpy)$_3$]$^{2+}$ in subsequent dark reactions. [*Cu(dmp)$_2$]$^+$ is a reductant [136, 137]. Also of considerable interest are excited-state porphyrin species which have been used to study reactions in biological systems. Further details of these reactions are presented in section 4.8.

$$[*Cr(bpy)_3]^{3+} + e^- \rightleftharpoons [Cr(bpy)_3]^{2+} \qquad (2.60)$$

$$[*Cu(dmp)_2]^+ + e^- \rightleftharpoons [Cu(dmp)_2]^{2+} \qquad (2.61)$$

2.8 NON-AQUEOUS MEDIA

A good number of investigations of self-exchange rates are now available for non-aqueous media. This extends the variety of reagents particularly to include reactions where one of the reagents is uncharged, and consequently electrostatic repulsion is less important. Nevertheless, salt effects due to ion-pairing effects with background electrolytes are more prevalent in non-aqueous media. Generally ion pairing reduces the reactivity of reagents, and considerable care is required in the application of Marcus Theory.

Selected self-exchange rates are presented in Table 2.11. The bulk of studies are performed in dipolar aprotic media, principally acetonitrile. For some reagents comparisons have been possible between different solvents and, although systematic approaches are not widespread, a few systems have been studied in some detail and are discussed in section 2.11.

Table 2.11. Selected self-exchange rates in non-aqueous media at 25.0°C

Reductant	Solvent	k $(M^{-1} s^{-1})$	$\Delta H^{\ddagger}$ $(kJ\,mol^{-1})$	$\Delta S^{\ddagger}$ $(J\,K^{-1}\,mol^{-1})$	$\Delta V^{\ddagger}$ $(cm^3\,mol^{-1})$	Ref
$[Ru_3O(OAc)_6(py)_3]^{+/0}$	$CH_2Cl_2{}^a$	1.1×10^8	18	−29		138
$[Mn(CNC_6H_4CH_3)_6]^{2+/+}$	$0.1, CD_3CN^b$	3.0×10^7	10	−67		139
$[Mn(CNCH_3)_6]^{2+/+}$	$0.1, CD_3CN^b$	2.1×10^7	7.5	−82	−2.4	140
$[Mn(CNC(CH_3)_3)_6]^{2+/+}$	$0.1, CD_3CN^b$	6.5×10^4	22.6	−77	−10.2	140
$[Mn(CNC_6H_{11})_6]^{2+/+}$	$0.08, CH_3CN$	4.38×10^5	14.2	−88	−17.4	141
$[Mn(bpyO_2)_3]^{3+/2+}$	$0.1, CH_3CN$	80				142
$[Fe(Cp)_2]^{+/0}$	$0.15, CH_3CN$	7.5×10^6	24	−38	−7	143,144
$[Co(Cp)_2]^{+/0}$	$0.15, CH_3CN$	3.8×10^7				145
$[Co(P(OMe)_3)_4]^{0/-}$	$0.07, THF$	9×10^3	28	−76		146
$[Ru(hfac)_3]^{-/0}$	$0.05, CH_3CN$	5.0×10^6	≈25			147

^{a}24°C. ^{b}26°C.

Volume of activation data cast interesting light on the reactions. Theory, described in section 2.12, predicts that in the self-exchange reactions of $[Mn(CNCH_3)_6]^{2+/+}$ and its derivatives $\Delta V^{\ddagger}$ values should be of the order of −5 cm^3 mol^{-1}, decreasing slightly as the ligand bulk increases. This is quite different from the observed trend which appears to be related to the flexibility of the ligands. Ligands with increased flexibility can distort to form a more compact electron transfer precursor complex, allowing for closer interaction of the donor and acceptor orbitals on the metal centers. This observation suggests that electron transfer is very sensitive to the Mn–Mn distance [148].

Where the Marcus relationship can be used, it holds reasonably well in non-aqueous media. It must be remembered that the reduction potentials of the complexes are solvent dependent and hence the driving force for each reaction must be known under the experimental conditions. A number of cross-reaction rate constants and activation parameters are presented in Table 2.12. The positive value for $\Delta V^{\ddagger}$ for the reaction of $[Cr(CNdipp)_6]^{2+}$ with $[Co(dpg)_3(BPh)_2]$ should be noted. Normally for electron transfer reactions $\Delta V^{\ddagger}$ is negative and in this case the unusual value is thought to be due to solvent release in attaining the transition state.

Table 2.12. Cross-reaction rates in non-aqueous solvents at 25.0°C

Oxidant and Reductant		Solvent	k_{AB} $(M^{-1} s^{-1})$	$\Delta H^{\ddagger}$ $(kJ\,mol^{-1})$	$\Delta S^{\ddagger}$ $(J\,K^{-1}\,mol^{-1})$	$\Delta V^{\ddagger}$ $(cm^3\,mol^{-1})$	Ref.
$[Co_2(CO)_8]$	$[Re(CO)_5]^-$	THF	13				149
$[Mn_2(CO)_{10}]$	$[Re(CO)_5]^-$	THF	1.4				149
$[Cp_2Mo_2(CO)_6]$	$[Re(CO)_5]^-$	THF	20				149
$[Fe(bpy)_3]^{3+}$	$[Os(bpy)_3]^{2+}$	CH_3CN	1.3×10^5				150

(continues)

72   The outer-sphere mechanism   [Ch. 2]

Table 2.12. *(continued)*

Oxidant and Reductant		Solvent	k_{AB} ($M^{-1} s^{-1}$)	$\Delta H^{\ddagger}$ (kJ mol^{-1})	$\Delta S^{\ddagger}$ (J K^{-1} mol^{-1})	$\Delta V^{\ddagger}$ (cm^3 mol^{-1})	Ref.
[Co(dmg)$_3$(BF)$_2$]$^+$	[Fe(Cp)$_2$]	CH$_3$CN	1.63×10^4	35	−46	−9	151,152
[Co(dmg)$_3$(BC$_6$H$_5$)$_2$]$^+$	[Fe(Cp)$_2$]	CH$_3$CN	4.6×10^3			−4	152,153
[Cr(CNdipp)$_6$]$^{2+}$	[Co(dpg)$_3$(BPh)$_2$]	CH$_3$CN	1.24×10^5	34	−33	10.8	154

A fascinating development has been the study of electron transfer self-exchange and cross-reaction rates in the gas phase by Fourier transform ion cyclotron resonance [155]. In this experiment a resonance current due to isotopically pure cation is dissipated when other isotopic forms of the neutral species are introduced into the sample chamber of the spectrometer. The rate of loss of the signal is directly related to the rate of electron transfer between the gas phase components. Results are most easily considered by comparing the electron transfer rate with the gas phase capture rate constant which yields a measure of the efficiency of the electron transfer within a gas phase encounter. This parameter is presented in Table 2.13. The low efficiency in the case of [Mn(Cp)$_2$]$^{+/0}$ is thought to be due to a substantial change in structure between the oxidized and reduced states.

Table 2.13. Electron transfer efficiencies of gas phase encounters

Reaction	Efficiency	Ref.
[Cr(Cp)$_2$]$^{+/0}$	0.48	156
[Mn(Cp)$_2$]$^{+/0}$	0.013	156
[Fe(Cp)$_2$]$^{+/0}$	0.27	156
[Fe(Cp)$_2$]$^{+/0}$	0.14	157
[Co(Cp)$_2$]$^{+/0}$	0.78	156
[Co(Cp)$_2$]$^{+/0}$	0.74	157
[Ni(Cp)$_2$]$^{+/0}$	0.65	156
[Ru(Cp)$_2$]$^{+/0}$	0.25	156
[Cr(CO)$_6$]$^{+/0}$	0.15	156

2.9   STEREOSELECTIVITY IN ELECTRON TRANSFER

A key question which arises in any discussion of outer-sphere electron transfer reactions is the relative proximity and orientation of the reactants during the time at which the electron is transferred. It has been implied from earlier discussions that the reactants are in intimate contact and that relative orientation is of little consequence but as yet no hard evidence for these points has been introduced. Attempts have been made to answer these questions by examining whether an optically active oxidant Δ-[A^{ox}] reacts at different rates with the Δ-[B^{red}] and Λ-[B^{red}] forms of a reductant (eqs (2.62) and (2.63)). It is thought that detection of significant stereo selectivity in the reaction indicates that the reactants are in intimate contact and from analysis of steric and electronic effects, orientation information might be obtained [158].

$$\Delta\text{-}[A^{ox}] + \Delta\text{-}[B^{red}] \xrightarrow{k_{\Delta\Delta}} \Delta\text{-}[A^{red}] + \Delta\text{-}[B^{ox}] \quad (2.62)$$

$$\Delta\text{-}[A^{ox}] + \Lambda\text{-}[B^{red}] \xrightarrow{k_{\Delta\Lambda}} \Delta\text{-}[A^{red}] + \Lambda\text{-}[B^{ox}] \quad (2.63)$$

The detection of stereoselectivity is not trivial and has provided a significant challenge. Some of the problems which arise can be seen in studies carried out on the reaction between $\Delta\text{-}[Co(phen)_3]^{3+}$ as a chiral oxidant and rac-$[Cr(phen)_3]^{2+}$ as a racemic reductant [159]. In this instance, the oxidant can be resolved but the reductant is too labile for resolution and, consequently, direct measurement of differences in the rates cannot be attempted. However, if there are differences in the rates for Δ- and $\Lambda\text{-}[Cr(phen)_3]^{2+}$, the substitution inert kinetic product, $[Cr(phen)_3]^{3+}$, might be expected to be optically active. In addition, if the chiral induction is small, product analysis is inherently more sensitive than kinetic methods because the errors on rate constants tend to be rather large [160].

Reaction of an equimolar mixture of $\Delta\text{-}[Co(phen)_3]^{3+}$ and rac-$[Cr(phen)_3]^{2+}$ does not lead to chiral induction in the $[Cr(phen)_3]^{3+}$ product [161]. The rate of the electron transfer reaction is very fast, 2×10^8 M^{-1} s^{-1} (eq. (2.64)) [162], whereas racemization of $[Cr(phen)_3]^{2+}$ (eq. (2.65)), is quite slow, 0.12 s^{-1} [163], so that equal amounts of both Δ- and $\Lambda\text{-}[Cr(phen)_3]^{3+}$ are produced. In other words for detection of optical activity in the kinetic product, the racemic reagent must remain a racemic mixture throughout the course of the reaction. The experiment can be changed to accommodate this requirement by using a large excess of rac-$[Cr(phen)_3]^{2+}$ over the oxidant so that any change in the concentrations of Δ- or $\Lambda\text{-}[Cr(phen)_3]^{2+}$ will be negligible. Unfortunately, under these conditions, racemization of optically active $[Cr(phen)_3]^{3+}$ will take place by the rapid self-exchange reaction (eq. (2.66)), which has a rate constant of 1×10^9 M^{-1} s^{-1} [164].

$$\Delta\text{-}[Co(phen)_3]^{3+} + \text{rac-}[Cr(phen)_3]^{2+} \rightarrow \Delta\text{-}[Co(phen)_3]^{2+} +$$
$$+ (\text{rac-}[Cr(phen)_3]^{3+}) \quad (2.64)$$

$$\Delta\text{-}[Cr(phen)_3]^{2+} \rightleftharpoons \Lambda\text{-}[Cr(phen)_3]^{2+} \quad (2.65)$$

$$\Delta\text{-}[Cr(phen)_3]^{3+} + \text{rac-}[Cr(phen)_3]^{2+} \rightleftharpoons \Delta\text{-}[Cr(phen)_3]^{2+}$$
$$+ \text{rac-}[Cr(phen)_3]^{3+} \quad (2.66)$$

$[Cr(phen)_3]^{2+}$ is a poor probe for the detection of electron transfer stereoselectivity. Since it is important that the racemic reagent be labile and that the kinetic product should have a low self-exchange rate with the racemic form, reagents such as $[Co(edta)]^{2-/-}$ and $[Co(en)_3]^{3+/2+}$ are ideal for this purpose and have been used extensively as stereoselectivity probes [158]. A number of results of these studies are listed in Table 2.14. The results are quoted as the enantiomeric excess of the dominant isomer in the product. In general, the stereoselectivity is small and would be difficult to detect by kinetic methods, amounting to 10–20% difference in rate constants. A 10% ΔΔ enantiomeric excess corresponds to $k_{\Delta\Delta}/k_{\Lambda\Lambda} = 55/45$ or 1.22. Nevertheless kinetic determinations have been made in suitable instances, and where comparisons are possible, they show good agreement with the results of product analysis [165]. Although small, the stereoselectivity is significant and it can be concluded that the reactants are in intimate contact during the reaction. In addition, the stereoselectivity is very sensitive to the structure of the reactants.

Table 2.14. Stereoselectivity in outer-sphere electron transfer reactions at 25°C

Reaction	μ (M)	k (M^{-1} s^{-1})	Stereoselectivity (% ee)	Ref.
$[Co(edta)]^- + [Co(en)_3]^{2+}$	1.0	5.3	9% ΔΛ	160
$[Co(edta)]^- + [Co(sep)]^{2+}$	0.1	6×10^4	17% ΔΛ	166
$[Co(edta)]^- + [Co((\pm)\text{-chxn})_3\text{-lel}_3]^{2+}$	0.1		24% ΔΛ	166
$[Co(ox)_3]^{3-} + [Co(en)_3]^{2+}$	0.1	3.9×10^2	8% ΔΛ	167
$[Co(ox)_3]^{3-} + [Co(sep)]^{2+}$	0.1		21% ΔΛ	167
$[Co(ox)_3]^{3-} + [Co((\pm)\text{-chxn})_3\text{-lel}_3]^{2+}$	0.1		38% ΔΛ	167
$[Co(ox)_3]^{3-} + [Co(phen)_3]^{2+}$	0.017	15.8	24% ΔΛ	28
$[Co(phen)_3]^{3+} + [Co(4,7\text{-Me}_2\text{phen})_3]^{2+}$	0.1	2×10^3	15% ΔΛ	55
$[Ru(bpy)_3]^{3+} + [Co(edta)]^{2-}$			<1% ΔΛ	160
$[Co(edta)]^- + [*Ru(bpy)_3]^{2+}$			7% ΔΛ	168

Information on the interactions comes from comparisons of stereoselectivity in the electron transfer reactions with stereoselectivity in ion-pair formation between analogues for the reactants. These ion pairs model the interactions which are expected in the formation of the electron transfer precursor complex. For example, the ion pair formed between $[Rh(ox)_3]^{3-}$ and $[Ru(phen)_3]^{2+}$ serves as a model for the precursor complex in the reaction between $[Co(ox)_3]^{3-}$ and $[Co(phen)_3]^{2+}$ (eq. (2.67)) [28]. Conductivity studies reveal that $K_{os\Delta\Lambda} = 206$ M^{-1} while $K_{os\Delta\Delta} = 195$ M^{-1}, giving chiral recognition $K_{\Delta\Lambda}/K_{\Delta\Delta} = 1.05$, smaller but with the same preference as the chiral induction in the electron transfer process where $k_{\Delta\Lambda}/k_{\Delta\Delta} = 1.7$. NMR investigations have been used to provide structural information on the ion-pairs and it is proposed that there is a dominant interaction along the C_3 axes of the complexes (Fig. 2.7).

$$[Co(ox)_3]^{3-} + [Co(phen)_3]^{2+} \underset{}{\overset{K_{os}}{\rightleftharpoons}} \{[Co(ox)_3]^{3-}, [Co(phen)_3]^{2+}\} \overset{k_{et}}{\rightarrow}$$

$$[Co(ox)_2]^{2-} + [Co(phen)_3]^{2+} + ox^{2-} \quad (2.67)$$

Stereoselectivity has also been detected in reactions between like-charged reagents where no strong ion-pairing interaction is to be expected. The evidence is strong

Sec. 2.10] Theoretical detals 75

that the reagents come into intimate contact and the implication is that orientation plays a significant role in electron transfer.

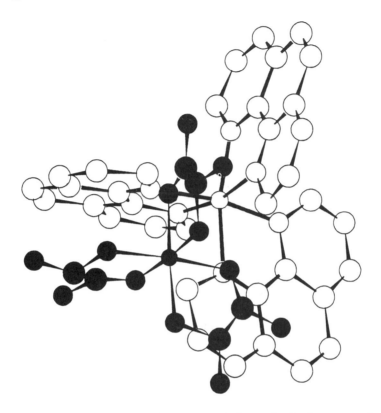

Fig. 2.7. Proposed orientation derived from nmr and other studies for the precursor in the reaction between $[Co(ox)_3]^{3-}$ (shaded) and $[Co(phen)_3]^{2+}$, [28].

2.10 THEORETICAL DETAILS

The remarkable success of Marcus Theory focuses attention on the self-exchange rates as characteristic quantities which express the dynamics of electron transfer, independent of driving force. There are some exceptions to this success, particularly among the metal aqua ions, but there does not seem to be any pronounced dependence on the electronic stucture of the metal ions or on the nature of the ligands. As indicated previously, adherence to the Marcus relationship for a reagent requires that the activation processes involved in electron transfer are independent of reaction partner. It does not necessarily imply that all the reactions are adiabatic, merely that the degree of adiabaticity is the same for the cross-reactions as it is for the self-exchange reaction. However, in itself this behavior argues strongly for adiabatic character since it unlikely that all cross-reactions examined will have a consistent degree of non-adiabaticity. In order to explore this argument and to rationalize the range of self-exchange rates in Tables 2.7 and 2.9, it is necessary to examine the electron transfer process in greater detail.

Within the series in Tables 2.7 and 2.9 there is a considerable variation in the self-exchange rates covering almost 20 orders of magnitude, and this variation shows some trends with the electronic structures of the reactants. Some understanding of the molecular basis for these trends is possible within the framework of classical dynamics with minor additions from quantum theory [4]. This semi-classical approach is by no means the only way to examine the trends which are observed and other detailed statistical and quantum mechanical treatments are now available [169–171]. However, at present their application is more limited and the simpler treatment has the distinct advantage that it works relatively well for reactions at ambient temperatures. At lower temperatures, quantum mechanical effects such as tunneling are much more important.

Consider the model for electron transfer presented in eqs (2.68) and (2.69) where there is initial formation of a precursor complex followed by the electron transfer. The rate constant for the reaction can be written as in eq. (2.70). In this expression, k_o is the diffusion rate constant and k_{et} is the rate constant for electron transfer within the precursor given by eq. (2.71), where υ_{eff} is the effective nuclear frequency, κ_{el} the electronic factor, and $\Delta G^{\ddagger}$ is the activation barrier. For completeness, another factor Γ, a tunneling factor, is added to the pre-exponential term. As noted in section 1.11, the tunneling factor originates from a quantum mechanical effect in which the reactants proceed to products by tunneling through the activation barrier for the reaction with less energy than it takes to go over the barrier. In most instances, at ambient temperatures, it is close to unity but for reactions with large barriers, it can become important. The various factors affecting the rate will now be examined in some detail. It must be remembered that the precursor assembly comprises of a variety of different structures at different distances and orientations and that each configuration will have its own value for the quantities υ_{eff}, κ_{el}, Γ, and $\Delta G^{\ddagger}$. However, for simplicity, a single 'averaged' configuration will be considered and quantities related to this will be evaluated.

$$[A^{ox}] + [A^{red}] \underset{k_{-o}}{\overset{k_o}{\rightleftharpoons}} \{[A^{red}],[A^{ox}]\} \qquad (2.68)$$

$$\{[A^{red}],[A^{ox}]\} \overset{k_{et}}{\rightarrow} [A^{red}] + [A^{ox}] \qquad (2.69)$$

$$k = k_o k_{et}/(k_{-o} + k_{et}) \qquad (2.70)$$

$$k_{et} = \upsilon_{eff}\,\kappa_{el}\,\Gamma\,\exp(-\Delta G^{\ddagger}/RT) \qquad (2.71)$$

2.11 THE ACTIVATION BARRIER $\Delta G^{\ddagger}$

Far from the diffusion limit, the activation barrier for the self-exchange process within the precursor assembly, $\Delta G^{\ddagger}$ is given by $\lambda/4$, and reflects the changes in the reactants and the solvent which are necessary to attain the compromise geometry of the transition state, Fig. 2.8. It is essential to separate λ into a component λ_{in}, derived from the reorganization of the bond lengths and angles within the precursor complex assembly, and a component, λ_{out}, reflecting the rearrangement of solvent as the charge is transferred (eq. (2.72)).

$$\lambda = \lambda_{in} + \lambda_{out} \tag{2.72}$$

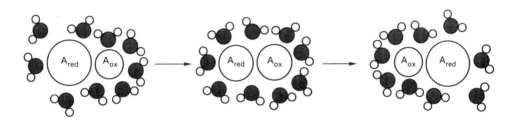

Fig. 2.8. Changes in the inner-coordination sphere, λ_{in}, indicated by the different sizes of the reagents, and the outer-coordination sphere, λ_{out}, indicated by the reorientation of the solvent, which are necessary to attain the compromise geometry of the transition state.

Contributions to the inner-sphere term, λ_{in}, arise from the reorganization of the complexes to overcome the Frank–Condon barriers associated with the differences in equilibrium bond lengths for the oxidized and reduced forms. For efficient electron transfer, the activated complex has an intermediate geometry and the activation energy can be approximated as a simple harmonic potential (eq. (2.73)), where f_i^{red} and f_i^{ox} are force constants for the ith vibrations of the reduced and oxidized species, d^{red} and d^{ox} are the equilibrium bond distances for the species and $d^{\ddagger}$ is the bond distance appropriate for the transition state in the reaction. There will also be contributions from angle deformations but these generally are less energetically important.

$$\lambda_{in} = 1/2 \sum f_i^{red} (d^{red} - d^{\ddagger})_i^2 + 1/2 \sum f_i^{ox} (d^{ox} - d^{\ddagger})_i^2 \tag{2.73}$$

The major changes in structure are generally found in the metal–ligand bond length and metal–ligand bond stretches vary with oxidation state. For example, the Co—N stretch in $[Co(NH_3)_6]^{2+}$ is 357 cm^{-1} and for $[Co(NH_3)_6]^{3+}$ it is 494 cm^{-1} and the force constants can be estimated from normal coordinate analysis.

A significant problem with the application of eq. (2.73) is that the geometry of the transition state must be known to partition the bond length change between the two oxidation states. This can be overcome by using the approximation (eq. (2.74)), where f_i is a reduced force constant $(2\, f^{red}\, f^{ox}/(f^{red} + f^{ox}))$ for the ith inner-sphere vibration of each reactant complex and $(d^{red} - d^{ox})_i$ is the corresponding difference in equilibrium bond length for the two oxidation states. Some representative estimates for reduced force constants are 170 N m^{-1} ([Co(NH$_3$)$_6$]$^{3+/2+}$), 220 N m^{-1} ([Ru(NH$_3$)$_6$]$^{3+/2+}$), 200 N m^{-1} ([Fe(H$_2$O)$_6$]$^{3+/2+}$), 100 N m^{-1} ([Co([9]aneS$_3$)$_2$]$^{3+/2+}$).

$$\lambda_{in} = 1/2 \sum f_i\, (d^{red} - d^{ox})_i^2 \qquad (2.74)$$

Although the magnitude of the inner-sphere component is dependent on all structural changes between the two oxidation states, the largest change is generally recorded in the metal–ligand bond lengths and this is generally the sole structural change which is considered. More sophisticated methods using molecular mechanics to examine all the important bond length and bond angle deformations have also been reported but the method is not in general use [172]. Additionally, experimental measurement of λ_{in} can be made from photoemission experiments [173, 174].

In Table 2.15, the metal–ligand bond length changes for a number of complexes are listed. Some of these have been obtained from x-ray structures of the oxidized and reduced complexes in the solid state; others have been obtained from EXAFS [175, 176] data directly on species in solution. In instances where both types of measurements have been carried out, there is good agreement between the methods.

Table 2.15. Correlation of bond-length changes with electron transfer rate constants

Reaction	Δd_o (Å)	k_{AA} (M^{-1} s^{-1})	Ref.
[Fe(H$_2$O)$_6$]$^{3+/2+}$	0.13	1.1	44
[Co(OH$_2$)$_6$]$^{3+/2+}$	0.21	5	44
[Cr(OH$_2$)$_6$]$^{3+/2+}$	0.20	<1.9 × 10^{-5}	44
[V(OH$_2$)$_2$]$^{3+/2+}$	0.13	1 × 10^{-2}	177, 178
[Ru(OH$_2$)$_6$]$^{3+/2+}$	0.09	20	179, 180
[Co(en)$_3$]$^{3+/2+}$	0.21	7.7 × 10^{-5}	72
[Ru(en)$_3$]$^{3+/2+}$	≈0.05	3.1 × 10^4	90
[Co(NH$_3$)$_6$]$^{3+/2+}$	0.22	2 × 10^{-8}	75
[Ru(NH$_3$)$_6$]$^{3+/2+}$	0.04	6.6 × 10^3	91
[Co(sep)]$^{3+/2+}$	0.17	5.1	56
[Co([9]aneN$_3$)$_2$]$^{3+/2+}$	0.18	0.19	181
[Ni([9]aneN$_3$)$_2$]$^{3+/2+}$	0.09	1.2 × 10^4	182, 183
[Fe([9]aneN$_3$)$_2$]$^{3+/2+}$	0.04	4.6 × 10^3	184–186
[Co([9]aneS$_3$)$_2$]$^{3+/2+}$	0.07	1.3 × 10^4	80,81

(continues)

Table 2.15 (continued)

Reaction	Δd_o (Å)	k_{AA} (M^{-1} s^{-1})	Ref.
[Mn(sar)]$^{3+/2+}$	0.11	39	63
[Fe(sar)]$^{3+/2+}$	0.04	7×10^5	63
[Ni(sar)]$^{3+/2+}$	0.09	4×10^3	63
[Ru(sar)]$^{3+/2+}$	0.01	6×10^5	78
[Co(sar)]$^{3+/2+}$	0.19	2.1	54
[Co(phen)$_3$]$^{3+/2+}$	0.19	12	55
[Co(bpy)$_3$]$^{3+/2+}$	0.19	5.7	28
[Co(bpy)$_3$]$^{2+/+}$	−0.02	5.7	187
[Ru(bpy)$_3$]$^{3+/2+}$	0.00	4×10^8	188
[Fe(phen)$_3$]$^{3+/2+}$	0.00	1.3×10^7	68
[Fe(bpy)$_3$]$^{3+/2+}$	0.00	3×10^8	189
[Cr(bpy)$_3$]$^{3+/2+}$	0.10	1×10^9	190
[Ni(bpy)$_3$]$^{3+/2+}$	0.12	1.5×10^3	191–193
[Co(terpy)$_2$]$^{3+/2+}$	0.13	4×10^2	44
[Fe(CN)$_6$]$^{3-/4-}$	0.03	2×10^4	70
[Ru(CN)$_6$]$^{3-/4-}$	0.04	8.3×10^3	20

Data for several series of compounds where structural data are available for both oxidized and reduced forms are listed. For each type of structurally similar complexes such as [M(bpy)$_3$]$^{3+/2+}$ or [M(sar)]$^{3+/2+}$, the data show a good correlation between the magnitude of the self-exchange rate and the structural rearrangement involved. Particular attention should be focused on the series of ruthenium complexes with amine donor atoms. In this series the Ru—N bond distance varies little between the oxidized and reduced forms and consistently the rates are faster than for structurally similar complexes with other metal ions. For this series, the approximation can be made that $\lambda_{in} \approx 0$ and hence that the rate of the electron transfer process will be largely dependent on λ_{out}.

For structurally similar complexes, some idea of the applicability of eq. (2.74) can be obtained. For example, the rate constant for the [M(sar)]$^{3+/2+}$ self-exchange can be written as eq. (2.75) and approximated as eq. (2.76) provided the pre-exponential factors are substantially constant for this series of complexes. Plots of $(\lambda_{in}/4RT)$ obtained from eq. (2.76) against $(d^{red} - d^{ox})^2$ for several series are shown in Fig. 2.9, and there is a plausible correlation despite the crudity of the model.

$$\ln k_{[M(sar)]} = \ln(K_o \upsilon_{eff} \Gamma \kappa_{el}) - (\lambda_{out}/4RT) - (\lambda_{in}/4RT) \quad (2.75)$$

$$\ln k_{[M(sar)]} = \ln k_{[Ru(sar)]} - (\lambda_{in}/4RT) \quad (2.76)$$

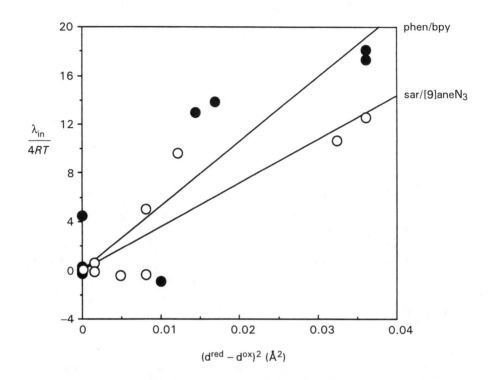

Fig. 2.9. Plot of ($\lambda_{in}/4RT$) obtained from eq. (2.76) against $(d^{red} - d^{ox})^2$ for $[M([9]aneN_3)_2]^{3+/2+}$ and $[M(sar)]^{3+/2+}$, open circles; and $[M(phen)_3]^{3+/2+}$ and $[M(bpy)_3]^{3+/2+}$, closed circles. The correlations are not particularly well defined. Differences in slope for the two sets of compounds indicate differences in the force constants associated with the distortions.

For the $[M(bpy)_3]^{3+/2+}$ and $[M(sar)]^{3+/2+}$ series, there are excellent correlations revealing no dependence on the different metal ion electronic structures involved and although the force constants vary within each series, this does not appear to be of great importance. However, for the $[M(H_2O)_6]^{3+/2+}$ series, the correlation is much poorer.

The self-exchange rates for reactions of ruthenium amine complexes where λ_{in} is small show a trend, decreasing with decreasing ligand bulk. In part, this might be ascribed to the electrostatic repulsion between the ions in the formation of the precursor complex but correction for this in a plot of $\log(k/K_o)$ against $1/a$ (Fig. 2.10), where a is the radius of the complex, shows that the general trend is maintained. The complexes considered have the same charges in the oxidized and reduced forms and hence the interaction with solvent will be dependent on a^{-1}. As with inner-sphere reorganization, the outer coordination sphere composed principally of solvent dipoles must also rearrange prior to the electron transfer process. A value for the outer-sphere term, λ_{out}, can be estimated from a classical electrostatic treatment for the movement of charge in a medium of continuous dielectric constant and is given by eq. (2.77).

$$\lambda_{out} = (\Delta e)^2 \ (1/2a_A + 1/2a_B - 1/r)(1/D_{op} - 1/D_s) \tag{2.77}$$

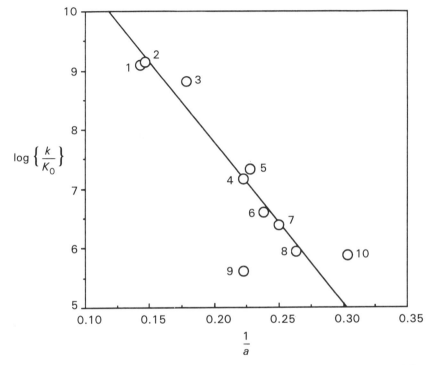

Fig. 2.10. Plot of $\log(k/K_o)$ against $1/a$ for ruthenium amine complexes; (1) $[Ru(bpy)_3]^{3+/2+}$, (2) $[Ru(phen)_3]^{3+/2+}$, (3) $[Ru(NH_3)_2(bpy)_2]^{3+/2+}$, (4) $[Ru(sar)]^{3+/2+}$, (5) $[Ru(NH_3)_4(bpy)]^{3+/2+}$, (6) $[Ru(en)_3]^{3+/2+}$, (7) $[Ru([9]aneN_3)_2]^{3+/2+}$, (8) $[Ru(NH_3)_5spy]^{3+/2+}$, (9) $[Ru(NH_3)_5(isonic)]^{3+/2+}$, (10) $[Ru(NH_3)_6]^{3+/2+}$, from Refs [57] and [93].

In this expression, the properties of the solvent are rather crudely represented by $(1/D_{op} - 1/D_s)$ where D_s is the static dielectric constant, and D_{op} is the optical dielectric constant equal to the square of the refractive index. The term $(1/D_{op} - 1/D_s)$ takes account of the slower frequency response of solvent orientation and vibration to instantaneous movement of charge, reflected in D_s, and the much faster response of the solvent electronic polarization, reflected in D_{op}. The expression eq. (2.77) is valid for moderately well separated ions where $r \gg (a_A + a_B)/2$. However, the commonly held view of electron transfer involves intimate contact between the reactants. Modification of eq. (2.77) to account for close contact leads to considerable complications [194] and, in practice, the simplicity of eq. (2.77) leads to its widespread application. Despite these limitations, eq. (2.77) has the same functional dependence on the radius of the ions as is observed in the experimental data.

A key test of the expression for λ_{out} is the prediction of the solvent dependence of the self-exchange rate. A plot of $\log k_{AA}$ against $(1/D_{op} - 1/D_s)$ for the self-exchange rate of $[Cr(C_6H_6)_2]^{+/0}$ in different solvents is shown in Fig. 2.11. The slope gives a value of 6.4 Å for the radius, a_A, of the complex, when $a_A = a_B$ and $r = 2a_A$ [195, 196]. Unfortunately for a number of bimolecular systems, correlation of the rate with $(1/D_{op} - 1/D_s)$ is poor even when the effects of precursor formation are taken into account. In some instances the correlation is in the opposite sense from that predicted [141]. Other approaches have been attempted. Part of the problem appears to be that specific interactions between the solvent and solute are not taken into account properly [197–199]. There has been an attempt to replace the expression with an empirical fit to the Taft parameters. These take account of the acidity, basicity and polarizability of the solvent [200], but the correlations are empirical and have little predictive value at present. Despite this failure of theory to predict the rate dependencies on solvent accurately, there is sufficient evidence to merit continued use of eq. (2.77).

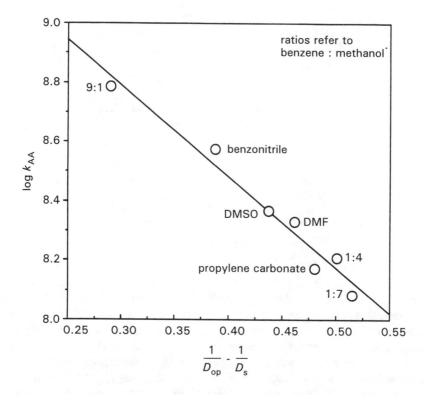

Fig. 2.11. Plot of $\log k_{AA}$ against $(1/D_{op} - 1/D_s)$ for the slef-exchange of $[Cr(C_6H_6)_2]^{+/0}$ in a number of different solvents, from Ref. [196].

2.12 ACTIVATION PARAMETERS AND ISOTOPE EFFECTS

At this point some consideration can be given to the computation of values for activation parameters. There is no well-defined prescription for the prediction of $\Delta H^{\ddagger}$ and $\Delta S^{\ddagger}$ for outer-sphere electron transfer [3]. It must be recalled that the rates for self-exchange reactions are generally first-order in both reactants so that the activation parameters are composite and reflect both precursor association and the electron transfer within the assembly (eqs (2.78) and (2.79)).

$$\Delta H^{\ddagger} = \Delta H^{\circ}_{os} + \Delta H^{\ddagger}_{et} \tag{2.78}$$

$$\Delta S^{\ddagger} = \Delta S^{\circ}_{os} + \Delta S^{\ddagger}_{et} \tag{2.79}$$

Qualitatively, it has been pointed out that for reactions between ions of similar charge, ΔH°_{os} and ΔS°_{os} are expected to be small and negative for formation of the assembly due to an increase in electrostricted water. For the electron transfer within the assembly, the activation process involves bond stretching and solvent rearrangement so that $\Delta H^{\ddagger}_{et}$ should be positive and directly related to the Frank–Condon barrier, depending on the magnitude of the structural rearrangement involved. It is expected that $\Delta S^{\ddagger}_{et}$ should be negative because the transition state has a higher charge and therefore is significantly more ordered than the reactants. Thus overall, $\Delta H^{\ddagger}$ should be small and positive while $\Delta S^{\ddagger}$ is negative and this is borne out by experiments.

The volume of activation, $\Delta V^{\ddagger}$, can be calculated from eq. (2.80) [31, 201, 202], where $\Delta V^{\ddagger}_{col}$ measures the coulombic interaction on formation of the transition state, $\Delta V^{\ddagger}_{out}$ is the contribution from solvent reorganization, $\Delta V^{\ddagger}_{in}$ is the contribution from changes in the metal–ligand bonds and $\Delta V^{\ddagger}_{dh}$ is a Debye–Hückel term dealing with changes in activity as a result of interactions with other ions in solution.

$$\Delta V^{\ddagger} = \Delta V^{\ddagger}_{col} + \Delta V^{\ddagger}_{out} + \Delta V^{\ddagger}_{in} + \Delta V^{\ddagger}_{dh} \tag{2.80}$$

The relative contributions from the various terms are collected in Table 2.16 for a number of reactions in aqueous solution where $\Delta V^{\ddagger}$ measurements have been made experimentally.

Table 2.16. Calculated and experimental values for $\Delta V^{\ddagger}$ (cm^3 mol^{-1})

	$\Delta V^{\ddagger}_{col}$	$\Delta V^{\ddagger}_{out}$	$\Delta V^{\ddagger}_{in}$	$\Delta V^{\ddagger}_{dh}$	$\Delta V^{\ddagger a}_{calc}$	$\Delta V^{\ddagger}_{expt}$	Ref.
$[Fe(H_2O)_6]^{3+/2+}$	−7.1	−5.7	+0.5	+3.8	−8.5	−11.0	66
$[Co(en)_3]^{3+/2+b}$	−8.1	−6.6	+0.6	+4.4	−9.5	−20.0	73
$[Co(sep)_3]^{3+/2+}$	−6.1	−4.9	+0.6	+3.8	−6.6	−6.4	76

[a]Ref. [31], [b]65 °C.

The agreement is mixed. $\Delta V^{\ddagger}_{in}$ is generally a minor term and the activation volumes are dominated by solvation terms, particularly $\Delta V^{\ddagger}_{out}$. A significant disagreement between $\Delta V^{\ddagger}_{expt}$ and $\Delta V^{\ddagger}_{calc}$ suggests that the reactions are not determined solely by

Frank–Condon effects and hence that contributions from the electronic factor may be important [203].

Hydrogen/deuterium isotope effects have been examined for a number of outer-sphere electron transfer reactions including the $[Fe(H_2O)_6]^{3+/2+}$ exchange [204] and the reduction of $[Co(NH_3)_6]^{3+}$ by $[Cr(bpy)_3]^{2+}$ [164]. It is concluded that the effect is small, consistent with a dominant role for metal–ligand bond stretching in the activation process. Calculation of the isotope effects at ambient temperatures provides good agreement [205, 206].

2.13 THE PRE-EXPONENTIAL FACTORS

Up to this point no account has been taken of the pre-exponential factors in eq. (2.71), and indeed from a practical point of view these factors are revealed in deviations from the simple model for dealing with λ_{in} and λ_{out}. The effective nuclear frequency, υ_{eff}, is the frequency with which the activated complex breaks down to give products. It is generally in the range 10^{12} s^{-1} to 10^{14} s^{-1} [4]. Where λ_{in} is large, internal rearrangements will dominate the activation process and the effective nuclear frequency will correspond to some combination of internal breathing modes of the complexes, typically around 400 cm^{-1} for a M—N bond stretch, corresponding to $\upsilon_{eff} \approx 10^{13}$ s^{-1}. Where internal rearrangements are less important, reorientation of solvent dipoles will dominate the energetics of the process. The relative contributions are weighted according to eq. (2.81).

$$\upsilon_{eff} = \left\{ \frac{\upsilon_{out}^2 \lambda_{out} + \upsilon_{in}^2 \lambda_{in}}{\lambda_{out} + \lambda_{in}} \right\}^{1/2} \tag{2.81}$$

Sutin has developed a useful model for examining the effective frequency for barrier crossing [46]. The molecular motions are separated into a low frequency (approximately 10^{12} s^{-1}) component from the solvent, and a high frequency intramolecular component from the reactants (Fig. 2.12). As the reactants proceed to products along the dotted line corresponding to the pathway of steepest descent from the activated complex, the low frequency mode is relatively more important in the initial stages of the reaction, whereas close to the transition state the high frequency mode is more important.

In instances where the activation energies are modest, solvent dynamics can play an important role in determining the barrier crossing frequency [207–209]. The solvent mode is inversely related to the longitudinal solvent relaxation time, τ_L, as shown in eq. (2.82).

$$\upsilon_{out} = \tau_L^{-1} \left\{ \frac{\lambda_{out}}{16\pi kT} \right\}^{1/2} \tag{2.82}$$

For water, $\tau_L^{-1} = 1.9 \times 10^{12}$ s^{-1}, but this is reduced considerably in less polar media, and the effects of solvent dynamics have now been investigated extensively for [Fe(Cp)$_2$]$^{+/0}$ and [Co(Cp)$_2$]$^{+/0}$ [145, 210, 211].

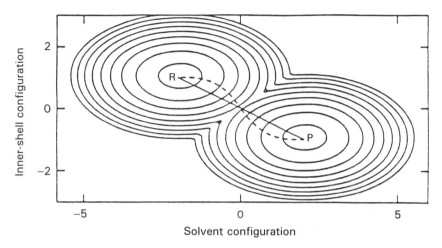

Fig. 2.12. Separation of molecular motions into a low-frequency component from the solvent and a high-frequency intramolecular component from the reactants. The dotted pathway gives the lowest energy route from the precursor potential energy well to the successor potential energy well, from Sutin, N.; Brunschwig, B.S.; Creutz, C.; Winkler J.R. *Pure Appl. Chem.* 1988, **60**, 1819 with permission.

The electronic factor, κ_{el}, has has no amenable classical approximation and determination of this parameter requires measurement or calculation by quantum mechanical treatment. However, as a starting point for the discussion of the electronic factor or electronic transmission coefficient, the metal complex reactants can be considered as hard spheres which come into contact in the precursor ion pair with sufficient orbital overlap to allow electron transfer with unit probability ($\kappa_{el} = 1$). This is the adiabatic limit where the electron travels smoothly from the reactant surface to the product surface, (Fig. 2.13), and represents a limiting form of the Landau–Zener expression (eq. (2.83)), where the electronic frequency, υ_{el}, is much greater than the effective nuclear frequency, υ_{eff}.

$$\kappa_{el} = \frac{2(1 - \exp(-\upsilon_{el}/2\upsilon_{eff}))}{2 - \exp(-\upsilon_{el}/2\upsilon_{eff})} \tag{2.83}$$

The electronic frequency is related to the extent of the overlap between donor and acceptor orbitals (eq. (2.84)), and can be estimated by calculation of the electronic matrix coupling element $H_{AB} = \langle \psi_A | H | \psi_B \rangle$ where ψ_A and ψ_B are the wavefunctions for the donor and acceptor orbitals respectively.

$$\upsilon_{el} = \frac{2 H_{AB}^2}{h} \left\{ \frac{\pi^3}{(\lambda_{out} + \lambda_{in})RT} \right\}^{1/2} \tag{2.84}$$

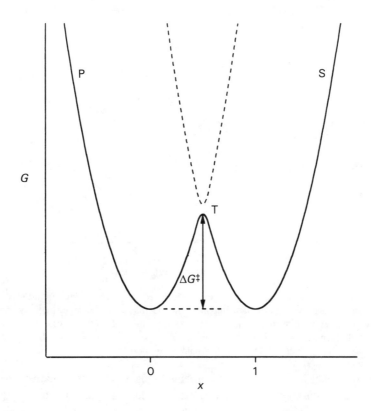

Fig. 2.13. Reaction coordinate diagram for a self-exchange reaction in the adiabatic limit where coupling between the precursor surface P and the successor surface S is weak. $\Delta G_{AA}^\circ = 0$ and $\Delta G_{AA}^\ddagger = \lambda_{AA}/4$.

The magnitude of the coupling has a strong dependence on the relative orientations of the complexes, as can be readily appreciated from consideration of Fig. 2.14. Results of such calculations and of estimates using alternative methods are available for a few systems (Table 2.17). Details of spectroscopically derived coupling constants are presented in section 4.2.

Also presented in Table 2.17 are values for the nuclear tunneling factors, Γ, estimated from eq. (2.85) [5]. Tunneling is appreciable at ambient temperatures only where the barrier heights are large. Interestingly, there is a compensation between the reduction of the electronic transmission coefficient and the tunneling factor which makes experimental evidence for this latter parameter hard to obtain.

$$\Gamma = \exp\left\{\frac{\lambda_{in}}{h\upsilon_{in}}\left(\tanh\frac{h\upsilon_{in}}{4kT} - \frac{h\upsilon_{in}}{4kT}\right)\right\} \qquad (2.85)$$

Sec. 2.13] The pre-exponential factors 87

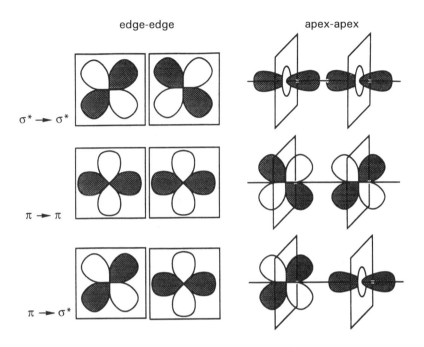

Fig. 2.14. The effects of reactant orientation on orbital overlap. Two orientations for an octahedral metal complex are illustrated. In the edge–edge orientation, the edges of the octahedra come in contract. While π–σ* changes are forbidden, π–π changes are favored over σ*–σ* changes. In the apexapex orientation, the apices of the octahdra come in contact. Again π–σ* changes are forbidden, but this time σ*–σ* changes are favored over π–π changes. Other mutual orientations will favor σ*–π.

Table 2.17. Values for transmission coefficients and tunneling constants for metal ion self-exchange reactions at 25 °C

Reaction	Δd (Å)	Γ	κ	Ref.
$[Fe(H_2O)_6]^{3+/2+}$	0.14	3.2	1×10^{-2}	44
$[Ru(H_2O)_6]^{3+/2+}$	0.01	1.5		179
$[Ru(NH_3)_6]^{3+/2+}$	0.04	1.0	1.0	44
$[Co(NH_3)_6]^{3+/2+}$	0.22	7.0	1×10^{-4}	44
$[Ru(bpy)_3]^{3+/2+}$	0.00	1.0	1.0	44
$[Co(bpy)_3]^{3+/2+}$	0.19	5.0	1×10^{-3}	44

With this framework one can begin to understand features of the self-exchange rates for metal ion complexes and to approach the manifestation of electronic and structural effects on electron transfer rates from an experimental point of view.

2.14 ELECTRONIC AND STRUCTURAL EFFECTS

The available structural and self-exchange rate data for ruthenium amine complexes indicate that internal rearrangement during the electron transfer process is minimal and hence $\lambda_{in} \approx 0$. These complexes therefore provide an important benchmark for comparisons with other metal ion species. The self-exchange rates span a considerable range from 10^3 M^{-1} s^{-1} for [Ru(NH$_3$)$_6$]$^{3+/2+}$ to 10^9 M^{-1}s^{-1}, close to the diffusion-controlled limit, for [Ru(bpy)$_3$]$^{3+/2+}$, and there is a good correlation with the size of the reagent which suggests that λ_{out} dominates the reaction energetics. The reactions are generally considered to be adiabatic with $\kappa_{el} \approx 1.0$ and the small structural rearrangement involved means that tunneling is unimportant. There is no evidence that increasing the bulk of the ligands decreases the probability for electron transfer, even though the donor and acceptor orbitals are predominantly metal-centered. This raises the point that direct overlap between metal-centered orbitals is not essential for adiabatic electron transfer. Suitable overlap may be maintained by delocalization over ligand orbitals, especially the π orbitals of oligopyridine ring systems. However, there is some evidence that extensive modification of the ligand periphery with insulating substituents can lead to a decrease in κ_{el} [212].

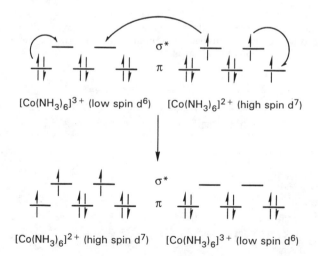

Fig. 2.15. Schematic diagram of the d orbitals of [Co(NH$_3$)$_6$]$^{3+/2+}$ showing the multiple electronic changes required in the self-exchange process.

In contrast to the behavior of the ruthenium(III)/(II) reagents, the corresponding cobalt(III)/(II) reagents present a more complex problem. The self-exchange rate for $[Co(NH_3)_6]^{3+/2+}$ ($k_{AA} \approx 2 \times 10^{-7}$ M^{-1} s^{-1} at 25.0 °C) [74] is ten orders of magnitude slower than that for $[Ru(NH_3)_6]^{3+/2+}$ and has provided one of the most studied problems in electron transfer chemistry. At the root of the problem is the fact that electron transfer from high-spin cobalt(II) ^4T ($\pi^5\sigma^{*2}$) to low-spin cobalt(III) ^1A (σ^{*6}) is spin-forbidden (Fig. 2.15). In fact this is a misnomer, the forbiddenness derives from the fact that electron transfer involves movement of three electrons [213, 214]. The major mechanisms whereby this process can become allowed involve either thermal population or spin–orbit coupling with the ^2E ($\pi^6\sigma^{*1}$) excited state of cobalt(II). Computation shows that the ^2E state requires a significantly shorter Co—N bond length (2.09 Å) than ^4T (2.19 Å). The excited state may be thermally populated as the bond contracts while the reagent approaches the geometry appropriate for the transition state for the reaction (2.14 Å). This provides a facile mechanism for electron transfer [215]. This explanation is unlikely since the ^2E lies too high in energy above the ground state for the two state surfaces to cross [216]. The alternative mechanism involves quantum mechanical mixing of the ^2E excited state with the ground state by spin–orbit coupling, and this provides a pathway of slightly lower energy [217–219]. The calculations reveal that the reaction is mildly non-adiabatic with $\kappa_{el} \approx 10^{-4}$ and that the coupling between the donor and acceptor orbitals shows a strong dependence on orientation. Apex–apex alignment of the reactants which gives overlap between the donor σ^* and acceptor σ^* orbitals is strongly preferred [220].

Although the spin-state change presents little intrinsic barrier to the electron transfer process, changes in spin state have important structural consequences. The structural differences between $[Co(NH_3)_6]^{3+}$ and $[Co(NH_3)_6]^{2+}$ are much larger than those in the corresponding ruthenium complex. Hence λ_{in} is much larger and is the dominant factor in determining the activation barrier. It has also been pointed out that large structural changes, with the attendant changes in vibrational frequencies, result in substantial changes in vibrational entropy which contribute to $\Delta S^\ddagger$ [221].

It must be emphasized that there is little experimental evidence for non-adiabaticity in reactions involving $[Co(NH_3)_6]^{3+/2+}$ [75]. In part this is the result of the compensating increase in tunneling which results from the substantial structural barrier. A series of interesting observations by Endicott and coworkers on the outer-sphere reduction of a series of cobalt(III) complexes by $[Co(sep)]^{2+}$ and $[Ru(NH_3)_6]^{2+}$ provides some of the most convincing evidence for non-adiabaticity in reactions of cobalt(III) complexes [222–224]. The reactions must necessarily be outer-sphere and a constant ratio for the rates k_{sep}/k_{ru} is expected (section 2.6). However, the experimental data reveal that this ratio varies over a wide range (Table 2.18), and the result has been interpreted to indicate that there is a change in the electronic coupling between the donor and acceptor orbitals. Reductions by $[Ru(NH_3)_6]^{2+}$ are adiabatic but those by $[Co(sep)]^{2+}$ are not. The ratio, k_{sep}/k_{ru}, expected for adiabatic reaction by both reductants is around 40 and so the retardation factor which approximates to κ_{el} for the series can be estimated. The trend is explained by an increase in intermolecular ligand–metal charge transfer between the reductant and the ligands coordinated to cobalt(III) along the series $(NH_2C_6H_{11}) > (NH_2C_6H_5)$ $(NH_2C_6H_4NO_2)$. This ligand–metal charge transfer lowers the energy of an electron transfer pathway involving spin–orbit coupling with a triplet excited state of the precursor complex.

Table 2.18. Rates of reduction of cobalt(III) complexes by $[Ru(NH_3)_6]^{2+}$ and $[Co(sep)]^{2+}$ at 25.0 °C, $\mu = 0.20$ M

Oxidant	k_{sep} $(M^{-1} s^{-1})$	k_{ru} $(M^{-1} s^{-1})$	$\frac{k_{sep}}{k_{ru}}$	κ
$[Co(en)_2(NH_2C_6H_{11})Cl]^{2+}$	0.82	45	0.018	0.00045
$[Co(en)_2(NH_2C_6H_5)Cl]^{2+}$	3.3	56	0.06	0.0015
$[Co(en)_2(NH_2C_6H_4NO_2)Cl]^{2+}$	92	81	1.1	0.025
$[Co(NH_3)_5(NH_2C_6H_{11})]^{3+}$	0.14	0.07	2	0.05
$[Co(NH_3)_5(NH_2C_6H_5)]^{3+}$	0.60	0.11	5.5	0.14
$[Co(NH_3)_5(NH_2C_6H_4NO_2)]^{3+}$	2.1	0.14	15	0.4
$[Co(NH_3)_6]^{3+}$	0.15	0.006	25	0.6

Similar observations have been made [225–227] in comparisons of the reductions of $[Co(phen)_3]^{3+}$ with $[Ru(NH_3)_6]^{2+}$ and $[Co(sep)]^{2+}$ where outer-sphere ion pair formation between $[Co(phen)_3]^{3+}$ and reducing anions enhances the coupling between donor and acceptor, a superexchange mechanism, and results in an increase in rate [228]. The effect is most obvious in the significantly non-adiabatic reaction between the two cobalt complexes. Specific anion effects have been noted in other reactions involving $[Co(phen)_3]^{3+/2+}$ [15].

Within the group of cobalt complexes, there is a considerable variation in self-exchange rate. Particular attention is drawn to the difference in self-exchange rate between $[Co(en)_3]^{3+/2+}$, $(7.7 \times 10^{-5}$ M^{-1} s^{-1}) and $[Co(sep)]^{3+/2+}$, (5 M^{-1} s^{-1}). Both complexes involve similar low-spin–high-spin changes. Molecular mechanics calculations reveal [214, 229] that the preferred Co—N bond length by the rigid cage structure, 2.10 Å, lies intermediate between the values for the Co^{III}—N (1.99 Å) and Co^{II}—N (2.16 Å) bonds and the relief of strain in attaining the activated complex lowers the activation barrier to electron transfer [56, 230]. This strain within the complex is reflected in a smaller change in bond length between the two oxidation states, again consistent with a lower barrier to electron transfer in the $[Co(sep)]^{3+/2+}$ complex.

There are a number of complexes for which the cobalt(II) is in the low-spin ^2E state and these show larger self-exchange rates. Examples are $[Co(terpy)_2]^{3+/2+}$, $[Co([9]aneS_3)_2]^{3+/2+}$, and $[Co(azacapten)]^{3+/2+}$ for which self-exchange rates range from 10^2 M^{-1} s^{-1} to 10^5 M^{-1} s^{-1}. Again, this is reflected in the structural changes involved in the electron transfer process (Table 2.15). In Fig. 2.9 a plot of $\{\Delta G^{\ddagger} - \lambda_{out}\}$ against Δr^2 is shown for a range of cobalt complexes, including those for which the cobalt(II) state is high-spin and low-spin. The correlation is excellent even though the variation in force constants is not taken into consideration [80].

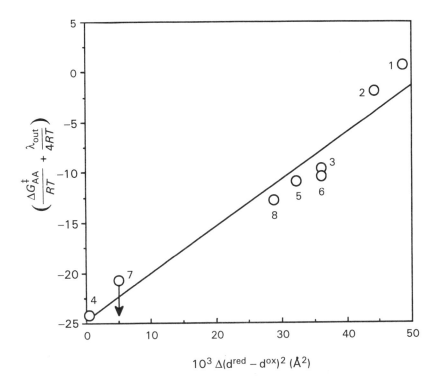

Fig. 2.16. Plot of $-(\Delta G^{\ddagger}_{AA} - \lambda_{out})$ against $\Delta(d^{red} - d^{ox})^2$ for a series of cobalt(III/II) amine complexes; (1) $[Co(NH_3)_6]^{3+/2+}$, (2) $[Co(en)_3]^{3+/2+}$, (3) $[Co(bpy)_3]^{3+/2+}$, (4) $[Co(bpy)_3]^{2+/+}$, (5) $[Co([9]aneN_3)_2]^{3+/2+}$, (6) $[Co(phen)_3]^{3+/2+}$, (7) $[Co([9]aneS_3)_2]^{3+/2+}$, (8) $[Co(sep)]^{3+/2+}$, from Ref. [80].

Self-exchange rate data for metal ions other than cobalt(III)/(II) and ruthenium(III)/(II) are insufficient in variety to support much comment. The rates follow the structural changes involved in the electron transfer process. It is notable that most nickel(III)/(II) self-exchange rates are in the range 10^3–10^4 M^{-1} s^{-1} while those for iron(III)/(II) tend to be somewhat higher.

Balzani and coworkers [129, 231] have proposed a more general method for examining deviations from adiabatic behavior in reactions of metal complexes. The equation for the cross-electron transfer reaction rate (eq. (2.86)) is recast in general steady-state form as eq. (2.87). The electron transfer rate constant, k_{et}, can be approximated as eq. (2.88) with $\upsilon_{eff} = kT/h$, the universal frequency of activation rate theory [49] so that eq. (2.87) can be rewritten as eq. (2.89).

$$[A^{ox}] + [B^{red}] \underset{k_{-o}}{\overset{k_o}{\rightleftharpoons}} \{[A^{ox}],[B^{red}]\} \underset{k_{-et}}{\overset{k_{et}}{\rightleftharpoons}} \{[A^{red}],[B_{ox}]\} \underset{k_o'}{\overset{k_{-o}'}{\rightleftharpoons}} [A^{red}] + [B^{ox}] \quad (2.86)$$

$$k_{AB} = \frac{k_o}{1 + \dfrac{k_{-o}}{k_{et}} + \dfrac{k_{-o}k_{-et}}{k'_o k_{et}}} \quad (2.87)$$

$$k_{et} = \frac{kT}{h}(\kappa_{el})_{AB} \exp(-\Delta G^{\ddagger}/RT) \quad (2.88)$$

$$k_{AB} = \frac{k_o}{1 + \dfrac{k_{-o}}{kT/h(\kappa_{el})_{AB} \exp(-\Delta G^{\ddagger}/RT)} + \dfrac{K'_o}{K_o K_{AB}}} \quad (2.89)$$

The activation free energy, $\Delta G^{\ddagger}_{AB}$, is approximated as eq. (2.90), though the more conventional Marcus expression, eq. (2.24), can also be used.

$$\Delta G^{\ddagger}_{AB} = \Delta G + \frac{\lambda_{AA} + \lambda_{BB}}{8\ln 2} \ln\left[1 + \exp\left(-\frac{8\Delta G \ln 2}{\lambda_{AA} = \lambda_{BB}}\right)\right] \quad (2.90)$$

In order to use eq. (2.90) in the analysis of reactivity, rate data with a homogeneous series of reactants spanning a range of potentials are required. Reagents such as $[M(bpy)_3]^{3+/2+}$ and $[M(phen)_3]^{3+/2+}$ where M is Fe, Ru, Cr and Os are ideal since the self-exchange rates are very similar, $k_{AA} \approx 1 \times 10^9$ M^{-1} s^{-1}, and $(\kappa_{el})_{AA} \approx 1.00$. The assumption is made that $(\kappa_{el})_{AB} \approx \sqrt{((\kappa_{el})_{AA}(\kappa_{el})_{BB})}$, and fits to experimental data are used to determine $(\kappa_{el})_{BB}$, the degree of adiabaticity in the self-exchange reaction of the reagent under study. Data for reactions of $[Ru(NH_3)_6]^{3+/2+}$ show an excellent fit with $(\kappa_{el})_{BB} = 1.0$ and the reaction with $[Fe(H_2O)_6]^{3+/2+}$ is mildly non-adiabatic with $(\kappa_{el})_{BB} \approx 10^{-3}$ although comparable fits can be obtained with $(\kappa_{el})_{BB} = 1.0$ (Fig. 2.17). However, for other reagents notably $[Eu_{aq}]^{3+/2+}$, there is a considerable discrepancy with $(\kappa_{el})_{BB} \approx 10^{-7}$. This approach has been criticized [232] as it assumes that the non-adiabaticity arises exclusively from the self-exchange process. There is poor agreement between the self-exchange rate calculated from the derived parameters and the experimentally determined value. Despite these limitations, the treatment highlights particular redox couples such as those for $[Eu_{aq}]^{3+/2+}$ where substantial deviations from adiabaticity can be expected. It must be stated that comparisons of experimentally observed and Marcus-calculated rate constants give comparable information.

One of the most important classes of reagents where significant deviations between experimentally observed and Marcus-calculated rate constants are found is the metal aqua complexes, $[M(H_2O)_6]^{3+/2+}$. Rates calculated by Marcus Theory are consistently larger than those obtained experimentally, particularly at large driving forces [1, 98]. This is reflected in comparisons of the experimental self-exchange rate data in Table 2.7 and the Marcus-calculated data in Table 2.9, particularly for $[Fe(H_2O)_6]^{3+/2+}$ and $[Co(H_2O)_6]^{3+/2+}$ where the values differ by 3 and 13 orders of magnitude, respectively. There have been several reasons advanced for this discrepancy. One is that the mechanisms for the self-exchange reactions are not exclusively outer-sphere but have considerable inner-sphere character [66]. Against this is the considerable evidence presented in Chapter 3 that H_2O will not function as a bridge in inner-sphere reactions since it is of insufficient basicity. The discrepancies are largest for $[Co(H_2O)_6]^{3+/2+}$, and to a lesser extent $[Fe(H_2O)_6]^{3+/2+}$ and this observation has prompted the proposal that an inner-sphere water-bridged mechanism competes only where the metal is strongly oxidizing and the barrier for metal–ligand bond homolysis is low [103]. There are alternative explanations. The metal aqua ions will be more strongly coupled to solvent water by hydrogen bonding and this may offer substantial differences between self-exchange reactions where both reactants are strongly hydrogen bonded and cross-reactions where this is absent [199]. Effects of hydrogen bonding have been treated by including a specific term for hydrogen bonding, λ_H, in the reorganizational energy (eq. (2.91)) [198]. However, a functional form for λ_H has not been elucidated. Nevertheless, statistical mechanical calculations point to some special stability for hydrogen bond-bridged outer-sphere transition states [233, 234]

$$\lambda = \lambda'_{in} + \lambda'_{out} + \lambda_H \tag{2.91}$$

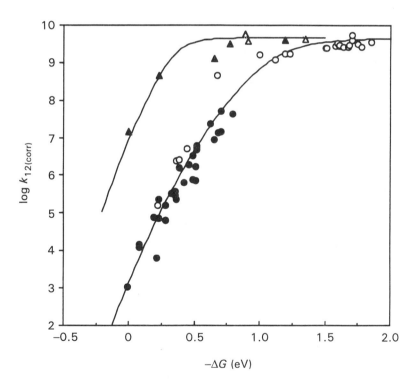

Fig. 2.17. Plot of log second-order rate constant (corrected for differences in encounter distance, charge product and ionic strength) against driving force for reaction $[M(phen)_3]^{3+/2+}$ and $[M(bpy)_3]^{3+/2+}$ with $[Ru(NH_3)_6]^{2+}$ closed triangles; $[Ru(NH_3)_6]^{3+}$ open triangles; $[Fe(H_2O)_6]^{2+}$ closed circles; and $[Fe(H_2O)_6]^{3+}$ open circles. The solid lines represent fits to eq. (2.89).

If this explanation is correct, outer-sphere cross-reactions between metal aqua complexes should show significantly better agreement with the predictions of Marcus Theory than cross-reactions with reagents which cannot form hydrogen bonds. Unfortunately data for this sort of correlation are rather difficult to obtain since base-catalyzed inner-sphere pathways tend to compete and the reactions are subject to substantial medium effects. A comparison of some available data is presented in Table 2.19, and included in the table are rate constants calculated from the Marcus expression. For $[Fe(H_2O)_6]^{3+/2+}$ and $[Co(H_2O)_6]^{3+/2+}$ the self-exchange rates used in the calculations are the calculated values of 1×10^{-3} M^{-1} s^{-1} and 1×10^{-12} M^{-1} s^{-1} respectively. Much poorer agreement results from use of the experimental self-exchange rates. This would seem to argue against a specific role for hydrogen bonding in explaining the specific discrepancy in reactions involving the aqua ions and an inner-sphere component in these reactions seems likely.

Table 2.19. Observed and calculated rate constants for reactions between aqua ion complexes at 25.0°C

	μ (M)	k_{expt} (M^{-1} s^{-1})	k_{calc} (M^{-1} s^{-1})	Ref.
$[Co(H_2O)_6]^{3+} + [Fe(H_2O)_6]^{2+}$	1.0	5.0×10	4	235
$[Co(H_2O)_6]^{3+} + [V(H_2O)_6]^{2+}$	3.0	9×10^5	2×10^6	236
$[Co(H_2O)_6]^{3+} + [Cr(H_2O)_6]^{2+}$	3.0	1.3×10^4	7×10^3	236
$[V(H_2O)_6]^{3+} + [Cr(H_2O)_6]^{2+}$	3.0	0.20	2×10^{-5}	237
$[Fe(H_2O)_6]^{3+} + [Cr(H_2O)_6]^{2+}$	1.0	5.7×10^2	6.8×10	238
$[Fe(H_2O)_6]^{3+} + [V(H_2O)_6]^{2+}$	1.0	1.8×10^4	3×10^4	235
$[Fe(H_2O)_6]^{3+} + [Ru(H_2O)_6]^{2+}$	1.0	2.3×10^3	1×10^3	239
$[Ru(H_2O)_6]^{3+} + [V(H_2O)_6]^{2+}$	1.0	2.8×10^2	2×10^3	239

The data in Table 2.19 are presented in Fig. 2.18 along with the experimental and Marcus-derived self-exchange rate data for the metal aqua ions. Agreement for both sets of data is excellent for rate constants greater than approximately 1 M^{-1} s^{-1}. Below this value, agreement is poor and it seems probable that other mechanisms, most likely inner-sphere, may be operative. Much depends on the position of a proton!

There is one particular well studied redox reagent which shows very poor agreement with the predictions of Marcus Theory and where there is no consistency in the evaluated self-exchange rates. This is the $[Eu_{aq}]^{3+/2+}$ system. Only an upper limit of 3×10^5 M^{-1} s^{-1} at 40°C and 2 M ionic strength for the self-exchange rate has been experimentally determined [240]. It has been argued that the rate should be no lower than that for $[Fe(H_2O)_6]^{3+/2+}$ since both λ_{in} and λ_{out} should be smaller on account of the larger size of $[Eu_{aq}]^{3+/2+}$ and that the low value points to non-adiabaticity. This is explained by poor overlap between the 4f donor and acceptor which are shielded by the 5s and 5p orbitals [241]. Values for the self-exchange rate evaluated from cross-reaction data range from 10^{-4} M^{-1} s^{-1} to 10^{-7} M^{-1} s^{-1}. However, a self-exchange rate of approximately 10^{-5} M^{-1} s^{-1} is generally accepted. Similar situations exist for the $[U_{aq}]^{4+/3+}$ couple where a self-exchange rate of approximately 10^{-6} M^{-1} s^{-1} is accepted [242, 243], $[Ce_{aq}]^{4+/3+}$ [97], $[UO_{2aq}]^{2+/+}$ [113], $[Np_{aq}]^{4+/3+}$ [244], and $[Y_{aq}]^{3+/2+}$ [245].

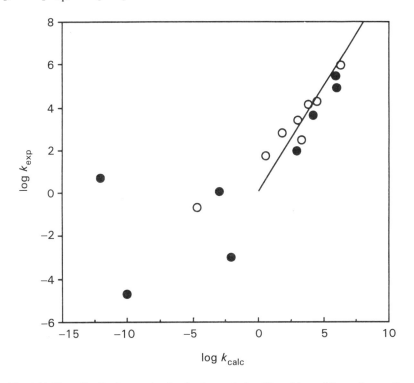

Fig. 2.18. Plot of $\log(k_{AB})_{exp}$ against $\log(k_{AB})_{calc}$ calculated from Marcus Theory for reactions of a series of metal–aqua ion complexes. Closed circles represent cross-reaction rates, open circles represent self-exchange rates. The solid line has a slope of unity. Note that the correlation is good for data with $k_{exp} \geq 1$ M^{-1} s^{-1} but below this value the experimentally determined rates are faster than those calculated for outer-sphere mechanisms leading to speculation that inner-sphere pathways are operating.

2.15 ELECTRON TRANSFER INVOLVING LARGE STRUCTURAL CHANGES

It has been pointed out that in a number of outer-sphere reactions, particularly those involving cobalt(III)/(II), the structural changes associated with the change in oxidation state can be substantial. However, in all these cases, the inner-coordination sphere remains substantially intact. There are complexes such as $[Co(edta)]^{-/2-}$ where there is a change in coordination as a result of the electron transfer process. The ligand is sexidentate in $[Co(edta)]^-$, but on reduction one of the carboxylate arms is substituted by solvent, $[Co(edta)(OH_2)]^{2-}$ [246]. This change in coordination can have a profound effect on the energetics of electron transfer [85].

96 The outer-sphere mechanism

Large structural changes are perhaps most readily examined in complexes with a predominantly square-planar geometry which lose or gain H_2O in the course of oxidation or reduction. There is a considerable amount of data on ligands which provide strong square-planar ligand fields in d^8 nickel(II) and copper(III) complexes with very weakly bound axial ligands. On oxidation of nickel(II) to nickel(III) or reduction of copper(III) to copper(II), distorted tetragonal geometries with more strongly bound axial ligands are preferred.

Outer-sphere reactions of the copper(III) peptide complexes are generally well-behaved and there is good agreement between self-exchange rates measured directly [59] and those derived with the Marcus linear free energy relationship, despite the change in axial coordination during the electron transfer process. The self-exchange rates are of the order of 10^4–10^6 M^{-1} s^{-1} [247, 248]. Activation parameters are also in good agreement with the predictions of Marcus Theory.

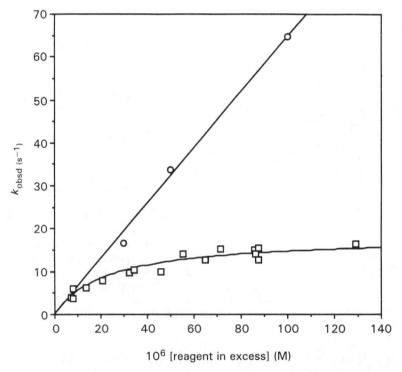

Fig. 2.19. Pseudo-first-order rate constants for the reduction of $[Cu^{III}(H_{-3}G_4)]^-$ by $[Cu^I(dpmp)_2]^{3-}$; oxidant in exces open circles; reductant in excess closed circles, from Ref. [249].

There is one exception [249]. In the reduction of the complex $[CuH_{-3}G_4]^-$ with the anionic copper(I) reagent $[Cu(dpsmp)_2]^{3-}$, (eqs (2.92)–(2.93)), limiting first-order behavior is detected (eq. (2.94)), with $k_1 = 18$ s^{-1} and $k_{-1}/k_2 = 2.3 \times 105$ M^{-1}, when the reaction is examined under pseudo-first-order conditions with an excess of the reductant. However, the reaction remains second order in the presence of an excess of [[CuH$_{-3}$G$_4$]$^-$] (Fig. 2.19). A mechanism involving a limiting first-order activation process on copper(III), possibly the addition of axial solvent, is proposed but it should be noted that inner-sphere reactions can be competitive with outer-sphere reactions with these oxidants [250–253].

$$[\text{CuH}_{-3}\text{G}_4]^- \underset{k_{-1}}{\overset{k_1}{\rightleftharpoons}} [*\text{CuH}_{-3}\text{G}_4]^- \qquad (2.92)$$

$$[*\text{CuH}_{-3}\text{G}_4]^- + [\text{Cu(dpsmp)}_2]^{3-} \xrightarrow{k_2} [\text{CuH}_{-3}\text{G}_4]^{2-} + [\text{Cu(dpsmp)}_2]^{2-} \qquad (2.93)$$

$$\frac{-d[[\text{Cu(dpsmp)}_2]^{3-}]}{dt} = \frac{k_1 k_2 [[\text{Cu(dpsmp)}_2]^{3-}][[\text{CuH}_{-3}\text{G}_4]^-]}{k_{-1} + k_2 [[\text{Cu(dpsmp)}_2]^{3-}]} \qquad (2.94)$$

With nickel(III) complexes, the behavior is more complex. Directly measured self-exchange rates [52] and those determined from cross-reactions with dissimilar nickel(III) complexes [254] are higher by a factor of 10^4–10^7 than those determined by reactions with well-defined outer-sphere reagents such as [Ru(NH$_3$)$_6$]$^{2+}$, where the apparent self-exchange rate is ≈ 0.1 M^{-1} s^{-1} [52]. This is another instance, like the [Co(H$_2$O)$_6$]$^{3+/2+}$ system, where water bridging is proposed. Again, the metal complex is strongly oxidizing. In contrast, reactions of the tetra-aza macrocyclic species such as [Ni[14]aneN$_4$]$^{3+}$ are well-behaved [255, 256] and there is good agreement with the Marcus relationship [51]. The situation is similar with the corresponding [Co([14]aneN$_4$)(H$_2$O)$_2$]$^{3+/2+}$ case [103].

The [Cu$_{aq}$]$^{2+}$ ion has a reduction potential of 0.15 V and is a modest oxidant. Aqueous [Cu$_{aq}$]$^+$ ion is thermodynamically unstable with respect to disproportionation but may be produced in acidic media as a metastable species [257]. Both the tetragonal oxidized and tetrahedral reduced forms are very labile but outer-sphere behavior can be examined in reactions with appropriate inert oxidants or reductants and a self-exchange rate around 2×10^{-4} M^{-1} s^{-1} has been calculated with use of the Marcus relationship [111, 258]. This low reactivity coupled with the high substitution lability and the coordination change on electron transfer ensures that inner-sphere mechanisms are favored unless they can be prevented. The self-exchange rate measured directly in concentrated HCl solutions is 5×10^7 M^{-1} s^{-1} [259] and is almost certainly an inner-sphere reaction.

Early work on the outer-sphere reactions of copper complexes centered on reactions of [Cu(bpy)$_2$]$^{2+/+}$, [Cu(phen)$_2$]$^{2+/+}$ and its derivative [Cu(dmp)$_2$]$^{2+/+}$. Again, both oxidation states are labile and while the oxidized form is flattened tetrahedral, [Cu(phen)$_2$]$^{2+}$ [260], or five-coordinate, [Cu(dmp)$_2$(OH$_2$)]$^{2+}$ [261], the copper(I) ions are much closer to a regular tetrahedral geometry [262, 263]. Reduction potentials (Table 2.20) reflect the tendency of the ligands to favor the lower coordination geometry of the copper(I) state. The lability of the complexes presents some problems in studying reactions, and conditions of excess chelating ligand must be used. For example in the reduction of $[Cu(dmp)_2]^{2+}$ by $[Ru(NH_3)_5py]^{2+}$, the general rate law is of the form of eq. (2.95), consistent with the mechanism in eqs (2.96)–(2.100) [258]. In practice, the dependence of the reaction rate on the concentration of [dmp] indicates that the bis complex, k_2, is the sole reductant.

Table 2.20. Reduction potentials for copper(II) complexes in aqueous solution at 25° C

Complex	$E°$ (V)
$[Cu_{aq}]^{2+/+}$	0.15
$[Cu(phen)_2]^{2+/+}$	0.17
$[Cu(bpy)_2]^{2+/+}$	0.12
$[Cu(dmp)_2]^{2+/+}$	0.61
$[Cu(dpsmp)_2]^{3-/-}$	0.59

$$\frac{d[[Cu(dmp)_2]^+]}{dt} = \frac{K_1 k_1 [dmp] + K_1 K_2 k_2 [dmp]^2}{1 + K_1 [dmp] + K_1 K_2 [dmp]^2} \quad (2.95)$$

$$[Cu_{aq}]^{2+} + dmp \underset{}{\overset{K_1}{\rightleftharpoons}} [Cu(dmp)]^{2+} \quad (2.96)$$

$$[Cu(dmp)]^{2+} + dmp \underset{}{\overset{K_2}{\rightleftharpoons}} [Cu(dmp)_2]^{2+} \quad (2.97)$$

$$[Cu(dmp)_2]^{2+} + [Ru(NH_3)_5py]^{2+} \xrightarrow{k_2} [Cu(dmp)_2]^+ + [Ru(NH_3)_5py]^{3+} \quad (2.98)$$

$$[Cu(dmp)]^{2+} + [Ru(NH_3)_5py]^{2+} \xrightarrow{k_1} [Cu(dmp)]^+ + [Ru(NH_3)_5py]^{3+} \quad (2.99)$$

$$[Cu(dmp)]^+ + dmp \underset{}{\overset{fast}{\rightleftharpoons}} [Cu(dmp)_2]^+ \quad (2.100)$$

Self-exchange rates evaluated from reactions which are assigned an outer-sphere mechanism in which the copper(I) forms are oxidized are higher than those evaluated from reactions in which copper(II) forms are reduced (Table 2.21), leading to the conclusion that this is a coupled reaction in which the geometry changes play an important role in determining the preferred reaction pathway. The self-exchange rate for $[Cu(dmp)_2]^{2+/+}$ has been measured directly [88] and has a value of 2.0×10^5 M^{-1} s^{-1}, in good agreement with estimates based on the geometric mean of the self-exchange rates from cross reactions [264] and electrochemical measurements [265].

Table 2.21. Rate and activation parameters for reduction and oxidation of copper complexes

Reaction	μ (M)	k (M^{-1} s^{-1})	$\Delta H^{\ddagger}$ (kJ mol^{-1})	$\Delta S^{\ddagger}$ (J K^{-1} mol^{-1})	k_{AA} (M^{-1} s^{-1})	Ref.
$[Cu_{aq}]^+ + [Ru(NH_3)_5py]^{3+}$	1.0	4.6×10			2×10^{-4}	258
$[Cu_{aq}]^+ + [Ru(NH_3)_5isn]^{3+}$	1.0	5.4×10^2			2×10^{-4}	258
$[Cu_{aq}]^+ + [Ru(NH_3)_4bpy]^{3+}$	1.0	3.8×10^3			2×10^{-4}	258
$[Cu_{aq}]^+ + [Ru(NH_3)_5py]^{3+}$	1.0	4.4×10^4			2×10^{-4}	258
$[Cu_{aq}]^+ + [Ru(phen)_3]^{3+}$	1.0	1.2×10^9			1×10^{-5}	266
$[Co(sep)]^{2+} + [Cu_{aq}]^{2+}$	0.5	5.0			5×10^{-7}	111
$[Cu(dmp)_2]^+ + [IrCl_6]^{2-}$	0.1	1.4×10^9			9×10^8	249
$[Ru(NH_3)_5py]^{2+} + [Cu(dmp)_2]^{2+}$	0.1	5.5×10^5	21.6	-64	4×10^4	258
$[Ru(NH_3)_5isn]^{2+} + [Cu(dmp)_2]^{2+}$	0.1	4.6×10^5	22.6	-59	4×10^4	258
$[Ru(NH_3)_4bpy]^{2+} + [Cu(dmp)_2]^{2+}$	0.1	3.9×10^5	23.7	-58	4×10^4	258
$[Co(phen)_3]^{2+} + [Cu(dmp)_2]^{2+}$	0.1	8.8×10^4	43.1	-6	4.4×10^4	267
$[Cu(phen)_2]^+ + [Co(acac)_3]$	0.25	7.56×10^2	33	-79		268
$[Cu(phen)_2]^+ + [Co(edta)]^-$	0.25	4.48×10^2	31	-88		268
$[Cu(phen)_2]^+ + [Co(edta)]^-$	0.5	3.12×10^2	26	-113	5×10^7	269
$[Cu(bpy)_2]^+ + [Co(edta)]^-$	0.5	2.59×10^2	19	-134	4×10^6	269

Rate-limiting behavior detected in the reduction of the $[Cu(dmp)_2]^{2+}$ derivative, $[Cu(dpsmp)_2OH_2]^{2-}$, with $k_1 = 230$ s^{-1} at pH 8.0 has been ascribed [270, 271] to loss of water from the five-coordinate copper(II) complex (eqs (2.101)–(2.102)).

$$[Cu(dpsmp)_2OH_2]^{2-} \underset{k_{-1}}{\overset{k_1}{\rightleftharpoons}} [Cu(dpsmp)_2]^{2-} + H_2O \tag{2.101}$$

$$[Cu(dpsmp)_2]^{2-} + [Fe(CN)_6]^{4-} \overset{k_2}{\rightarrow} [Cu(dpsmp)_2]^{3-} + [Fe(CN)_6]^{3-} \tag{2.102}$$

However, later work has shown [272] that this is the result of ionization of the coordinated water molecule with a pK_a = 8.3 and that $k_1 \geq 330$ s^{-1}. There is no evidence for a rate-limiting change of coordination in these reactions.

The series of copper thiaether macrocycles has attracted a great deal of interest. The complexes are moderately oxidizing with reduction potentials for $[Cu[14]aneS_4]^{2+}$ and $[Cu[15]aneS_5]^{2+}$ of 0.58 V and 0.68 V respectively at 25.0°C and 0.10 M ionic strength [273]. A 'square scheme' is used to describe the behavior where the reduction of the square pyramidal $[Cu[15]aneS_5(O)]^{2+}$ to the thermodynamically stable form of $[Cu[15]aneS_5(R)]^+$, which is tetrahedral with a non-bonded sulfur donor [274] can take place through a metastable reduced five-coordinate intermediate, $[Cu[15]aneS_5(P)]^+$ pathway A, or a metastable oxidized tetrahedral intermediate, $[Cu[15]aneS_5(Q)]^{2+}$ pathway B (Scheme 2.4). Pathway A is favored in reactions of $[Cu[15]aneS_5]^{2+}$ with reductants, whereas pathway B is found in reactions of $[Cu[15]aneS_5]^+$ with oxidants, explaining the discrepancies in the self-exchange rates evaluated from different studies (Table 2.22).

The self-exchange rate for $[Cu[15]aneS_5]^{2+/+}$ determined directly by NMR is 2×10^5 M^{-1} s^{-1} with $\Delta H^{\ddagger} = 14.0$ kJ mol^{-1} and $\Delta S^{\ddagger} = -103$ J K^{-1} mol^{-1} [275]. This value is similar to the self-exchange rate evaluated from reduction of $[Cu[15]aneS_5]^{2+}$ by $[Co(Me_4[14]tetraeneN_4)(OH_2)_2]^{2+}$ and implies that the pathway involving the metastable copper(I) species (A in Scheme 2.4) is preferred.

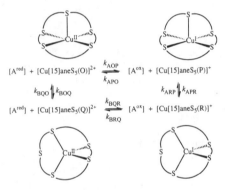

Scheme 2.4.

Table 2.22. Rates of electron transfer reactions of copper thiaether complexes in aqueous methanol at 25.0 °C

Reaction	μ (M)	k (M^{-1} s^{-1})	$\Delta H^{\ddagger}$ (kJ mol^{-1})	$\Delta S^{\ddagger}$ (J K^{-1} mol^{-1})	k_{AA} (M^{-1} s^{-1})	Ref.
$[Cu[14]aneS_4]^+$ + $[Fe(4,7-Me_2phen)_3]^{3+}$	0.1	9.9×10^6	33.9	2.5	2.2	276
$[Cu[15]aneS_5]^+$ + $[Fe(4,7-Me_2phen)_3]^{3+}$	0.1	4.6×10^7	19.7	−29	7.5×10^3	276
$[Co(Me_4[14]tetraeneN_4)(OH_2)_2]^{2+}$ + $[Cu[14]aneS_4]^{2+}$	0.1	7.0			1.6×10^3	276
$[Co(Me_4[14]tetraeneN_4)(OH_2)_2]^{2+}$ + $[Cu[15]aneS_5]^{2+}$	0.1	5.88×10^2			3.5×10^5	276

This is an example of 'gated' electron transfer where the pathway used is dependent on the properties of the reaction partner. The dynamics of this type of process have been considered in detail [277, 278]. It is concluded that reactive product intermediates (pathway A for reduction and pathway B for oxidation) will participate under all reaction conditions but that reactive reactant intermediates (pathway B for reduction and pathway A for oxidation) will participate only where the driving force is small such as in a self-exchange process. This is in substantial agreement with the experimental findings.

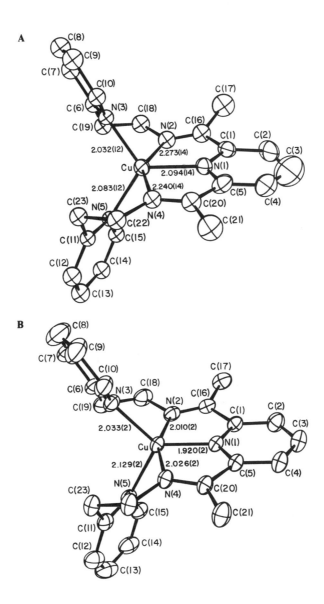

Fig. 2.20. Ortep views of the [Cu^I(py)$_2$DAP]$^+$, A, and [Cu^{II}(py)$_2$DAP]$^{2+}$, B, cations showing the retention of the coordination environment around the metal center, from Goodwin, J. A.; Stanbury, D. M.; Wilson, L. J.; Eigenbrot, C. W.; Scheidt, W. R. *J. Am. Chem. Soc.* 1987, **109**, 2979, with permission

More recently efforts have been extended to examine the copper(II)/(I) couple in ligands where the coordination number remains invariant between copper(II) and copper(I). A variety of five-coordinate ligands have been devised. While the coordination number remains invariant, there are substantial changes going from a trigonal environment with two additional but weaker donors in the reduced form to a square pyramidal structure in the oxidized form (Fig. 2.20) [279]. The substitution rate of the pyridine arm in $[Cu(py)_2DAP]^+$ is quite rapid, 310 s^{-1}, but it is likely that the reactions are outer-sphere in nature [280]. Self-exchange rates are presented in Table 2.23, and limited cross-reaction data are in reasonable agreement.

Table 2.23. Self-exchange rate constants for copper(II)/(I) in non-aqueous solvents

Reaction	Solvent (M^{-1}s^{-1})	$E°$ (V)	k_{AA} (M^{-1}s^{-1})	$\Delta H^\ddagger$ (kJ mol^{-1})	$\Delta S^\ddagger$ (J K^{-1} mol^{-1})	Ref.
$[Cu(TAAB)]^{2+/+}$	CD$_3$OD		5×10^{5c}			60
$[Cu(2,2'-(Im)_2biph)_2]^{2+/+}$	CH$_3$CN	0.11V	$<1 \times 10^2$			281
$[Cu(py)_2DAP]^{2+/+}$	0.05, CH$_3$CN	$-0.136V^a$	1.7×10^3			279
$[Cu(imidH)_2DAP]^{2+/+}$	0.038, CH$_3$CN	$-0.269V^a$	$<13 \times 10^2$			280
$[Cu(5-MeimidH)_2DAP]^{2+/+}$	0.025, CH$_3$CN		3.5×10^{4b}	16.2	-103	282

^{a}vs SCE. ^{b}20 °C. ^{c}22 °C.

Electron transfer reactions involving metal ions in higher oxidation states are frequently coupled to changes in the degree of hydrolysis of the complex and form a particularly important class of electron transfer processes with large structural changes. Some of the most widely studied examples of single-electron transfer reactions involve $[VO_{2aq}]^+/[VO_{aq}]^{2+}$ and $[TiO_{aq}]^{2+}/[Ti(H_2O)_6]^{3+}$. The coupling of electron and H$^+$ transfer in these reactions is of particular importance and key attention must be paid to the pH dependence of the rate and the elucidation of H$^+$ dependent pathways.

One of the simplest examples of this behavior is the outer-sphere reduction of hydrolyzed metal aqua complexes such as $[Co(H_2O)_5OH]^{2+}$ and $[Fe(H_2O)_5OH]^{2+}$. These species normally prefer inner-sphere pathways but outer-sphere reactions can be induced with appropriate choice of reductant. Reactions of the thermodynamically more powerful hexa-aqua-oxidants compete, but the sluggish outer-sphere behavior of $[Co(H_2O)_6]^{3+}$ allows investigations of reactions of $[Co(H_2O)_5OH]^{2+}$. Reductions of $[Co_{aq}]^{3+}$ in aqueous acid solution by the outer-sphere reductants $[Ni([9]aneN_3)_2]^{2+}$ and $[Ni([10]aneN_3)_2]^{2+}$ show two pathways, one independent of [H$^+$] corresponding to reaction of $[Co(H_2O)_6]^{3+}$ and the other [H$^+$]$^{-1}$ dependent and corresponding to outer-sphere reduction of $[Co(H_2O)_5OH]^{2+}$. This latter pathway shows a large solvent isotope effect but reaction with the N–D complex occurs at the same rate as the N–H reductant indicating that H$^+$ coupling is unlikely. A self-exchange rate of approximately 3 M^{-1} s^{-1} has been evaluated for $[Co(H_2O)_5OH]^{2+/+}$ with a reduction potential of 1.44 V [283].

A substantial rearrangement in the coordination sphere is required in the reduction of $[VO_2(OH_2)_4]^+$ (eq. (2.103)), which has a potential of 1.03 V in 1.0 M $HClO_4$. The protic equilibria for vanadium(V) and vanadium(IV) (eqs (2.104) and (2.105)) give a value of 0.76 V for the potential for the acid-independent reduction of $[VO(OH)(OH_2)_4]^{2+}$.

$$[VO_2(OH_2)_4]^+ + 2H^+ + e^- \rightleftharpoons [VO(OH_2)_5]^{2+} \tag{2.103}$$

$$[VO(OH)(OH_2)_4]^{2+} \rightleftharpoons [VO_2(OH_2)_4]^+ + H^+ \quad K_a = 30 \text{ M} \tag{2.104}$$

$$[VO(OH_2)_5]^{2+} \rightleftharpoons [VO(OH)(OH_2)_4]^+ + H^+ \quad K_a = 8.5 \times 10^{-7} \text{ M} \tag{2.105}$$

The kinetics of outer-sphere reduction of vanadium(V) by $[Os(bpy)_3]^{2+}$ are representative and show an acid dependence (eq. (2.106)) consistent with the mechanism (eqs (2.107)–(2.109)), with rate constants 2.7 M^{-1} s^{-1} and 225 M^{-1} s^{-1} at 1.0 M ionic strength [284].

$$-\frac{d[[Os(bpy)_3]^{2+}]}{dt} = \left\{ k_1 + \frac{k_2[H^+]}{K_a} \right\} [[VO_2(OH_2)_4]^+][[Os(bpy)_3]^{2+}] \tag{2.106}$$

$$[VO(OH)(OH_2)_4]^{2+} \xrightleftharpoons{K_a} [VO_2(OH_2)_4]^+ + H^+ \quad K_a = 4.5 \text{ M} \tag{2.107}$$

$$[VO_2(OH_2)_4]^+ + [Os(bpy)_3]^{2+} \xrightarrow{k_1} [VO_2(OH_2)_4]^{2+} + [Os(bpy)_3]^{3+} \tag{2.108}$$

$$[VO(OH)(OH_2)_4]^{2+} + [Os(bpy)_3]^{2+} \xrightarrow{k_2} [VO(OH)(OH_2)_4]^+ + [Os(bpy)_3]^{3+}$$

$$\tag{2.109}$$

Oxidation of $[VO(OH_2)_5]^{2+}$ by $[Ni(bpy)_3]^{3+}$ follows the rate law (eq. (2.110)), with rate constants 61 M^{-1} s^{-1} and 3.6×10^6 M^{-1} s^{-1} for pathways (2.112) and (2.113) respectively at 1.0 M ionic strength. It is noteworthy that the most facile pathway in both reactions involves coupling to a single proton where the structural rearrangement required is smallest. A self-exchange rate for $[VO(OH)(OH_2)_4]^{+/2+}$ is calculated to be approximately 10^{-3} M^{-1} s^{-1} from a range of outer-sphere rate data [285–287].

$$-\frac{d[[Ni(bpy)_3]^{3+}]}{dt} = \left\{ k_1 + \frac{k_2 K_a}{[H^+]} \right\} [[VO(OH_2)_5]^{2+}][[Ni(bpy)_3]^{3+}] \tag{2.110}$$

$$[VO(OH_2)_5]^{2+} \xrightleftharpoons{K_a} [VO(OH)(OH_2)_4]^+ + H^+ \quad K_a = 8.5 \times 10^{-7} \text{ M} \tag{2.111}$$

$$[VO(OH_2)_5]^{2+} + [Ni(bpy)_3]^{3+} \xrightarrow{k_1} [VO(OH_2)_5]^{3+} + [Ni(bpy)_3]^{2+} \quad (2.112)$$

$$[VO(OH)(OH_2)_4]^+ + [Ni(bpy)_3]^{3+} \xrightarrow{k_2} [VO(OH)(OH_2)_4]^{2+} + [Ni(bpy)_3]^{2+}$$

$$(2.113)$$

A similar situation exists in the reactions of $[TiO_{aq}]^{2+}$ and $[Ti(H_2O)_6]^{3+}$. The reduction potential of $[TiO_{aq}]^{2+}$ is -0.016 V in 1.0 M HCl (eq. (2.114)) [288], and is strongly pH-dependent owing to the change in the state of hydrolysis of the metal ion. The hydrolysis constants for both oxidation states are not known but reasonable estimates set the reduction potentials of $[Ti_{aq}]^{4+}$ and $[Ti(H_2O)_5OH]^{3+}$ at $+0.07$ V and $+0.1$ V respectively [289]. Outer-sphere mechansims have been assigned from linear free energy relationships [290], and cross-reactions of $[Ti(H_2O)_6]^{3+}$ with outer-sphere oxidants show a two-term rate law (eq. (2.115)), consistent with the mechanism in eqs (2.116)–(2.118), where $K_a \approx 2.3 \times 10^{-3}$ M and in this case $k_1 = 3.4 \times 10^5$ M^{-1} s^{-1} and $k_2 = 1.7 \times 10^7$ M^{-1} s^{-1} respectively. Estimates for the self-exchange rates are $>10^{-2}$ M^{-1} s^{-1} and $\geq 3 \times 10^{-4}$ respectively for $[Ti(H_2O)_5OH]^{3+/2+}$ and $[Ti_{aq}]^{4+/3+}$ and there is general consistency for a number of studies [288, 291].

$$[TiO_{aq}]^{2+} + 2H^+ + e^- \rightleftharpoons [Ti(H_2O)_6]^{3+} \quad (2.114)$$

$$-\frac{d[[Os(bpy)_3]^{3+}]}{dt} = \left\{\frac{k_1[H^+] + k_2 K_a}{[H^+] + K_a}\right\} [[Ti(H_2O)_6]^{3+}]_T [[Os(bpy)_3]^{3+}] \quad (2.115)$$

$$[Ti(H_2O)_6]^{3+} \underset{}{\overset{K_a}{\rightleftharpoons}} [Ti(H_2O)_5OH]^{2+} + H^+ \quad (2.116)$$

$$[Os(bpy)_3]^{3+} + [Ti(H_2O)_6]^{3+} \xrightarrow{k_1} [Os(bpy)_3]^{2+} + [TiO_{aq}]^{2+} + 2H^+ \quad (2.117)$$

$$[Os(bpy)_3]^{3+} + [Ti(H_2O)_5OH]^{2+} \xrightarrow{k_2} [Os(bpy)_3]^{2+} + [TiO_{aq}]^{2+} + H^+ \quad (2.118)$$

In these reactions, several generalities can be uncovered. Firstly, the dominant pathways for both oxidation and reduction tend to be those in which transfer of the electron is coupled to the smallest number of protons. The available evidence is that electron transfer from the metal center is quite distinct from H$^+$ transfer from the ligand and that electron transfer precedes H$^+$ transfer in the sequence of reaction steps. The lack of close coupling between the two processes may be related to the thermodynamic driving force which is favorable for both electron and H$^+$ transfer.

A rather different but particularly important case is illustrated by the comproportionation reaction of [Ru(bpy)$_2$(py)OH$_2$]$^{2+}$ with [Ru(bpy)$_2$(py)O]$^{2+}$ (eq. (2.119)) [292, 293]. Electron transfer is thermodynamically favorable by 0.11V, but only when coupled to the transfer of H$^+$ [294]. The rate is first-order in both [[Ru(bpy)$_2$(py)OH$_2$]$^{2+}$] and [[Ru(bpy)$_2$(py)O]$^{2+}$] and the second-order rate constant, k (eq. (2.120)), shows a strong dependence on pH consistent with the mechanism (eqs (2.121)–(2.123)), where k_{H_2O} = 2.15 × 10^5 M^{-1} s^{-1}, k_{OH} = 2.5 × 10^4 M^{-1} s^{-1} and pK_b = 3.20. An inner-sphere mechanism can be discounted on the basis of the product analysis but the reaction is much slower than might be expected for a simple outer-sphere process.

$$[\text{Ru(bpy)}_2(\text{py})\text{OH}_2]^{2+} + [\text{Ru(bpy)}_2(\text{py})\text{O}]^{2+} \rightleftharpoons 2\,[\text{Ru(bpy)}_2(\text{py})\text{OH}]^{2+} \tag{2.119}$$

$$k = \frac{k_{OH} + k_{H_2O}K_b/[\text{OH}^-]}{1 + K_b/[\text{OH}^-]} \tag{2.120}$$

$$[\text{Ru(bpy)}_2(\text{py})\text{OH}_2]^{2+} + \text{OH}^- \overset{K_b}{\rightleftharpoons} [\text{Ru(bpy)}_2(\text{py})\text{OH}]^{2+} \tag{2.121}$$

$$[\text{Ru(bpy)}_2(\text{py})\text{OH}_2]^{2+} + [\text{Ru(bpy)}_2(\text{py})\text{O}]^{2+} \overset{k_{H_2O}}{\rightleftharpoons} 2\,[\text{Ru(bpy)}_2(\text{py})\text{OH}]^{2+} \tag{2.122}$$

$$[\text{Ru(bpy)}_2(\text{py})\text{OH}]^+ + [\text{Ru(bpy)}_2(\text{py})\text{O}]^{2+} \overset{k_{OH}}{\rightleftharpoons} [\text{Ru(bpy)}_2(\text{py})\text{OH}]^{2+} + [\text{Ru(bpy)}_2(\text{py})\text{O}]^+ \tag{2.123}$$

A study of the reaction kinetics in D$_2$O is most revealing (Fig. 2.21). Both pathways show a solvent isotope effect but while k_H/k_D for the k_{OH} pathway is 1.5, k_H/k_D for the dominant reaction, k_{H_2O}, is 16.1. In this case a mechanism in which electron transfer precedes H$^+$ transfer can be ruled out since the initial electron transfer would be thermodynamically disfavored by at least 0.55 V and should therefore have a rate constant of no more than 3 × 10^3 M^{-1} s^{-1}. The mechanism proposed involves simultaneous transfer of an electron and H$^+$; a hydrogen atom transfer reaction with a transition state as shown in eq. (2.124).

$$\begin{array}{c} Ru^{IV}=O\text{-}\text{-}H\text{-}O\text{-}Ru^{II} \\ | \\ H \end{array} \qquad (2.124)$$

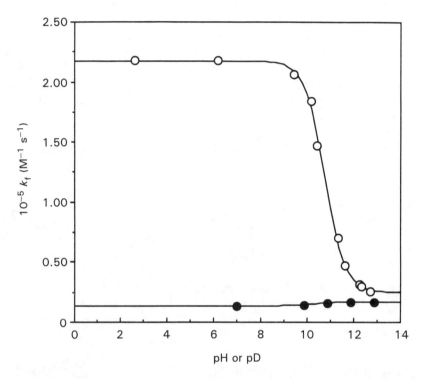

Fig. 2.21. pH dependence (open circles) and pD dependence (closed circles) of the rate costant for the comproportionation reaction between $[Ru^{IV}(bpy)_2(py)(O)]^{2+}$ and $[Ru^{II}(bpy)_2(py)(OH_2)]^{2+}$ at $\mu = 0.10 M$ (Na$_2$So$_4$) and 25°C, from Ref. [293].

Although the reaction is termed hydrogen atom transfer, there is no formation of free H·. The reduction potential of H$^+$ is estimated to be -2.31 V and the energetics of the reaction clearly preclude any involvement of this process. What is happening is that there is a very strong coupling between H$^+$ and electron transfer.

A keener appreciation of this coupling can be obtained by considering the two-dimensional reaction coordinate plot or More O'Ferral diagram (Fig. 2.22) [295]. In this figure, electron transfer is depicted from left to right and H$^+$ transfer top to bottom. The reactants are at the upper left corner and the products at the bottom right. At the upper right corner, the high energy product of electron transfer without the associated H$^+$ transfer is located, while at the lower left corner, the high energy product of H$^+$ transfer without electron transfer is located. A consecutive electron transfer followed by H$^+$ transfer would take the pathway involving the upper right corner as an intermediate, whereas the preferred, coupled H⁺ and electron transfer, the atom transfer, is depicted by the diagonal and represents a lower energy pathway.

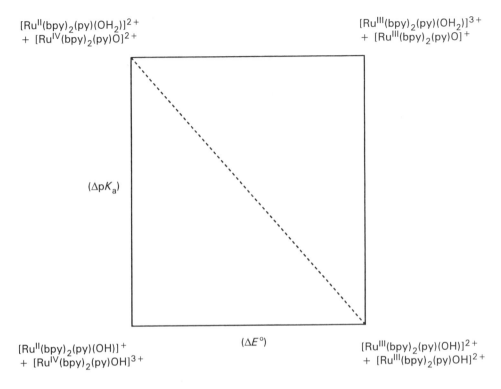

Fig. 2.22. More O'Farral plot for the coupled electron and H⁺ transfer in the reaction of $[Ru^{IV}(bpy)_2(py)(O)]^{2+}$ with $[Ru^{II}(bpy)_2(py)(OH_2)]^{2+}$. Horizontal changes represent electron transfer while vertical changes represent H⁺ transfer. The upper-right and lower-left reagents are thermodynamically at much higher energy than either the reactants or the products and the atom transfer pathway, represented by the dashed diagonal, is the preferred mechanism.

Like electron transfer, H⁺ transfer reactions follow strong a linear free energy relationship, the Brönsted relationship, which can be derived in a manner similar to that for electron transfer in section 2.4 [296, 297]. In this instance, the driving force is measured by the difference in pK_a values between the H⁺ donor and the H⁺ acceptor. The pathway followed in Fig. 2.22 will depend on the free energy differences for electron transfer and H⁺ transfer, and the degree to which atom transfer participates will be directly related to these changes. There is a series of reactions shown in Table 2.24 where the thermodynamic products require both H⁺ and electron transfer with differing driving forces. No large isotope effects are detectable, which suggests that they are best described as outer-sphere electron transfer reactions with coupled H⁺ transfer. However, the rates for reaction are in excess of those calculated for the thermodynamically less favorable simple electron transfer by the Marcus relationship,

108 The outer-sphere mechanism [Ch. 2]

and the ratio $\log(k_{expt}/k_{calc})$ is linearly related to $\Delta E°_{atom} - \Delta E°_{electron}$, suggesting that atom transfer provides a contribution to the rate.

Table 2.24. Rates for coupled H^+ and electron transfer at 25°C, $\mu = 0.1$ M as a function of the driving force for simple electron transfer, $\Delta E°_{electron}$, and the H^+ coupled process, $\Delta E°_{atom}$ [298, 299]

Reaciton	$\Delta E°_{electron}$ (V)	$\Delta E°_{atom}$ (V)	k_{expt} (M^{-1} s^{-1})	k_{calc} (M^{-1} s^{-1})
$[Ni^{III}(Me_2LH)]^{2+} + [Fe^{II}(Me_2LH)]^+$ $\rightarrow [Ni^{II}(Me_2LH_2)]^{2+} + [Fe^{III}(Me_2L)]^+$	0.07	0.38	3×10^6	5×10^4
$[Ni^{III}(Me_2L)]^+ + [Fe^{II}(Me_2LH)]^+$ $\rightarrow [Ni^{II}(Me_2LH)]^{2+} + [Fe^{III}(Me_2L)]^+$	−0.15	0.27	8×10^5	2×10^3
$[Ni^{IV}(Me_2L)]^{2+} + [Ni^{II}(Me_2LH_2)]^{2+}$ $\rightarrow [Ni^{III}(Me_2LH)]^{2+} + [Ni^{III}(Me_2LH)]^{2+}$	−0.42	0.13	4×10^2	4×10^2

At this point, with the coupling of atom and electron transfer, it is appropriate to consider the inner-sphere mechanism.

QUESTIONS

2.1 The rate constant for reduction of $[Co(edta)]^-$ by $[Ru(NH_3)_5bpy]^{2+}$ is 195 M^{-1} s^{-1} at 25.0°C and 0.10 M ionic strength. Addition of β-cyclodextrin, a cyclic sugar which forms an inclusion complex with aromatic residues, affects the rate constants as shown in the figure. Provide an explanation for the data.
(Johnson, M. D.; Reinsborough, V. C.; Ward, S. *Inorg. Chem.* 1992, **31**, 1085–1087.)

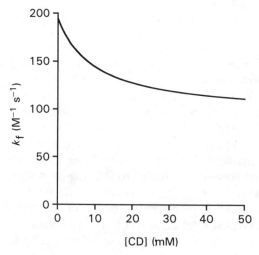

2.2 Use the data in Tables 2.7 and 2.9 to predict rate constants for electron transfer by an outer-sphere mechanism for the following reactions:

$$[Fe(dipic)_2]^- + [Co(bpy)_3]^{2+} \rightarrow$$

$$[Ru(bpy)_3]^{3+} + [Co(sar)]^{2+} \rightarrow$$

$$[Co([9]aneS_3)_2]^{3+} + [Ru(en)_3]^{2+} \rightarrow$$

2.3 The rate constant for the reduction of $[Co(edta)]^-$ by $[Co(sep)]^{2+}$, $[Fe(edta)]^{2-}$, and $[Ru(en)_3]^{2+}$ are 8.3×10^4 M^{-1} s^{-1}, 4.4 M^{-1} s^{-1} and 9×10^3 M^{-1} s^{-1} respectively at 25 °C and 0.10 M ionic strength. Assuming an outer-sphere mechanism, calculate a self-exchange rate for $[Co(edta)]^{-/2-}$ with data from Tables 2.7 and 2.9, using (a) eq. (2.26), and (b) eq. (2.28) with electrostatics corrections. (Hint: an iterative procedure is required. In general, one iteration is sufficient.)

2.4 Use the second-order rate constants in the table to discuss mechanisms for the reduction of the cobalt(III) complexes by $[Co(en)_3]^{2+}$ and $[Cr(bpy)_3]^{2+}$.

Table. Second-order rate constants (M^{-1} s^{-1}) at 25.0 °C and 0.10 M ionic strength

	Reductant	
Oxidant	$[Cr(bpy)_3]^{2+}$	$[Co(en)_3]^{2+}$
$[Co(ox)_3]^{3-}$	—	390
$[Co(gly)(ox)_2]^{2-}$	—	20
$[Co(edta)]^-$	2.0×10^6	18
$cis(\beta)$-$[Co(edda)(ox)]^-$	1.1×10^5	0.87
C_1-$cis(N)$-$[Co(gly)_2(ox)]^-$	1.7×10^5	1.50
$cis(\alpha)$-$[Co(edda)(ox)]^-$	2.1×10^5	0.37
C_2-$cis(N)$-$[Co(gly)_2(ox)]^-$	2.4×10^5	0.44
$trans(N)$-$[Co(gly)_2(ox)]^-$	1.65×10^5	0.28
$[Co(ox)_2(en)]^-$	8.5×10^4	0.16
$[Co(en)_2(ox)]^+$	1.0×10^3	0.0011

(Warren, R. M. L.; Tatehata, A.; Lappin, A. G. *Inorg. Chem.* 1993, **32**, 1191–1196.)

2.5 Discuss the following statement:
'Experimental evidence for non-adiabatic behavior in bimolecular outer-sphere electron transfer reactions is difficult to obtain as a result of nuclear tunneling.'

2.6 Dielectric continuum theory has proved to be of considerable importance in describing the effect of changing solvent on charge transfer processes. Indicate what limitations there might be on the application of this theory to (a) self-exchange reactions involving charged complexes, and (b) cross-reactions between dissimilar complexes.

2.7 Describe the experimental criteria necessary for the detection of stereoselectivity in outer-sphere electron transfer reactions. Indicate two ways in wich stereoselectivity in the reaction between [Co(edta)]$^-$ and [Co(sep)]$^{2+}$ might be investigated. Note that the cobalt(II) complex, [Co(sep)]$^{2+}$, ca be isolated in eantiomeric forms.

REFERENCES

[1] Chou, M.; Creutz, C.; Sutin, N. *J. Am. Chem. Soc.* 1977, **99**, 5615–5623.
[2] Marcus, R. A. *Ann. Rev. Phys. Chem.* 1964, **15**, 155–196.
[3] Hush, N. S. *Trans. Faraday Soc.* 1961, **57**, 557–580.
[4] Sutin, N. *Acc. Chem. Res.* 1982, **15**, 275–282.
[5] Marcus, R. A.; Sutin, N. *Biochim. Biophys. Acta* 1985, **811**, 265–322.
[6] Zahir, K.; Espenson, J. H.; Bakac, A. *Inorg. Chem.* 1988, **27**, 3144–3146.
[7] Bock, C. R.; Meyer, T. J.; Whitten, D. G. *J. Am. Chem. Soc.* 1974, **96**, 4704–4712.
[8] McArdle, J. V.; Yocom, K.; Gray, H. B. *J. Am. Chem. Soc.* 1977, **99**, 4141–4145.
[9] Braun, P.; van Eldik, R. *J. Chem. Soc., Chem. Commun.* 1985, 1349–1350.
[10] McArdle, J. V.; Yocom, K.; Gray, H. B. *J. Am. Chem. Soc.* 1977, **99**, 4141–4145.
[11] Endicott, J. F.; Taube, H. *J. Am. Chem. Soc.* 1964, **86**, 1686–1691.
[12] Ronco, S.; Ferraudi, G. *J. Chem. Soc., Dalton Trans.* 1990, 887–889.
[13] Brown, G. M.; Krentzien, H. J.; Abe, M.; Taube, H. *Inorg. Chem.* 1979, **18**, 3374–3379.
[14] Bänsch, B.; Martinez, P.; van Eldick, R. *J. Phys. Chem.* 1992, **96**, 234–238.
[15] Przystas, T. J.; Sutin, N. *J. Am. Chem. Soc.* 1973, **95**, 5545–5555.
[16] Macartney, D. H.; Sutin, N. *Inorg. Chem.* 1983, **22**, 3530–3534.
[17] Toma, H. E.; Lellis, F. T. P. *Polyhedron* 1985, **4**, 993–997.
[18] Wilkins, R. G.; Yelin, R. *J. Am. Chem. Soc.* 1970, **92**, 1191–1194.
[19] Grossman, B.; Wilkins, R. G. *J. Am. Chem. Soc.* 1967, **89**, 4230–4232.
[20] Hoddenbagh, J. M. A.; Macartney, D. H. *Inorg. Chem.* 1990, **29**, 245–251.
[21] Ehighaokhuo, J. O.; Ojo, J. F.; Olubuyide, O. *J. Chem. Soc., Dalton Trans.* 1985, 1665–1667.
[22] Haim, A.; Sutin, N. *Inorg. Chem.* 1976, **15**, 476–478.
[23] Miralles, A. J.; Armstrong, R. E.; Haim, A. *J. Am. Chem. Soc.* 1977, **99**, 1416–1420.
[24] Gaswick, D.; Haim, A. *J. Am. Chem. Soc.* 1971, **93**, 7347–7348.
[25] Miralles, A. J.; Szecsy, A. P.; Haim, A. *Inorg. Chem.* 1982, **21**, 697–699.
[26] Gaus, P. L.; Villaneuva, J. L. *J. Am. Chem. Soc.* 1980, **102**, 1934–1938.
[27] Kremer, E.; Cha, G.; Morkevicius, M.; Seaman, M.; Haim, A. *Inorg. Chem.* 1984, **23**, 3028–3030.
[28] Warren, R. M. L.; Lappin, A. G.; Tatehata, A. *Inorg. Chem.* 1992, **31**, 1566–1574.

[29] Huck, H.-M.; Wieghardt, K. *Inorg. Chem.* 1979, **18**, 1799–1807.
[30] Yokoyama, H.; Kon, H. *J. Phys. Chem.* 1991, **95**, 8956–8963.
[31] Stranks, D. R. *Pure Appl. Chem.* 1974, **38**, 303–323.
[32] Krack, I.; van Eldik, R. *Inorg. Chem.* 1986, **25**, 1743–1747.
[33] Krack, I.; van Eldik, R. *Inorg. Chem.* 1989, **28**, 851–855.
[34] van Eldik, R.; Kelm, H. *Inorg. Chim. Acta* 1983, **73**, 91–94.
[35] Sasaki, Y.; Endo, K.; Nagasawa, A.; Saito, K. *Inorg. Chem.* 1986, **25**, 4845–4847.
[36] Kanesato, M.; Ebihara, M.; Sasaki, Y.; Saito, K. *J. Am. Chem. Soc.* 1983, **105**, 5711–5713.
[37] Krack, I.; van Eldik, R. *Inorg. Chem.* 1990, **29**, 1700–1704.
[38] Fuoss, R. M. *J. Am. Chem. Soc.* 1958, **80**, 5059–5061.
[39] Hirata, F.; Friedman, H. L.; Holz, M.; Hertz, H. G. *J. Chem. Phys.* 1980, **73**, 6031–6038.
[40] Marusak, R. A.; Lappin, A. G. *J. Phys. Chem.* 1989, **93**, 6856–6859.
[41] Tatehata, A.; Fujita, M.; Ando, K.; Asaba, Y. *J. Chem. Soc. Dalton Trans.* 1987, 1977–1982.
[42] Niederhoffer, E. C.; Martell, A. E.; Rudolph, P.; Clearfield, A. *Cryst. Struct. Commun.* 1982, **11**, 1951–1954.
[43] Boys, D.; Escobar, C.; Wittke, O. *Acta Cryst.* 1984, **C40**, 1359–1362.
[44] Brunschwig, B. S.; Creutz, C.; Macartney, D. H.; Sham, T.-K.; Sutin, N. *Faraday Discuss. Chem. Soc.* 1982, **74**, 113–127.
[45] Stynes, H. C.; Ibers, J. A. *Inorg. Chem.* 1971, **10**, 2304–2308.
[46] Sutin, N.; Brunschwig, B. S.; Creutz, C.; Winkler, J. R. *Pure Appl. Chem.* 1988, **60**, 1817–1830.
[47] Ratner, M. A.; Levine, R. D. *J. Am. Chem. Soc.* 1980, **102**, 4898–4900.
[48] Marcus, R. A.; Sutin, N *Inorg. Chem.* 1975, **14**, 213–216.
[49] Espenson, J. H. *Chemical Kinetics and Reaction Mechanisms.* McGraw-Hill: New York, 1981; pp. 50–55.
[50] Meyer, T. J.; Taube, H. *Inorg. Chem.* 1968, **11**, 2369–2379.
[51] McAuley, A.; Macartney, D. H.; Oswald, T. *J. Chem. Soc., Chem. Commun.* 1982, 274–275.
[52] Wang, J.-F.; Kumar, K.; Margerum, D. W. *Inorg. Chem.* 1989, **28**, 3481–3488.
[53] Dwyer, F. P.; Gyarfas, E. C. *Nature* 1950, **166**, 481.
[54] Creaser, I. I.; Sargeson, A. M.; Zanella, A. W. *Inorg. Chem.* 1983, **22**, 4022–4029.
[55] Warren, R. M. L.; Lappin, A. G.; Mehta, B. D.; Neumann, H. M. *Inorg. Chem.* 1990, **29**, 4185–4189.
[56] Creaser, I. I.; Harrowfield, J. M.; Herlt, A. J.; Sargeson, A. M.; Springborg, J.; Geue, R. J.; Snow, M. R. *J. Am. Chem. Soc.* 1982, **104**, 6016–6025.
[57] Brown, G. M.; Sutin, N. *J. Am. Chem. Soc.* 1979, **101**, 883–892.
[58] McAuley, A.; Xu, C. *Inorg. Chem.* 1988, **27**, 1204–1209.
[59] Koval, C. A.; Margerum, D. W. *Inorg. Chem.* 1981, **20**, 2311–2318.
[60] Pulliam, E. J.; McMillin, D. R. *Inorg. Chem.* 1984, **23**, 1172–1175.
[61] Krishnamurty, K. V.; Wahl, A. C. *J. Phys. Chem.* 1958, **80**, 5921–5924.

[62] Anderson, A.; Bonner, N. A. *J. Am. Chem. Soc.* 1954, **76**, 3826–3830.
[63] Bernhard, P.; Sargeson, A. M. *Inorg. Chem.* 1987, **26**, 4122–4125.
[64] Spiccia, L.; Swaddle, T. W. *J. Chem. Soc., Chem. Commun.* 1985, 67–68.
[65] Spiccia, L.; Swaddle, T. W. *Inorg. Chem.* 1987, **26**, 2265–2271.
[66] Jolley, W. H.; Stranks, D. R.; Swaddle, T. W. *Inorg. Chem.* 1990, **29**, 1948–1951.
[67] Doine, H.; Swaddle, T. W. *Can. J. Chem.* 1990, **68**, 2228–2233.
[68] Doine, H.; Swaddle, T. W. *Can. J. Chem.* 1988, **66**, 2763–2767.
[69] Campion, R. J.; Deck, C. F.; King, P.; Wahl, A. C. *Inorg. Chem.* 1967, **6**, 672–681.
[70] Takagi, H.; Swaddle, T. W. *Inorg. Chem.* 1992, **31**, 4669–4673.
[71] Habib, H. S.; Hunt, J. P. *J. Am. Chem. Soc.* 1966, **88**, 1668–1671.
[72] Dwyer, F. P.; Sargeson, A. M. *J. Phys. Chem.* 1961, **65**, 1892–1894.
[73] Jolley, W. H.; Stranks, D. R.; Swaddle, T. W. *Inorg. Chem.* 1990, **29**, 385–389.
[74] Hammershøi, A.; Geselowitz, D.; Taube, H. *Inorg. Chem.* 1984, **23**, 979–982.
[75] Geselowitz, D.; Taube, H. *Advances in Inorganic and Bioinorganic Mechanisms,* 1982, **1**, 391–407.
[76] Doine, H.; Swaddle, T. W. *Inorg. Chem.* 1991, **30**, 1858–1862.
[77] Hammershøi, A.; Sargeson, A. M.; *Inorg. Chem.* 1983, **22**, 3554–3561.
[78] Dubs, R. V.; Gahan, L. R.; Sargeson, A. M. *Inorg. Chem.* 1983, **22**, 2523–2527.
[79] Osvath, P; Sargeson, A. M.; Skelton, B. W.; White, A. H. *J. Chem. Soc., Chem. Commun.* 1991, 1036–1038.
[80] Küppers, H.-J.; Neves, A.; Pomp, C.; Ventur, D.; Wieghardt, K.; Nuber, B.; Weiss, J. *Inorg. Chem.* 1986, **25**, 2400–2408.
[81] Küppers, H.-J.; Wieghardt, K..; Steenken, S.; Nuber, B.; Weiss, J. *Z. anorg. allg. Chem.* 1989, **573**, 43–62.
[82] Chandrasekhar, S.; McAuley, A. *Inorg. Chem.* 1992, **31**, 480–487.
[83] Im, Y. A.; Busch, D. H. *J. Am. Chem. Soc.* 1961, **83**, 3357–3362.
[84] Wilkins, R. G.; Yelin, R. E. *Inorg. Chem.* 1968, **7**, 2667–2669.
[85] Jolley, W. H.; Stranks, D. R.; Swaddle, T. W. *Inorg. Chem.* 1992, **31**, 507–511.
[86] Rasmussen, P. G.; Brubaker, C. H. *Inorg. Chem.* 1964, **3**, 977–980.
[87] Marusak, R. A.; Sharp, C.; Lappin, A. G. *Inorg. Chem.* 1990, **29**, 4453–4456.
[88] Doine, H.; Yano, Y.; Swaddle, T. W. *Inorg. Chem.* 1989, **28**, 2319–2322.
[89] Bernhard, P.; Helm, L.; Ludi, A.; Merbach, A. E. *J. Am. Chem. Soc.* 1985, **107**, 312–317.
[90] Beattie, J. K.; Smolenaers, P. J. *J. Phys. Chem.* 1986, **90**, 3684–3686.
[91] Smolenaers, P. J.; Beattie, *Inorg. Chem.* 1986, **25**, 2259–2262.
[92] Taube, H. *Adv. Chem.* 1977, **162**, 127–144.
[93] Bernhard, P.; Sargeson, A. M. *Inorg. Chem.* 1988, **27**, 2582–2587.
[94] Young, R. C.; Keene, F. R.; Meyer, T. J. *J. Am. Chem. Soc.* 1977, **99**, 2468–2473.
[95] Foucher, D. A.; Macartney, D. H. *J. Chem. Res.* 1992, 346–347.
[96] Hurwitz, P.; Kustin, K. *Trans. Faraday Soc.* 1966, **62**, 427–432.
[97] Duke, F. R.; Parchen, F. R. *J. Phys. Chem.* 1956, **78**, 1540–1543.
[98] Weaver, M. J.; Yee, E. L. *Inorg. Chem.* 1980, **19**, 1936–1945.
[99] Ferraudi, G. J.; Endicott, J. F. *Inorg. Chim. Acta* 1979, **37**, 219–223.
[100] Williams, N. H.; Yandell, J. K.; 1983, **36**, 2377–2386.

[101] Lannon, A. M.; Lappin, A. G.; Segal, M. G. *J. Chem. Soc., Dalton Trans.* 1986, 619–624.
[102] Toma, H. E.; Malin, J. M. *Inorg. Chem.* 1974, **13**, 1772–1774.
[103] Endicott, J. F.; Durham, B.; Kumar, K. *Inorg. Chem.* 1982, **21**, 2437–2444.
[104] Farina, R.; Wilkins, R. G. *Inorg. Chem.* 1968, **7**, 514–518.
[105] Hendry, P.; Ludi, A. *J. Chem. Soc., Chem. Commun.* 1987, 891–892.
[106] Beattie, J. K.; Binstead, R. A.; Broccardo, M. *Inorg. Chem.* 1978, **17**, 1822–1826.
[107] Holwerda, R. A.; Knaff, D. B.; Gray, H. B.; Clemmer, J. D.; Crowley, R.; Smith, J. M.; Mauk, A. G. *J. Am. Chem. Soc.* 1980, **102**, 1142–1146.
[108] Mauk, A. G.; Bordignon, E.; Gray, H. B. *J. Am. Chem. Soc.* 1982, **104**, 7645–7657.
[109] Ellis, R. M.; Quilligan, J. D.; Williams, N. H.; Yandell, J. K. *Aust. J. Chem.* 1989, **42**, 1–7.
[110] McAuley, A.; Norman, P. R.; Olubuyide, O. *Inorg. Chem.* 1984, **23**, 1938–1943.
[111] Sisley, M. J.; Jordan, R. B. *Inorg. Chem.* 1992, **31**, 2880–2884.
[112] Mentasti, E.; Kirschenbaum, L. J. *Inorg. Chim. Acta* 1987, **129**, 99–102.
[113] Howes, K. R.; Bakac, A.; Espenson, J. H. *Inorg. Chem.* 1988, **27**, 791–794.
[114] Toppen, D. L.; Link, R. G. *Inorg. Chem.* 1971, **10**, 2635–2636.
[115] Chen, J. C.; Gould, E. S. *J. Am. Chem. Soc.* 1973, **95**, 5539–5544.
[116] Gould, E. S. *Inorg. Chem.* 1979, **18**, 900–901.
[117] Scott, K. L.; Sykes, A. G. *J. Chem. Soc., Dalton Trans.* 1972, 1832–1837.
[118] Fan, F.-R. F.; Gould, E. S. *Inorg. Chem.* 1974, **13**, 2647–2651.
[119] Stranks, D. R. *Discuss. Faraday Soc.* 1961, 73–79.
[120] Lytle, F. E.; Hercules, D. M. *J. Am. Chem. Soc.* 1969, **91**, 253–257.
[121] Gafney, H. D.; Adamson, A. W. *J. Am. Chem. Soc.* 1972, **94**, 8238–8239.
[122] Demas, J. N.; Adamson, A. W. *J. Am. Chem. Soc.* 1973, **95**, 5159–5168.
[123] Navon, G.; Sutin, N. *Inorg. Chem.* 1974, **13**, 2159–2164.
[124] Creutz, C.; Sutin, N. *Inorg. Chem.* 1976, **15**, 496–499.
[125] Bolletta, F.; Maestri, M.; Moggi, L.; Balzani, V.; *J. Am. Chem. Soc.* 1973, **95**, 7864–7866.
[126] Mok, C.-Y.; Zanella, A. W.; Creutz, C.; Sutin, N. *Inorg. Chem.* 1984, **23**, 2891–2897.
[127] Rybak, W.; Haim, A.; Netzel, T. L.; Sutin, N. *J. Phys. Chem.* 1981, **85**, 2856–2860.
[128] Lin, C.-T.; Böttcher, W.; Chou, M.; Creutz, C.; Sutin, N. *J. Am. Chem. Soc.* 1976, **98**, 6536–6544.
[129] Balzani, V.; Scandola, F.; Orlandi, G.; Sabbatini, N.; Indelli, M. T. *J. Am. Chem. Soc.* 1981, **103**, 3370–3378.
[130] Berkoff, R.; Krist, K.; Gafney, H. D. *Inorg. Chem.* 1980, **19**, 1–7.
[131] Creutz, C.; Sutin, N. *J. Am. Chem. Soc.* 1977, **99**, 241–243.
[132] Lay, P. A.; Mau, A. W. H.; Sasse, W. H. F.; Creaser, I. I.; Gahan, L. R.; Sargeson, A. M. *Inorg. Chem.* 1983, **17**, 2347–2349.
[133] Zahir, K.; Böttcher, W.; Haim, A. *Inorg. Chem.* 1985, **24**, 1966–1968.

[134] Shaoyung, L.; Bakac, A.; Espenson, J. H. *Inorg. Chem.* 1989, **28**, 1367–1369.
[135] Serpone, N.; Jamieson, M. A.; Sriram, R.; Hoffman, M. Z. *Inorg. Chem.* 1981, **20**, 3983–3988.
[136] Ahn, B.-T.; McMillin, D. R. *Inorg. Chem.* 1978, **17**, 2253–2258.
[137] Gamache, R. E.; Rader, R. A.; McMillin, D. R. *J. Am. Chem. Soc.* 1985, **107**, 1141–1146.
[138] Walsh, J. L.; Baumann, J. A.; Meyer, T. *J. Inorg. Chem.* 1980, **19**, 2145–2151.
[139] Nielson, R. M.; Wherland, S. *Inorg. Chem.* 1986, **25**, 2437–2440.
[140] Nielson, R. M.; Wherland, S. *J. Am. Chem. Soc.* 1985, **107**, 1505–1510.
[141] Nielson, R. M.; Wherland, S. *Inorg. Chem.* 1984, **23**, 1338–1344.
[142] Macartney, D. H.; Thompson, D. W. *J. Chem. Res.* (S) 1992, 344–345.
[143] Nielson, R. M.; McManis, G. E.; Safford, L. K.; Weaver, M. *J. Phys. Chem.* 1989, **93**, 2152–2157.
[144] Kirchner, K.; Dang, S.-Q.; Stebler, M.; Dodgen, H. W.; Wherland, S.; Hunt, J. P. *Inorg. Chem.* 1989, **28**, 3604–3608.
[145] Nielson, R. M.; McManis, G. E.; Golovin, M. N.; Weaver, M. J. *J. Phys. Chem.* 1988, **92**, 3441–3450.
[146] Protasiewicz, J. D.; Theopold, K. H.; Schulte, G. *Inorg. Chem.* 1988, **27**, 1133–1136.
[147] Chan, M.-S.; Wahl, A. C. *J. Phys. Chem.* 1982, **86**, 126–130.
[148] Nielson, R. M.; Hunt, J. P.; Dodgen, H. W.; Wherland, S. *Inorg. Chem.* 1986, **25**, 1964–1968.
[149] Corraine, M. S.; Atwood, J. D. *Inorg. Chem.* 1989, **28**, 3781–3782.
[150] Braga, T. G.; Wahl, A. C. *J. Phys. Chem.* 1985, **89**, 5822–5828.
[151] Borchardt, D.; Pool, K.; Wherland, S. *Inorg. Chem.* 1982, **21**, 93–97.
[152] Murguia, M. A.; Wherland, S. *Inorg. Chem.* 1991, **30**, 139–144.
[153] Borchardt, D.; Wherland, S. *Inorg. Chem.* 1986, **25**, 901–905.
[154] Anderson, K. A.; Wherland, S. *Inorg. Chem.* 1991, **30**, 624–629.
[155] Marshall, A. G. *Acc. Chem. Res.* 1985, **18**, 316–322.
[156] Richardson, D. E.; Christ, C. S.; Sharpe, P.; Eyler, J. R. *J. Am. Chem. Soc.* 1987, **109**, 3849–3902.
[157] Phelps, D. K.; Gord, J. R.; Freiser, B. S.; Weaver, M. J. *J. Phys. Chem.* 1991, **95**, 4338–4342.
[158] Lappin, A. G.; Marusak, R. A. *Coord. Chem. Rev.* 1991, **109**, 125–180.
[159] Sutter, J. H.; Hunt, J. B. *J. Am. Chem. Soc.* 1969, **91**, 3107–3108.
[160] Geselowitz, D. A.; Taube, H. *J. Am. Chem. Soc.* 1980, **102**, 4525–4526.
[161] Kane-Maguire, N. A. P.; Tollison, R. M.; Richardson, D. E. *Inorg. Chem.* 1976, **15**, 499–500.
[162] Khurram, Z.; Espenson, J. H.; Bakac, A. *Inorg. Chem.* 1988, **27**, 3144–3146.
[163] Blinn, E. L.; Wilkins, R. G. *Inorg. Chem.* 1976, **15**, 2952.
[164] Zwickel, A. M.; Taube, H. *Discuss. Faraday Soc.* 1960, **29**, 42–48.
[165] Ficke, J. T.; Pladziewicz, J. R.; Sheu, E. C.; Lappin, A. G. *Inorg. Chem.* 1991, **30**, 4282–4285.
[166] Osvath, P.; Lappin, A. G. *Inorg. Chem.* 1987, **26**, 195–202.

[167] Marusak, R. A.; Osvath, P.; Kemper, M.; Lappin, A. G. *Inorg. Chem.* 1989, **28**, 1542–1548.
[168] Kaizu, Y.; Mori, T.; Kabayashi, H. *J. Phys. Chem.* 1985, **89**, 332–335.
[169] Newton, M. D.; Sutin, N. *Ann. Rev. Phys. Chem.* 1984, **35**, 437–480.
[170] Brunschwig, B. S.; Sutin, N. *Comments Inorg. Chem.* 1987, **6**, 209–235.
[171] Newton, M. D. *Chem. Rev.* 1991, **91**, 767–792.
[172] Brubaker, G. R.; Johnson, D. W. *Coord. Chem. Rev.* 1984, **53**, 1–36.
[173] Tunuli, M. S. *J. Phys. Chem.* 1986, **90**, 1983–1984.
[174] Tunuli, M. S.; Khan, S. U. M. *J. Phys. Chem.* 1987, **91**, 3474–3478.
[175] Sham, T. K.; *Acc. Chem. Res.* 1986, **19**, 99–104.
[176] Sakane, H.; Miyanaga, T.; Wanatabe, I.; Yokoyama, Y. *Chem. Letters,* 1990, 1623–1626.
[177] Cotton, F. A.; Fair, C. K.; Lewis, G. E.; Mott, G. N.; Ross, F. K.; Schulz, A. J.; Williams, J. M. *J. Am. Chem. Soc.* 1984, **106**, 5319–5323.
[178] Holt, D. G.; Larkworthy, L. F.; Povey, D. C.; Smith, G. W.; Leigh, G. J. *Inorg. Chim. Acta* 1990, **169**, 201–205.
[179] Bernhard, P.; Bürgi, H.-B.; Hauser, J.; Lehmann, H.; Ludi, A. *Inorg. Chem.* 1982, **21**, 3936–3941.
[180] Bernhard, P.; Ludi, A. *Inorg. Chem.* 1984, **23**, 870–872.
[181] Ventur, D.; Wieghardt, K.; Nuber, B.; Weiss, J. *Z. anorg. allg. Chem.* 1987, **551**, 33–60.
[182] Zompa, L. J.; Margulis, T. N. *Inorg. Chim. Acta* 1978, **28**, L157–L162.
[183] Wieghardt, K.; Walz, W.; Nuber, B.; Weiss, J.; Ozarowski, A.; Stratemeier, H.; Reinen, D. *Inorg. Chem.* 1986, **25**, 1650–1654.
[184] Boeyens, J. C. A.; Forbes, A. G. S.; Hancock, R. D.; Wieghardt, K. *Inorg. Chem.* 1985, **24**, 2926–2931.
[185] Wieghardt, K.; Schmidt, W.; Herrmann, Küppers, H.-J. *Inorg. Chem.* 1983, **22**, 2953–2956.
[186] Pohl, K.; Wieghardt, K.; Kaim, W.; Steenken, S. *Inorg. Chem.* 1988, **27**, 440–447.
[187] Szalda, D. J.; Creutz, C.; Mahajan, D.; Sutin, N. *Inorg. Chem.* 1983, **22**, 2372–2379.
[188] Biner, M.; Bürgi, H.-B.; Ludi, A.; Röhr, C. *J. Am. Chem. Soc.* 1992, **114**, 5197–5203.
[189] Ruff, I.; Zimonyi, M. *Electrochim. Acta* 1973, **18**, 515–516.
[190] Brunschwig, B. S.; Sutin, N. *J. Am. Chem. Soc.* 1978, **100**, 7568–7577.
[191] Wada, A.; Sakabe, N.; Tanaka, J. *Acta Crystallogr., Sect. B* 1976, **B32**, 1121–1127.
[192] Wada, A.; Katayama, C.; Tanaka, J. *Acta Crystallogr., Sect. B* 1976, **B32**, 3194–3199.
[193] Szalda, D. J.; Macartney, D. H.; Sutin, N. *Inorg. Chem.* 1984, **23**, 3473–3479.
[194] Cannon, R. D. *Electron Transfer Reaction,* Butterworth, 1980.
[195] Li, T. T.-T.; Brubaker, C. H. *J. Organomet. Chem.* 1981, **216**, 223–234.
[196] Li, T. T.-T.; Weaver, M. J.; Brubaker, C. H. *J. Am. Chem. Soc.* 1982, **104**, 2381–2386.

[197] Hupp, J. T.; Weaver, M. J. *J. Phys Chem.* 1985, **89,** 1601–1608.
[198] Lay, P. A. *J. Phys. Chem.* 1986, **90,** 878–885.
[199] Lay, P. A.; McAlpine, N. S.; Hupp, J. T.; Weaver, M. J.; Sargeson, A. M. *Inorg. Chem.* 1990, **29,** 4322–4328.
[200] Abbott, A. P.; Rusling, J. F. *J. Phys. Chem.* 1990, **94,** 8910–8912.
[201] Wherland, S. *Inorg. Chem.* 1983, **22,** 2349–2350.
[202] van Eldik, R.; Asano, T.; Le Noble, W. *J. Chem. Rev.* 1989, **89,** 549–688.
[203] Swaddle, T. W. *Inorg. Chem.* 1990, **29,** 5017–5025.
[204] Hudis, J.; Dodson, R. W. *J. Am. Chem. Soc.* 1956, **78,** 911–913.
[205] Jortner, J. *J. Chem. Phys.* 1976, **64,** 4860–4867.
[206] Buhks, E.; Bixon, M.; Jortner, J. *J. Phys. Chem.* 1981, **85,** 3763–3766.
[207] Weaver, M. J.; McManis, G. E. *Acc. Chem. Res.* 1990, **23,** 294–300.
[208] Weaver, M. J. *Chem. Rev.* 1992, **92,** 463–480.
[209] Maroncelli, M.; MacInnis, J.; Fleming, G. R. *Science,* 1989, **243,** 1674–1681.
[210] Nielson, R. M.; McManis, G. E.; Weaver, M. J. *J. Phys. Chem.* 1989, **93,** 4703–4706.
[211] McManis, G. E.; Nielson, R. M.; Gochev, A.; Weaver, M. J. *J. Am. Chem. Soc.* 1989, **111,** 5533–5541.
[212] Koval, C. A.; Pravata, R. L. A.; Reidsema, C. M. *Inorg. Chem.* 1984, **23,** 545–553.
[213] Geselowitz, D. A. *Inorg. Chim. Acta* 1988, **154,** 225–228.
[214] Endicott, J. F.; Brubaker, G. R.; Ramasami, T.; Kumar, K.; Dwarakanath, K.; Cassel, J.; Johnson, D. *Inorg. Chem.* 1983, **22,** 3745–3762.
[215] Larsson, S.; Ståhl, K.; Zerner, M. C. *Inorg. Chem.* 1986, **25,** 3033–3037.
[216] Geselowitz, D. A. *Inorg. Chim. Acta* 1989, **163,** 79–86.
[217] Buhks, E.; Bixon, M.; Jortner, J.; Navon, G. *Inorg. Chem.* 1979, **18,** 2014–2018.
[218] Newton, M. D. *J. Phys. Chem.* 1986, **90,** 3734–3739.
[219] Newton, M. D. *J. Phys. Chem.* 1991, **95,** 30–38.
[220] Newton, M. D. *Chem. Rev.* 1991, **91,** 767–792.
[221] Richardson, D. E.; Sharpe, P. *Inorg. Chem.* 1991, **30,** 1412–1414.
[222] Ramasami, T.; Endicott, J. F. *Inorg. Chem.* 1984, **23,** 2918–2920.
[223] Ramasami, T.; Endicott, J. F. *J. Am. Chem. Soc.* 1985, **107,** 389–396.
[224] Ramasami, T.; Endicott, J. F. *J. Phys. Chem.* 1986, **90,** 3740–3747.
[225] Endicott, J. F.; Ramasami, T. *J. Am. Chem. Soc.* 1982, **104,** 5252–5254.
[226] Endicott, J. F.; Ramasami, T.; Gaswick, D. C.; Tamilarasan, R.; Heeg, M. J.; Brubaker, G. R.; Pyke, S. C. *J. Am. Chem. Soc.* 1983, **105,** 5301–5310.
[227] Ramasami, T.; Endicott, J. F. *Inorg. Chem.* 1984, **23,** 3324–3333.
[228] Endicott, J. F. *Acc. Chem. Res.* 1988, **21,** 59–66.
[229] Hancock, R. D. *Progress Inorg. Chem.* 1989, **37,** 187–291.
[230] Geselowitz, D. *Inorg. Chem.* 1981, **20,** 4457–4459.
[231] Sandrini, D.; Gandolfi, M. T.; Maestri, M.; Bolletta, F.; Balzani, V. *Inorg. Chem.* 1984, **23,** 3017–3023.
[232] Fürholz, U.; Haim, A. *Inorg. Chem.* 1985, **24,** 3094–3095.
[233] Kuharski, R. A.; Bader, J. S.; Chandler, D.; Sprik, M.; Klein, M. L.; Impey, R. W. *J. Chem. Phys.* 1988, **89,** 3248–3257.

[234] Bader, J. S.; Kuharski, R. A.; Chandler, D. *J. Chem. Phys.* 1990, **93**, 230–236.
[235] Bennett, L. E.; Sheppard, J. *J. Phys. Chem.* 1962, **66**, 1275–1279.
[236] Hyde, M. R.; Davies, R.; Sykes, A. G. *J. Chem. Soc., Dalton Trans.* 1972, 1838–1843.
[237] Rotzinger, F. P. *Inorg. Chem.* 1986, **25**, 4570–4572.
[238] Dulz, G.; Sutin, N. *J. Am. Chem. Soc.* 1964, **86**, 829–832.
[239] Böttcher, W.; Brown, G. M.; Sutin, N. *Inorg. Chem.* 1979, **18**, 1447–1451.
[240] Meier, D. J.; Garner, C. S. *J. Phys. Chem.* 1952, **56**, 853–857.
[241] Taube, H. *Adv. Chem.* 1977, **162**, 127–144.
[242] Wang *J. Am. Chem. Soc.* 1971, **93**, 380–386.
[243] Lavallee, C.; Lavallee, D. K.; Deutsch, E. A. *Inorg. Chem.* 1978, **17**, 2217–2222.
[244] Lavallee, C.; Lavallee, D. K. *Inorg. Chem.* 1977, **16**, 2601–2605.
[245] Christensen, R. J.; Espension, J. H.; Butcher, A. B. *Inorg. Chem.* 1973, **12**, 564–569.
[246] Harada, S.; Funaki, Y.; Yasunaga, T. *J. Am. Chem. Soc.* 1980, **102**, 136–139.
[247] Anast, J. M.; Hamburg, A. W.; Margerum, D. W. *Inorg. Chem.* 1983, **22**, 2139–2145.
[248] DeKorte, J. M.; Owens, G. D.; Margerum, D. W. *Inorg. Chem.* 1979, **18**, 1538–1542.
[249] Lappin, A. G.; Youngblood, M. P.; Margerum, D. W. *Inorg. Chem.* 1980, **19**, 407–413
[250] Owens, G. D.; Chellappa, K. L.; Margerum, D. W. *Inorg. Chem.* 1979, **18**, 960.
[251] Owens, G. D.; Margerum, D. W. *Inorg. Chem.* 1981, **20**, 1446–1453.
[252] Anast, J. M.; Margerum, D. W. *Inorg. Chem.* 1982, **21**, 3494–3501.
[253] Owens, G. D.; Phillips, D. A.; Czarnecki, J. J.; Raycheba, J. M. T.; Margerum, D. W. *Inorg. Chem.* 1984, **23**, 1345–1353.
[254] Murray, C. K.; Margerum, D. W. *Inorg. Chem.* 1983, **22**, 463–469.
[255] McAuley, A.; Norman, P. R.; Olubuyide, O. *J. Chem. Soc., Dalton Trans.* 1984, 1501–1505.
[256] Fairbank, M. G.; Norman, P. R.; McAuley, A. *Inorg. Chem.* 1985, **24**, 2639–2644.
[257] Espenson, J. H.; Shaw, K.; Parker, O. J. *J. Am. Chem. Soc.* 1967, **89**, 5730–5731.
[258] Davies, K. M. *Inorg. Chem.* 1983, **22**, 615–619.
[259] McConnell, H. M.; Weaver, H. E. *J. Chem. Phys.* 1956, **25**, 307–311.
[260] Foley, J.; Tyagi, S.; Hathaway, B. J. *J. Chem. Soc., Dalton Trans.* 1984, 1–5.
[261] Burke, P. J.; Henrick, K.; McMillin, D. R. *Inorg. Chem.* 1982, **21**, 1881–1886.
[262] Dobson, J. F.; Green, B. E.; Healy, P. C.; Kennard, C. H. L.; Pakawatchai, C.; White, A. H. *Aust. J. Chem.* 1984, **37**, 649–659.
[263] Munakata, M.; Kitagawa, S.; Asahara, A.; Masuda, H. *Bull. Chem. Soc. Japan* 1987, **60**, 1927–1929.
[264] Lee, C.-W.; Anson, F. C. *J. Phys. Chem.* 1983, **87**, 3360–3362.
[265] Lee, C.-W.; Anson, F. C. *Inorg. Chem.* 1984, **23**, 837–844.
[266] Hoselton, M. A.; Lin, C.-T.; Schwarz, H. A.; Sutin, N. *J. Am. Chem. Soc.* 1978, **100**, 2383–2388.

[267] Augustin, M. A.; Yandell, J. K. *Inorg. Chem.* 1979, **18**, 577–583.
[268] de Araujo, M. A.; Hodges, H. L. *Inorg. Chem.* 1982, **21**, 3167–3172.
[269] Yoneda, G. S.; Blackmer, G. L.; Holwerda, R. A. *Inorg. Chem.* 1977, **16**, 3376–3378.
[270] Al-Shatti, N.; Lappin, A. G.; Sykes, A. G. *Inorg. Chem.* 1981, **20**, 1466–1469.
[271] Leupin, P.; Al-Shatti, N.; Sykes, A. G. *J. Chem. Soc., Dalton Trans.* 1982, 927–930.
[272] Allan, A. E.; Lappin, A. G.; Laranjeira, M. C. M. *Inorg. Chem.* 1984, **23**, 477–482.
[273] Bernardo, M. M.; Heeg, M. J.; Schroeder, R. R.; Ochrymowycz, L. A. Rorabacher, D. D. *Inorg. Chem.* 1992, **31**, 191–198.
[274] Corfield, P. W. R.; Ceccarelli, C.; Glick, M. D.; Moy, I. W.-Y.; Ochrymowycz, L. A.; Rorabacher, D. B. *J. Am. Chem. Soc.* 1985, **107**, 2399–2404.
[275] Linde, A. M. Q. V.; Juntunen, K. L.; Mols, O.; Ksebati, M. B.; Ochrymowycz, L. A.; Rorabacher, D. B. *Inorg. Chem.* 1991, **30**, 5037–5042.
[276] Martin, M. J.; Endicott, J. F.; Ochrymowycz, L. A.; Rorabacher, D. B. *Inorg. Chem.* 1987, **26**, 3012–3022.
[277] Hoffman, B. M.; Ratner, M. R. *J. Am. Chem. Soc.* 1987, **109**, 6237–6243.
[278] Brunschwig, B. S.; Sutin, N. *J. Am. Chem. Soc.* 1989, **111**, 7454–7465.
[279] Goodwin, J. A.; Stanbury, D. M.; Wilson, L. J.; Eigenbrot, C. W.; Scheidt, W. R. *J. Am. Chem. Soc.* 1987, **109**, 2979–2991.
[280] Goodwin, J. A.; Wilson, L. J.; Stanbury, D. M.; Scott, R. A. *Inorg. Chem.* 1989, **28**, 42–50.
[281] Knapp, S.; Keenan, T. P.; Zhang, X.; Fikar, R.; Potenza, J. A.; Schugar, H. J. *J. Am. Chem. Soc.* 1990, **112**, 3452–3464.
[282] Coggin, D. K.; González, J. A.; Kook, A. M., Bergman, C.; Brennan, T. D.; Scheidt, W. R.; Stanbury, D. M.; Wilson, L. J. *Inorg. Chem.* 1991, **30**, 1125–1134.
[283] Fairbank, M. G.; McAuley, A.; Norman, P. R.; Olubuyide, O. *Can. J. Chem.* 1985, **63**, 2983–2989.
[284] Macartney, D. H. *Inorg. Chem.* 1986, **25**, 2222–2225.
[285] Macartney, D. H.; McAuley, A.; Olubuyide, O. A. *Inorg. Chem.* 1985, **24**, 307–312.
[286] Birk, J. P. *Inorg. Chem.* 1977, **16**, 1381–1383.
[287] Birk, J. P.; Weaver, S. V. *Inorg. Chem.* 1972, **11**, 95–98.
[288] Braunschwig, B. S.; Sutin, N. *Inorg. Chem.* 1979, **18**, 1731–1736.
[289] Huston, P.; Espenson, J. H.; Bakac, A. *Inorg. Chem.* 1989, **28**, 3671–3674.
[290] Thompson, G. A. K.; Sykes, A. G. *Inorg. Chem.* 1976, **15**, 638–642.
[291] McAuley, A.; Olubuyide, O.; Spencer, L.; West, P. R. *Inorg. Chem.* 1984, **23**, 2594–2599.
[292] Binstead, R. A.; Moyer, B. A.; Samuels, G. J.; Meyer, T. J. *J. Am. Chem. Soc.* 1989, **103**, 2897–2899.
[293] Binstead, R. A.; Meyer, T. J. *J. Am. Chem. Soc.* 1987, **109**, 3287–3297.
[294] Moyer, B. A.; Meyer, T. J. *Inorg. Chem.* 1981, **20**, 436–444.
[295] Jencks, W. P. *Chem. Rev.* 1972, **72**, 705.

[296] Bell, R. P. *The Proton in Chemistry,* Chapman and hall, 1973.
[297] Alberry, W. J. *Ann. Rev. Phys. Chem.* 1980, **31**, 227–263.
[298] Lappin, A. G.; Martone, D. P.; Osvath, P. *Inorg. Chem.* 1985, **24**, 4187–4191.
[299] Lappin, A. G.; Laranjeira, M. C. M.; Marusak, R. A. unpublished data.

3

The inner-sphere mechanism

3.1 INTRODUCTION

The reduction of $[Co(NH_3)_5Cl]^{2+}$ by $[Cr(H_2O)_6]^{2+}$ takes place with a second-order rate constant of 6.0×10^5 M^{-1} s^{-1} (25.0°C, $\mu = 0.10$ M, pH = 1) [1] which is faster than the rate of substitution of Cl$^-$ on $[Co(NH_3)_5Cl]^{2+}$ but slower than the normal rates of substitution on the labile $[Cr(H_2O)_6]^{2+}$. More importantly, the initial identifiable oxidized product is a substitutionally inert kinetic product, $[Cr(H_2O)_5Cl]^{2+}$. The thermodynamic product, $[Cr(H_2O)_6]^{3+}$, is formed very much more slowly from this species [2, 3]. Tracer studies in the presence of radioactive ^{36}Cl$^-$ reveal that >99% of the chloride incorporated into the product arises from the $[Co(NH_3)_5Cl]^{2+}$ oxidant. It is therefore deduced that the electron transfer reaction also involves transfer of a Cl$^-$ ion, and that at some point along the reaction coordinate, a precursor intermediate, $[(NH_3)_5Co^{III}ClCr^{II}(H_2O)_5]^{4+}$, is formed in which the Cl$^-$ bridges the two metal centers (eq. (3.1)). After the electron transfer (eq. (3.2)), in the acidic conditions, the successor complex decomposes by cleavage of the more labile Co^{II}—Cl bond to yield the chromium(III) containing kinetic product (eq. (3.3)).

$$[Co(NH_3)_5Cl]^{2+} + [Cr(H_2O)_6]^{2+} \underset{k_{-1}}{\overset{k_1}{\rightleftharpoons}} [(NH_3)_5Co^{III}ClCr^{II}(H_2O)_5]^{4+} \qquad (3.1)$$

$$[(NH_3)_5Co^{III}ClCr^{II}(H_2O)_5]^{4+} \overset{k_2}{\rightarrow} [(NH_3)_5Co^{II}ClCr^{III}(H_2O)_5]^{4+} \qquad (3.2)$$

$$[(NH_3)_5Co^{II}ClCr^{III}(H_2O)_5]^{4+} \rightleftharpoons [Cr(H_2O)_5Cl]^{2+} + [Co(H_2O)_6]^{2+}$$
$$5NH_4^+ \qquad (3.3)$$

This is an ideal example which has all the essential requirements of an inner-sphere mechanism where electron transfer takes place through some bridging atom or group of atoms. It is particularly well defined because the product $[Cr(H_2O)_5Cl]^{2+}$ is isolable. In the presence of an excess of the reductant $[Cr(H_2O)_6]^{2+}$, a competing reaction, eq. (3.4), has a rate constant of 9 M^{-1} s^{-1} (at 0°C) [4] and fortunately also proceeds by an inner-sphere mechanism with transfer of a Cl$^-$ ion, otherwise the chloro complex would be more difficult to isolate. Incorporation of the small amount of ^{36}Cl$^-$ into the product proceeds by a related pathway in which [Cr(H$_2$O)$_5$^{36}Cl]$^+$ is the reductant.

$$[Cr(H_2O)_5Cl]^{2+} + [Cr(H_2O)_6]^{2+} \; [Cr(H_2O)_5Cl]^{2+} + [Cr(H_2O)_6]^{2+} \qquad (3.4)$$

The [Cr(H$_2$O)$_6$]$^{2+}$ ion is very special in that it has a high substitution lability ($k \approx 10^9$ s^{-1}) and a low rate for outer-sphere self-exchange resulting in a strong preference for an inner-sphere pathway. The added advantage of detection of kinetic products has enhanced the use of this reagent in the study of inner-sphere reactions. Even when kinetic products are too short-lived for detection, provided other criteria for inner-sphere reaction are met, mechanistic assignment can be relatively straightforward.

3.2 BRIDGING LIGANDS

Inner-sphere reactions are generally suggested by the following criteria:

(i) When the rate of electron transfer is equal to or slower than the rates of substitution of the reactants. This is a corollary of the requirement for outer-sphere reactions, that a reaction must be outer-sphere if it takes place faster than the rate of substitution at the metal centers. However, outer-sphere reactions are found with labile reactants.

(ii) When the rates do not fit with the common tests for an outer-sphere mechanism such as the Marcus relationship. In general the reactions must be faster than predicted by the Marcus relationship. However, as has been observed, outer-sphere reactions need not conform to the Marcus relationship and a number of inner-sphere reactions show relationships similar to those for outer-sphere reactions.

(iii) The presence of a suitable bridge on one of the reactants. Minimal requirements are that the bridging group must have an available lone pair of sufficient basicity to coordinate to the labile coordination position of the reaction partner.

Transfer of an electron and a chloride ion is formally equivalent to chlorine atom transfer. However, chlorine atom formation can be excluded on energetic grounds. The reduction potential of Cl in aqueous solution is estimated to be approximately 2.44 V, whereas the cobalt(III) complex has a potential around 0 V. As will be seen, transfer of the bridging group is not a requirement for the inner-sphere mechanism, but the coupling between electron transfer and bridge formation is an important component of the mechanism. This can be understood by consideration of the orbitals involved in the electron transfer process (eq. (3.5)). Along the unique z-axis, the electron is transferred from a d_{z^2} orbital (σ^*) on chromium(II) to the d_{z^2}

orbital (σ^*) in cobalt(III) which is involved in the bonding to the bridging atom. The electron is transferred as a result of the orbital overlap (resonance transfer), which is modified by the presence of the bridge orbital. The energy of the electron donor orbital is directly related to the strength of the Co—Cl binding interaction, and so electron transfer will be strongly coupled to the Co—Cl stretch. Further, the group *trans* to the bridge will also have a major effect on the energy of the orbital. Weakening the Co—*trans* ligand bond will favor electron transfer. Strong field ligands raise the energy of the orbital and inhibit electron transfer, weak field ligands enhance the electron transfer [5]. Thus in addition to the Frank–Condon barriers to transfer of the electron, discussed in Chapter 2, special consideration must be given to formation of the bridge and the lowering of the transition state energy by the interaction of the metal-centered and bridge orbitals. Isotopic substitution of ^{14}N by ^{15}N in $[Co(NH_3)_5Cl]^{3+}$ has a minimal effect on the rate [6].

$$\text{Co} \text{—} \text{Cl} \text{—} \text{Cr} \tag{3.5}$$

The inner-sphere mechanism is complex and each of the steps — formation of the precursor complex, electron transfer within the precursor complex, and decomposition of the successor complex — can be rate-limiting. However, like outer-sphere reactions, inner-sphere reactions most commonly conform to a simple second-order rate law (eq. (3.6)).

$$\text{Rate} = k\,[[Co(NH_3)_5Cl]^{2+}][[Cr(H_2O)_6]^{2+}] \tag{3.6}$$

In this case, the rate constant is slower than the rate expected for substitution at $[Cr(H_2O)_6]^{2+}$ and it may be concluded that the rate-limiting step is the electron transfer itself, although in some faster reductions by $[Cr(H_2O)_6]^{2+}$ the rate is thought to approach the diffusion-controlled limit for reaction between two cationic species, approximately 10^7 M^{-1} s^{-1}. Application of the steady-state approximation to the mechanism gives $k = k_1 k_2/(k_{-1} + k_2)$ which reduces to $K_1 k_2$ when $k_{-1} \gg k_2$. It is unlikely that stoichiometric quantities of the binuclear intermediate will be formed in this reaction since both of the reacting species are positively charged and K_1 is not expected to be large (no greater than 1). Similar rate laws and mechanisms are proposed for the reactions with other bridging groups, and examples are presented in Table 3.1.

Table 3.1. Rate constants and activation parameters for reduction of cobalt(III) complexes by $[Cr(H_2O)_6]^{2+}$ at 25.0°C

Oxidant	μ (M)	k (M^{-1} s^{-1})	$\Delta H^{\ddagger}$ (kJ mol^{-1})	$\Delta S^{\ddagger}$ (J K^{-1} mol^{-1})	Ref.
$[Co(NH_3)_5F]^{2+}$	0.1	2.5×10^5			1
$[Co(NH_3)_5Cl]^{2+}$	0.1	6.0×10^5			1
$[Co(NH_3)_5Br]^{2+}$	0.1	1.4×10^6			1
$[Co(NH_3)_5I]^{2+}$	0.1	3.0×10^6			1
$[Co(NH_3)_5OH]^{2+}$	1.0	1.5×10^6			7
$[Co(NH_3)_5OH]^{2+}$	1.0	< 0.1			7
$[Co(NH_3)_5CN]^{2+}$	1.0	$3.6 \times 10(R)^a$			8
$[Co(NH_3)_5SCN]^{2+}$	1.0	$8.0 \times 10^4(A)^a$			9
$[Co(NH_3)_5SCN]^{2+}$	1.0	$1.9 \times 10^5(R)^a$			9
$[Co(NH_3)_5NCS]^{2+}$	1.0	$19(R)^a$	29	-121	10
$[Co(NH_3)_5N_3]^{2+}$	1.0	$\approx 3 \times 10^5(R)^a$			10
$[Co(NH_3)_5O_2CH]^{2+}$	1.0	7.2	35	-113	11
$[Co(NH_3)_5O_2CCH_3]^{2+}$	1.0	0.35	34	-138	11
$[Co(NH_3)_5O_2CC(CH_3)_3]^{2+}$	1.0	0.007	46	-130	11
$[Co(NH_3)_5O_2CCH_2Cl]^{2+}$	1.0	0.12	37	-138	11
$[Co(NH_3)_5O_2CCHCl_2]^{2+}$	1.0	0.075	34	-151	11
$[Co(NH_3)_5O_2CCF_3]^{2+}]$	0.2	0.017	39	-146	11
$[Co(NH_3)_5O_2CCH_2NH_3]^{3+}$	1.0	0.064	32	-159	12
$[Co(NH_3)_5O_2CCH_2N(CH_3)_3]^{3+}$	1.0	0.016	32	-172	12
$[Co(NH_3)_5O_2CCH_2CH_2NH_3]^{3+}$	1.0	0.098			13
$[Co(NH_3)_5NHCHO]^{2+}$	1.0	1.74	50	-71	14
$[Co(NH_3)_5O_2CNH_2]^{2+}$	1.0	2.42	48	-78	15
$[Co(NH_3)_5OC(S)NHCH_3]^{2+}$	1.0	68			16
$[Co(NH_3)_5SC(O)NHCH_3]^{2+}$	1.0	6.5×10^4			16

^{a}(R) signifies remote and (A) signifies adjacent attack (see text).

For the reactions with halide bridges, there is a trend in reaction rate with I > Br > Cl > F which parallels the ordering of these reagents with outer-sphere reductants and reflects the thermodynamic driving force. This ordering is known as the normal order for the halide complex reductions. The reaction of $[Co(NH_3)_5OH_2]^{3+}$ with $[Cr(H_2O)_6]^{2+}$ conforms to a two-term rate law (eq. (3.7)) [7].

$$-\frac{d[[Co(NH_3)_5OH_2]^{3+}]}{dt} = \left\{ k + \frac{k'}{[H^+]} \right\} [[Co(NH_3)_5OH_2]^{3+}] [[Cr(H_2O)_6]^{2+}] \quad (3.7)$$

The dominant term has an inverse [H$^+$] dependence and is interpreted in terms of reaction of $[Co(NH_3)_5OH]^{2+}$ while the acid-independent term, k, is negligible. The K_a of $[Co(NH_3)_5OH_2]^{3+}$ is 6.1×10^{-7} (25°C, 0.3 M) [17] and in the acidic conditions of the reaction $k' = K_a k_{OH}$. Hydroxide ion is an effective bridging ligand, but evidence for water acting as a bridging ligand is not well established. While a measurable rate is obtained for the $[Co(NH_3)_5OH_2]^{3+}$ pathway in $NaClO_4$–$HClO_4$ media [18], it becomes negligible in $LiClO_4$–$HClO_4$ media and is most likely a medium effect. Evidence [19] from product analysis that H_2O is transferred in this reaction must therefore be reinterpreted. However, the substitution of ^{18}O for ^{16}O in the reaction causes the rate to be lowered by a factor of 1.046 which indicates significant Co—O bond stretching in the reaction [20]. It can be deduced that the requirements for the bridging atom are that it has a lone pair of electrons of sufficient basicity available for donation to the incoming metal center. The basicity of the lone pairs on cobalt-bound water is too low to coordinate to the incoming chromium(II) center.

Examples of inner-sphere reactions are not confined to a monatomic ligand bridge. The thermodynamically unfavored isomer $[Cr(H_2O)_5NC]^{2+}$ is formed in the reduction of $[Co(NH_3)_5CN]^{2+}$ by $[Cr(H_2O)_6]^{2+}$, evidence that the electron can be transferred through two atoms [21]. In the reaction of $[Co(NH_3)_5SCN]^{2+}$ with $[Cr(H_2O)_6]^{2+}$, two products, $[Cr(H_2O)_5SCN]^{2+}$ and $[Cr(H_2O)_5NCS]^{2+}$ are detected [9]. The former product results from attack by the reductant (bridge formation) on the S-atom *adjacent* to the cobalt(III) center while the latter product results from attack at the *remote* N-atom. Both bridge-forming atoms have available a lone pair for coordination to the reductant. Interestingly, reduction of $[Co(NH_3)_5NCS]^{2+}$ by $[Cr(H_2O)_6]^{2+}$ results in $[Cr(H_2O)_5SCN]^{2+}$ as the sole product [10]. Remote attack is required because the coordinated nitrogen has no available lone pair for bridge formation.

Reactions with carboxylate complexes of the type $[Co(NH_3)_5O_2CCH_3]^{2+}$ are also accompanied by transfer of the bridge group to give $[Cr(H_2O)_5O_2CCH_3]^{2+}$ as product [11]. The rate law (eq. (3.8)) reveals inhibition by H^+ as a result of protonation of the bound carboxylate with $K_a = 0.25$ M at 25.0°C. Rate constants for the dominant pathway, k_1, for a series of substituted carboxylate complexes are substantially slower than those for the halogens. Electron-withdrawing substituents such as chlorine and fluorine retard the rates and there is evidence for steric retardation also, but the effects are relatively small. The product may be the result of adjacent attack on the bound carboxylate oxygen or remote attack on the unbound carbonyl oxygen (eq. (3.9)). The fact that the reaction of $[Co(NH_3)_5O_2CCH_3]^{2+}$ is inhibited by protonation suggests competition of $[H^+]$ with the binding of $[Cr(H_2O)_6]^{2+}$.

$$\frac{-d[[Co(NH_3)_5O_2CCH_3]^{2+}]}{dt} =$$

$$\frac{k_1 + k_2 K_a[H^+]}{1 + K_a[H^+]} [[Co(NH_3)_5O_2CCH_3]^{2+}] [[Cr(H_2O)_6]^{2+}] \quad (3.8)$$

$$[(NH_3)_5Co-O\overset{\overset{\text{remote}}{\nwarrow}}{\underset{\underset{\text{adjacent}}{\nwarrow}}{\overset{O}{C}}}-CH_3]^{2+} \quad (3.9)$$

The question of the position of attack has been answered to some extent [22, 23] in reactions of the acetate-bridged binuclear complex (eq. (3.10)), which shows no inner-sphere reactivity, supporting the proposal that attack takes place on the carbonyl oxygen atom. However, it is conventional to refer to attack at either the bound carboxylate oxygen of the unbound carbonyl as adjacent attack to avoid confusion with remote attack through attached organic units. The effect of increasing the charge on the oxidant has been elucidated by comparisons of complexes of the type $[Co(NH_3)_5O_2CCH_2CH_3]^{2+}$ with those of the type $[Co(NH_3)_5O_2CCH_2NH_3]^{3+}$ [13]. Rate differences are quite small, generally less than an order of magnitude decrease in rate for the more positively charged complex.

$$[(NH_3)_3Co(OH)(OH)Co(NH_3)_3]^{3+} \text{ bridged by benzoate}$$

(3.10)

Special note should be taken of the reduction of the formamido complex, $[Co(NH_3)_5NH_2CHO]^{3+}$ [14]. The rate law indicates that formation of the conjugate base is required for electron transfer to give the O-bonded product, $[Cr(H_2O)_5(OCHNH_2)]^{3+}$. Deprotonation of the amide enhances conjugation with the amide oxygen, the point of inner-sphere coordination to the reductant, and it is concluded that inner-sphere electron transfer cannot take place through a coordinated amino group. Reductions of the isomer $[Co(NH_3)_5OCHNH_2]^{3+}$ and of the urea complex $[Co(NH_3)_5OC(NH_2)_2]^{3+}$ are outer-sphere, but the carbamate complex, $[Co(NH_3)_5O_2CNH_2]^{2+}$, involves an inner-sphere pathway to give an O-bonded rather than an N-bonded chromium(III) product [15]. Comparisons have been made of the O- and S-bonded thiocarbamate complexes, $[Co(NH_3)_5OC(S)NHCH_3]^{2+}$ and $[Co(NH_3)_5SC(O)NHCH_3]^{2+}$ [16]. Both take place by inner-sphere mechanisms and give the same S-bonded chromium(III) product. In $[Co(NH_3)_5OC(S)NHCH_3]^{2+}$, the point of attack is considered to be the S-atom giving the product directly, whereas with $[Co(NH_3)_5SC(O)NHCH_3]^{2+}$, attack at the O-atom is most likely, followed by rapid isomerization of the product, although adjacent attack at the S-atom cannot be ruled out. The rate enhancement in the S-bonded complex is thought to be the result of a *trans*-effect where the *trans* Co—N bond is lengthened, thereby lowering the energy of the acceptor orbital.

Patterns of reactivity in these inner-sphere reactions are difficult to interpret because the rate constant is a composite quantity and reflects both the strength of precursor association and the effectiveness of the bridge in promoting electron transfer. Stabilization of the electron transfer precursor complex by chelate formation is effective in enhancing the overall rate. In Table 3.2, rate constants for a number of reductions involving chelate formation are given. Where the chelating group is a carboxylate, the reaction rate is acid dependent with a rate law given by eq. (3.11) where $k = 1 \times 10^2$ M^{-1} s^{-1} and $k' = 4.0 \times 10^2$ s^{-1} for [Co(NH$_3$)$_5$O$_2$CCO$_2$H]$^{2+}$.

$$-\frac{d[Co(III)]}{dt} = \left\{k + \frac{k'}{[H^+]}\right\} [[Co(NH_3)_5O_2CCO_2H]^{2+}] [[Cr(H_2O)_6]^{2+}] \quad (3.11)$$

The K_a of the carboxylate is 8.7×10^{-3} M, substantially lower than that for the free acid so that the rate for reduction of the protonated carboxylate complex, k, is little enhanced over that for acetate ion. However, the rate for the deprotonated species, k'/K_a, is two orders of magnitude faster. A chelated precursor complex, (3.12), is proposed. Notice that the electron transfer takes place through the adjacent bridge.

$$[(NH_3)_5Co-O\cdots C(=O)\cdots O\cdots Cr(H_2O)_4]^{4+} \quad (3.12)$$

In the reactions of the alanine complex, although the reaction product is [(H$_2$O)$_5$CrO$_2$CCH$_2$NH$_3$]$^{3+}$, indicating that a bridged ligand complex is formed, the reaction is likely to be outer-sphere, the so-called bridged outer-sphere mechanism (eq. (3.13)), since the coordinated amine has no lone pair for bridge formation and electron transfer through the saturated ligand backbone is unlikely [24]. Similar conclusions are reached in the reactions of the cyanoacetate complex [(NH$_3$)$_5$CoNCCH$_2$CO$_2$]$^{2+}$ [16].

$$[(NH_3)_5Co-NH_2-C(=O)-O-Cr(H_2O)_5]^{4+} \quad (3.13)$$

Table 3.2. Rate constants and activation parameters for reduction of Cobalt(III) complexes capable of chelation by [Cr(H$_2$O)$_6$]$^{2+}$ at 25.0°C

Oxidant	μ (M)	k (M^{-1} s^{-1})	$\Delta H^{\ddagger}$ (kJ mol^{-1})	$\Delta S^{\ddagger}$ (J K^{-1} mol^{-1})	Ref.
[Co(NH$_3$)$_5$O$_2$CCH$_2$OH]$^{2+}$	1.0	3.06	38	−109	25
[Co(NH$_3$)$_5$O$_2$CCO$_2$H]$^{2+}$	1.0	1.0×10^2			26
[Co(NH$_3$)$_5$O$_2$CCO$_2$]$^+$	1.0	4.6×10^4	10	−84	26
[Co(NH$_3$)$_5$O$_2$CCH$_2$N(CH$_2$CO$_2$)$_2$]	1.0	1.5×10^{10a}			27
[Co(NH$_3$)$_5$O$_2$CCH$_2$NHCH$_2$CO$_2$]$_2$]$^+$	1.0	9.2×10^6			28
[Co(NH$_3$)$_5$O$_2$CCH$_2$NHCH$_2$CO$_2$H]$^{2+}$	1.0	4.3×10^{-2}			28
[Co(NH$_3$)$_5$NH$_2$CH$_2$CO$_2$H]$^{3+}$	1.0	1.2×10^{-3}			24
[Co(NH$_3$)$_5$NH$_2$CH$_2$CO$_2$]$^{2+}$	1.0	2.5			24
[(NH$_3$)$_5$CoNCCH$_2$CO$_2$]$^{2+}$	1.0	2.1	52	−64	29

[a]Nominal rate: the reaction is acid dependent and protonated species may be the major reactants.

Effects of non-bridging ligands have been examined for a number of complexes and results are presented in Table 3.3. These data are difficult to interpret because the driving force for the reaction varies with changing the coordination of the oxidant. The reaction rates show fairly small variations although changing the *trans* ligand from NH_3 to Cl^- causes a rate enhancement, consistent with a reduction in the ligand field. Activation parameters are unusual with small and even negative $\Delta H^{\ddagger}$ values that are ascribed to a favorable enthalpy of formation of the precursor complexes. The data are limited [29].

Table 3.3. Rate constants and activation parameters for non-bridging group effects in chloride bridged electron transfer

Oxidant	μ (M)	k ($M^{-1}\,s^{-1}$)	$\Delta H^{\ddagger}$ (kJ mol^{-1})	$\Delta S^{\ddagger}$ (J K^{-1} mol^{-1})	Ref.
cis-[Co(en)$_2$(NH$_3$)Cl]$^{2+}$	0.1	2.5×10^5	-8	-172	30
cis-[Co(en)$_2$Cl$_2$]$^+$	0.1	7.7×10^5	-25	-213	30
trans-[Co(en)$_2$Cl$_2$]$^+$	0.1	5×10^6	-33	-222	30
cis-[Co(en)$_2$FCl]$^{2+}$	0.1	9×10^5	-42	-272	30

3.3 DOUBLE BRIDGE FORMATION

In the case where the *cis*-ligand is also capable of bridge formation, a double ligand bridge is possible and a number of important examples are presented in Table 3.4. In the diazido-case, the doubly bridged product, *cis*-[Cr(H$_2$O)$_4$(N$_3$)$_2$]$^+$ and the singly bridged product are detected in the ratio $0.6 : 1$ at $0\,^\circ\mathrm{C}$. Reduction of the diformato-complex, *cis*-[Co(en)$_2$(O$_2$CH)$_2$]$^+$, is inhibited by H^+ with a rate law shown in eq. (3.14), where $K_a = 0.44\,M^{-1}$. Product analysis reveals both *cis*-[Cr(H$_2$O)$_4$(O$_2$CH)$_2$]$^+$ and [Cr(H$_2$O)$_5$(O$_2$CH)]$^{2+}$ again indicating the presence of a doubly bridged pathway and a singly bridged pathway in the ratio $3 : 1$. The corresponding *trans*-[Co(en)$_2$(O$_2$CH)$_2$]$^+$ gives a singly bridged pathway and reacts an order of magnitude more slowly with $k = 10.2\,M^{-1}\,s^{-1}$.

$$\frac{-d[cis\text{-}[Co(en)_2(O_2CH)_2]^+]}{dt} = \frac{k_o}{1 + K_a[H^+]} [[Cr(H_2O)_6]^{2+}] [cis\text{-}[Co(en)_2(O_2CH)_2]^+] \quad (3.14)$$

The chelated product, [Cr(H$_2$O)$_4$(ox)]$^+$, is found in the reduction of [Co(NH$_3$)$_4$(ox)]$^+$ and [Co(ox)$_3$]$^{3-}$ and it is proposed that a symmetric double bridge is formed (eq. (3.15)). Although a singly bridged pathway followed by rapid ring closure in the resulting monodentate product is also possible, sufficient examples of these chelated double-bridged structures have been noted to suggest that ring closure is not a factor.

$$[(NH_3)_4Co\underset{O\quad\quad O}{\overset{O\quad\quad O}{\diagup\diagdown\diagdown\diagup}}Cr(H_2O)_4]^{3+} \quad\quad (3.15)$$

In all the instances of double bridging, attack by the reductant occurs at groups remote from the cobalt(III) center. The expected double bridge product, $[Cr(OH_2)_4Cl_2]^+$, is not detected in the reduction of cis-$[Co(NH_3)_4Cl_2]^+$, and only $[Cr(OH_2)_5Cl]^{2+}$ is isolated. However, this does not necessarily rule out double bridge formation since the rate of hydrolysis of $[Cr(OH_2)_4Cl_2]^+$ exceeds the rate of electron transfer. There are no documented examples of double bridge formation with two single atom bridges but the reasons for this are not well understood [19, 31].

Table 3.4. Rate and activation parameters for reactions which show double bridges

Oxidant	μ (M)	k ($M^{-1} s^{-1}$)	$\Delta H^{\ddagger}$ (kJ mol^{-1})	$\Delta S^{\ddagger}$ (J K^{-1} mol^{-1})	Ref.
cis-$[Co(NH_3)_4(N_3)_2]^+$	1.0	$>10^3$			32
cis-$[Co(en)_2(N_3)_2]^+$	1.0	$>10^3$			32
cis-$[Co(en)_2(O_2CH)_2]^+$	1.0	434	15.5	-146	33
$[Co(NH_3)_4(ox)]^+$	1.0	2×10^5			34
$[Co(ox)_3]^{3-}$	1.0^a	4.0×10^6			35

^{a}20 °C.

It is also possible for a single bridge to form part of a chelate ring on the oxidant (Table 3.5). In the first three entries of the table, the carboxylate forms the bridging group and the products show monodentate coordination of the ligand through the carboxylate group. Rates are similar to those observed for non-chelated carboxylate groups. In $[Co(en)_2(SCH_2CO_2)]^{2+}$, the product is monodentate, bonded through the sulfur which forms the bridging group and significant rate enhancement results.

Table 3.5. Electron transfer reactions where the bridge is part of a chelating ligand at 25.0 °C

Oxidant	μ (M)	k ($M^{-1} s^{-1}$)	$\Delta H^{\ddagger}$ (kJ mol^{-1})	$\Delta S^{\ddagger}$ (J K^{-1} mol^{-1})	Ref.
$[Co(en)_2(NH_2CH_2CO_2)]^{2+}$	1.0	2.2	37	-133	36
$[Co(NH_3)_4(NH_2CH_2CO_2)]^{2+}$	1.0	6.4	30	-130	37
$[Co(en)_2(OCH_2CO_2)]^{2+}$	1.0	9.9×10^2			38
$[Co(en)_2(SCH_2CO_2)]^{2+}$	1.0	$>2 \times 10^6$			38

Thus far the reactions which have been examined have been restricted to reductions of cobalt(III) complexes by $[Cr(H_2O)_6]^{2+}$. There is no direct detection of precursor complex formation but the mechanism is well established from product analysis. Precursor complex formation is inferred from unusual activation parameters and in some rate law variations [39]. Reductions of other oxidant series by $[Cr(H_2O)_6]^{2+}$ have been examined, and these provide some valuable insights and some differences in trends.

3.4 DEPENDENCE ON OXIDANT

In Table 3.6 are presented the rate and activation parameters for reduction of complexes of chromium(III). Reactions of $[Cr(NH_3)_5Cl]^{2+}$ and its derivatives with $[Cr(H_2O)_6]^{2+}$ are equivalent to the chromium(II)-catalyzed aquation of the complex. However, the product is $[Cr(H_2O)_5Cl]^{2+}$ and can be distinguished from the $[Cr(NH_3)_5(OH_2)]^{2+}$ obtained by the uncatalyzed reaction. The rates parallel in great measure the reactions of the corresponding $[Co(NH_3)_5Cl]^{2+}$ complexes. However, they are much slower and the trends are much larger, supporting the idea that in the corresponding reactions with $[Co(NH_3)_5Cl]^{2+}$, the rates approach substitution control. For example, with halide bridges, the rate constants for $[Cr(NH_3)_5X]^{2+}$ reduction span a factor of 2×10^3 while those for $[Co(NH_3)_5X]^{2+}$ reduction span a factor of only 10. Non-bridging ligand effects in particular are also much more pronounced. A special place is reserved for reactions of $[Cr(H_2O)_5X]^{2+}$ with $[Cr(H_2O)_6]^{2+}$. These involve no net chemical reaction and are the equivalent of an inner-sphere self-exchange process discussed in section 3.8. Rates have been determined by isotope-exchange methods.

Table 3.6. Rate and activation parameters for the reduction of $[Cr(NH_3)_5X]^{2+}$ and $[Cr(H_2O)_5X]^{2+}$ by $[Cr(H_2O)_6]^{2+}$ at 25.0 °C

Oxidant	μ (M)	k ($M^{-1} s^{-1}$)	$\Delta H^{\ddagger}$ (kJ mol^{-1})	$\Delta S^{\ddagger}$ (J K^{-1} mol^{-1})	Ref.
$[Cr(NH_3)_5F]^{2+}$	1.0	2.7×10^{-3}	56	-125	40
$[Cr(NH_3)_5Cl]^{2+}$	1.0	5.1×10^{-2}	46	-96	40
$[Cr(NH_3)_5Br]^{2+}$	1.0	3.2×10^{-1}	36	-138	40
$[Cr(NH_3)_5I]^{2+}$	1.0	5.5			40
cis-$[Cr(NH_3)_4(H_2O)Cl]^{2+}$	1.5	1.3×10^{-1}	44	-117	41
trans-$[Cr(NH_3)_4(H_2O)Cl]^{2+}$	1.5	1.3	38	-117	41
$[Cr(NH_3)_3(H_2O)_2Cl]^{2+}$	1.5	2.2	39	-109	41
$[Cr(NH_3)_2(H_2O)_3Cl]^{2+}$	1.5	6.9	36	-113	41
$[Cr(NH_3)(H_2O)_4Cl]^{2+}$	1.5	1.9×10^1	33	-113	41
$[Cr(H_2O)_5Cl]^{2+}$	1.5	3.3×10^1			41
$[Cr(H_2O)_5Cl]^{2+}$	1.0^a	9.0			42
$[Cr(H_2O)_5F]^{2+}$	1.0	2×10^{-2}			42
$[Cr(H_2O)_5NCS]^{2+}$	1.0	1.2×10^{-4}			42
$[Cr(OH_2)_5Br]^{2+}$	1.0	>60			42
$[Cr(OH_2)_5OH]^{2+}$		2.3			43
$[Cr(OH_2)_5OH_2]^{2+}$		≈ 0			43
$[Cr(H_2O)_4(N_3)_2]^{2+}$	0.5^a	$\approx 60^b$			44

^{a}0 °C.
^{b}doubly bridged.

With ruthenium(III) oxidants, the reduced from, ruthenium(II), is less labile than cobalt(II) or chromium(II) and this introduces some mechanistic differences. In the reaction of $[Cr(H_2O)_6]^{2+}$ with $[Ru(NH_3)_5Cl]^{2+}$, the rate law shows deviations from second-order behavior (eq. (3.16)) with $K = 70$ M^{-1} and $k = 3.2 \times 10^4$ M^{-1} s^{-1}, indicating the formation of a reaction intermediate in equilibrium with the reactants (eqs (3.17)–(3.18)) [45].

$$\frac{d[[Ru(NH_3)_5Cl]^{2+}]}{dt} = \left\{\frac{k\,[Cr(II)]}{1 + K\,[Cr(II)]}\right\}[[Ru(NH_3)_5Cl]^{2+}] \qquad (3.16)$$

$$[Ru(NH_3)_5Cl]^{2+} + [Cr(H_2O)_6]^{2+} \rightleftharpoons [(NH_3)_5RuClCr(H_2O)_6]^{4+} \qquad (3.17)$$

$$[(NH_3)_5RuClCr(H_2O)_6]^{4+} \rightarrow [Ru(NH_3)_5(OH_2)]^{2+} + [Cr(H_2O)_5Cl]^{2+} \qquad (3.18)$$

At first sight this reaction appears to provide an ideal system for the study of electron transfer within a binuclear intermediate. However, the intermediate binuclear complex is ascribed the formulation of a successor complex, $[(NH_3)_5Ru^{II}ClCr^{III}(H_2O)_6]^{4+}$, rather than of a precursor since the magnitude of the formation constant is much larger than expected for the precursor. Spectroscopic characterization of the intermediate which differs significantly from the reactants is also in agreement with this assignment of oxidation states. Electron transfer is rapid, commensurate with the rate of formation of the binuclear species and the limiting first-order rate is the decomposition of the successor complex. The ruthenium(II) center is inert; however, it is the Ru—Cl bond which cleaves and the reaction proceeds with transfer of the chloride ion, showing that this is indeed the successor complex and that an alternative mechanism in which the intermediate serves as a dead-end complex and electron transfer takes place by an outer-sphere reaction can be ruled out. Examples of rate parameters for inner-sphere reductions of a number of ruthenium(III) complexes are presented in Table 3.7. Note that in these reactions with the change in the rate-limiting step, the order of reactivity for the halide bridge has changed. This feature, the so-called abnormal order, has attracted much attention in terms of its mechanistic significance [46, 47]. Interpretation of the data for reduction of $[Ru(NH_3)_5OH_2]^{3+}$ deserves some comment. While it appears that this represents an inner-sphere reaction with H_2O as a bridging group, a mechanism in which decomposition of a binuclear hydroxy-bridged intermediate is acid catalyzed is preferred.

Table 3.7. Rate and activation parameters for the reduction of ruthenium(III) complexes by $[Cr(H_2O)_6]^{2+}$ at 25.0°C

Oxidant	μ (M)	k ($M^{-1}\,s^{-1}$)	$\Delta H^{\ddagger}$ (kJ mol^{-1})	$\Delta S^{\ddagger}$ (J K^{-1} mol^{-1})	Ref.
$[Ru(NH_3)_5Cl]^{2+}$	0.11	3.5×10^4	5.4	−138	48
cis-$[Ru(NH_3)_4Cl_2]^{2+}$	0.10^a	7.2×10^4			45
trans-$[Ru(NH_3)_4Cl_2]^{2+}$	0.10	1.3×10^2			45
$[Ru(NH_3)_5Br]^{2+}$	0.11	2.2×10^3	11.7	−142	48
$[Ru(NH_3)_5I]^{2+}$	0.11	2.5×10^2			48
$[Ru(NH_3)_5OH_2]^{3+}$	0.10	$\leq 5.0 \times 10^2$			48
$[Ru(NH_3)_5OH]^{2+}$	0.10	3.5×10^6			48
$[Ru(NH_3)_5O_2CCF_3]^{2+}$	0.12	1.6×10^3	5.4	−163	48
$[Ru(NH_3)_5O_2CCH_3]^{2+}$	0.12	3.2×10^4	4.2	−142	48
$[Ru(NH_3)_5O_2CH]^{2+}$	0.12^b	1.7×10^5			48

$^a K = 465\,M^{-1}$.
^{b} 10°C.

Careful analysis of the products of the reduction of $[IrCl_6]^{2-}$ by $[Cr(H_2O)_6]^{2+}$ reveals that this reaction takes place by parallel outer-sphere and inner-sphere pathways (eqs (3.19) and (3.20)), with $k_{OS}/k_{IS} = 2.45$ [49, 50]. A significant outer-sphere rate is not surprising since the electron transfer reaction has a large driving force and the rate of the outer-sphere reaction calculated by the Marcus expression is approximately 10^9 M^{-1} s^{-1}, comparable with the rate of substitution of anions at the chromium(II) center.

$$[IrCl_6]^{2-} + [Cr(H_2O)_6]^{2+} \xrightarrow{k_{OS}} [IrCl_6]^{3-} + [Cr(H_2O)_6]^{3+} \quad (3.19)$$

$$[IrCl_6]^{2-} + [Cr(H_2O)_6]^{2+} \xrightarrow{k_{IS}} [Cl_5Ir^{III}ClCr^{II}(H_2O)_5] \quad (3.20)$$

The kinetic product of the inner-sphere pathway, $[Cl_5IrClCr(H_2O)_5]$, can be isolated and has been shown to decay by both Ir—Cl (40%) and Cr—Cl (60%) bond cleavage roughly in proportion to the expected rates of hydrolysis of the metal centers. The result of this is that some of the reaction proceeds with transfer of the ligand and some does not.

In summary, the high substitution lability of $[Cr(H_2O)_6]^{2+}$ and low rate of self-exchange for outer-sphere electron transfer combine to make the inner-sphere mechanism dominant in the chemistry of this reagent. There are many subtleties in the mechanism, effects of chelation and the formation of double bridges. The evidence for the mechanism, principally from analysis of the kinetic products, is definitive. Rate laws are less informative and rate comparisons are of limited use since in many instances the rates approach the substitution limit. However, there are a few criteria which allow suggestion of inner-sphere reactions from rate comparisons: the low reactivity of OH_2 bridges compared with OH^- bridges, the high rate of reactivity with N_3^- bridges compared with NCS^- bridges. Other metal ion complex reductants show inner-sphere mechanisms but it is not always possible to obtain such definitive evidence for an inner-sphere mechanism and these rate comparisons become more important in the assignment of mechanism.

3.5 AQUA-ION REDUCTANTS

The $[V(H_2O)_6]^{2+}$ ion is a frequently used reagent whose behavior contrasts significantly with that of $[Cr(H_2O)_6]^{2+}$. It has moderate reducing power, $E = -0.26$ V and low substitution lability, approximately 10^2 s^{-1} for water exchange, characteristic of a π^3 electronic configuration. This latter aspect appears to dominate inner-sphere reactivity since reactions which are faster than the substitution rate must follow an outer-sphere mechanism. For reactions which occur at rates slower than 40 M^{-1} s^{-1} [51], ambiguities in the mechanism exist, and since the self-exchange rate for the outer-sphere process is much higher than for chromium(III)/(II), mixtures of outer-sphere and inner-sphere pathways are frequently proposed. The product vanadium(III) complexes are labile with a characteristic solvent exchange rate of approximately 10^2 s^{-1} and so detection of kinetic products, though possible, is not common. Assignment of mechanism is generally based on deviations from the correlation of rates with those for $[Ru(NH_3)_6]^{2+}$ as reductant, a test for outer-sphere mechanism, and on the reaction rates being slower than the expected rate of substitution. Rates of reactions of $[V(H_2O)_6]^{2+}$ with a number of oxidants are presented in Table 3.8 together with their mechanistic assignments.

Table 3.8. Rates of reduction by $[V(H_2O)_6]^{2+}$ at 25.0°C

Oxidant	μ (M)	k (M^{-1} s^{-1})	$\Delta H^\ddagger$ (kJ mol^{-1})	$\Delta S^\ddagger$ (J K^{-1} mol^{-1})	Designation	Ref.
$[Ru(NH_3)_5Cl]^{2+}$	0.11	3.0×10^3	15.9	−125	OS	48
$[Ru(NH_3)_5Br]^{2+}$	0.11	5.1×10^3	15.9	−142	OS	48
$[Ru(NH_3)_5OAc]^{2+}$	0.5	1.3×10^3			OS	48
$[Co(NH_3)_5F]^{2+}$	1.0	2.6			OS	10
$[Co(NH_3)_5Cl]^{2+}$	1.0	7.6			OS	52
cis-$[Co(en)_2Cl_2]^+$	1.0	10.1			OS	52
trans-$[Co(en)_2Cl_2]^+$	1.0	128			OS	52
$[Co(NH_3)_5Br]^{2+}$	1.0	25			OS	10
$[Co(NH_3)_5I]^{2+}$	1.0	1.2×10^2			OS	10
$[Co(NH_3)_5OAc]^{2+}$	1.0	1.15	48.5	−7.9	IS?	11
$[Co(NH_3)_5O_2CH]^{2+}$	1.0	3.63	58.1	−54	IS?	11
cis-$[Co(en)_2(N_3)_2]^+$	1.0	33			IS	53
$[(NH_3)_5Co(oxH)]^{2+}$	1.0	1.0×10^2			IS	26
$[(NH_3)_5Co(ox)]^+$	1.0	4.6×10^4	9.6	−84	IS+OS	26
$[(en)Co(ox)_2]^-$	1.0	1.08×10^2	38.9	−75	IS+OS	54
$[Co(ox)_3]^{3-}$	1.0	2.04×10^4	9.2	−132	IS+OS	54
$[Cr(H_2O)_5NCS]^{2+}$	3.0	4.41×10^{-5}	96.6	−4	IS	55
$[Cr(H_2O)_5SCN]^{2+}$	3.0	16.2	51.9	−48	IS	55

Reductions of ruthenium complexes such as $[Ru(NH_3)_5Cl]^{2+}$ clearly exceed the rates commonly expected for substitution at $[V(H_2O)_6]^{2+}$ and are assigned an outer-sphere mechanism. A plot of log k_V against log k_{Ru} is shown in Fig. 3.1. There is a good correlation for the outer-sphere data and the reactions of $[Co(NH_3)_5Cl]^{2+}$ and derivatives are assigned an outer-sphere mechanism on this basis [10]. The reduction of $[Cr(H_2O)_5SCN]^{2+}$ shows a rate which is close to rates of substitution at $[V(H_2O)_6]^{2+}$ and has been shown to be inner-sphere with the detection of $[V(H_2O)_5NCS]^{2+}$ as a transient product. In the case of reduction of $[Cr(H_2O)_5NCS]^{2+}$, the rate is much slower than the substitution rate and is tentatively assigned as inner-sphere where there is rate-limiting electron transfer on the basis of comparisons with the rates of the corresponding reductions by $[Cr(H_2O)_6]^{2+}$ [55]. Note that for inner-sphere reactions, $\Delta H^\ddagger$ is larger and $\Delta S^\ddagger$ less negative than for outer-sphere reactions and more in line with parameters for substitution.

In a few other instances it has proved possible to design experiments where the vanadium(III) kinetic product of inner-sphere electron transfer has sufficient longevity for detection. Reduction of cis-$[Co(en)_2(N_3)_2]^+$ by $[V(H_2O)_6]^{2+}$ leads to the formation of a transient $[V(H_2O)_5N_3]^{2+}$, absorbing at 350 nm, and the subsequent aquation of this complex is sluggish (10 s^{-1}) [53]. Similarly the chelate product, $[V(H_2O)_4(ox)]^+$, has been detected in the inner-sphere reduction of $[(NH_3)_5Co(oxH)]^{2+}$, $[(en)Co(ox)_2]^-$, and $[Co(ox)_3]^{3-}$ [26, 54]. In the latter two reactions, $[V(H_2O)_6]^{3+}$ is also detected as an initial reaction product consistent with parallel inner-sphere and outer-sphere pathways (eqs (3.21)–(3.22)). The proportions of the inner-sphere pathways are limited by substitution at the $[V(H_2O)_6]^{2+}$ center.

$$[Co(ox)_3]^{3-} + [V(H_2O)_6]^{2+} \xrightarrow{k_{IS}} [Co(ox)_2] + [V(H_2O)_4(ox)]^+ \quad (3.21)$$

$$[Co(ox)_3]^{3-} + [V(H_2O)_6]^{2+} \xrightarrow{k_{OS}} [Co(ox)_3]^{4-} + [V(H_2O)_6]^{3+} \quad (3.22)$$

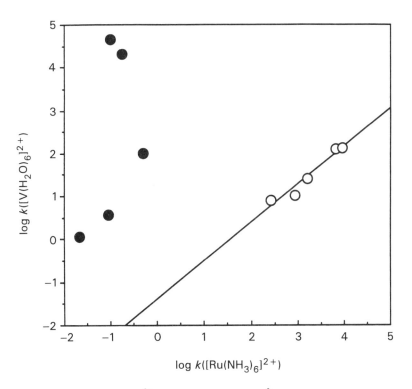

Fig. 3.1. Plot of log k($[V(H_2O)_6]^{2+}$) against log k($[Ru(NH_3)_6]^{2+}$) for the reduction of complexes shown in Table 3.8. Open circles are assigned an outer-sphere mechanism for both reagents. In the case of the closed circles the $[V(H_2O)_6]^{2+}$ reactions are faster than predicted for outer-sphere reaction and have some inner-sphere character. Note that the inner-sphere rates must be less than $\approx 10^2$ s^{-1}.

Where the rates for inner-sphere reductions by $[V(H_2O)_6]^{2+}$ lie in the range 1–40 M^{-1} s^{-1}, the activation parameters are generally similar to those for substitution at $[V(H_2O)_6]^{2+}$. However, for reactions slower than the substitution-controlled rate, electron transfer is rate-limiting, and there is no requirement for activation parameters to reflect those for substitution.

The reagent $[Fe(H_2O)_6]^{2+}$ is a poor reductant and is very labile with a solvent exchange rate of approximately 10^6 s^{-1}. Like $[V(H_2O)_6]^{2+}$, the oxidized form, $[Fe(H_2O)_6]^{3+}$, is significantly less labile than $[Fe(H_2O)_6]^{2+}$ with a solvent exchange rate of approximately 10^2 s^{-1} and detection of kinetic products is possible although isolation is not. Representative rate data are presented in Table 3.9. Reduction of $[Co(NH_3)_5F]^{2+}$ and derivatives is assigned an inner-sphere mechanism on the basis of large rate discrimination between $[Co(NH_3)_5N_3]^{2+}$ and $[Co(NH_3)_5NCS]^{2+}$ [56], and on the basis of the observation that the inverse order of reduction of the halide complexes is followed [57]. Volumes of activation have also been determined for a number of the reactions with $[Fe(H_2O)_6]^{2+}$ as reductant [58, 59]. The values are positive, consistent with the exclusion of solvent in the precursor complex formation.

Table 3.9. Rate constants and activation parameters for reduction by $[Fe(H_2O)_6]^{2+}$ at 25.0°C

Oxidant	μ (M)	k (M^{-1} s^{-1})	$\Delta H^{\ddagger}$ (kJ mol^{-1})	$\Delta S^{\ddagger}$ (J K^{-1} mol^{-1})	$\Delta V^{\ddagger}$ (cm^3 mol^{-1})	Ref.
$[Co(NH_3)_5F]^{2+}$	1.0	6.6×10^{-3}	57	−96	11	56
$[Co(NH_3)_5Cl]^{2+}$	1.0	1.35×10^{-3}	52	−125	9	56
$[Co(NH_3)_5Br]^{2+}$	1.0	7.3×10^{-4}	56	−117	6	56
$[Co(NH_3)_5OAc]^{2+}$	1.0	$>5 \times 10^{-5}$				56
$[Co(NH_3)_5NCS]^{2+}$	1.0	$>3 \times 10^{-5}$				56
$[Co(NH_3)_5N_3]^{2+}$	1.0	8.8×10^{-3}	63	−74	12	60
$[Co(NH_3)_5SCN]^{2+}$	1.0	1.2×10^{-1}				61
$[Co(ox)_3]^{3-}$	1.0	33				62
$[Co(en)(ox)_2]^{-}$	1.0	3.15×10^{-3}	66	−76		54
$[Co(en)_2(ox)]^{+}$	1.0	2.28×10^{-5}	80	−65		63
$[Co(dien)(en)Cl]^{2+}$	1.0	6.0×10^{-6}				64
$[Co(tetren)Cl]^{2+}$	1.0	$< 1 \times 10^{-8}$				64
cis-$[Co(en)_2(NH_3)Cl]^{2+}$	1.0	1.8×10^{-5}				65
cis-$[Co(en)_2pyCl]^{2+}$	1.0	7.9×10^{-4}				66
cis-$[Co(en)_2(NCS)Cl]^{+}$	1.0	1.7×10^{-4}				65
cis-$[Co(en)_2Cl_2]^{+}$	1.0	1.6×10^{-3}				65
cis-$[Co(en)_2(H_2O)Cl]^{2+}$	1.0	4.6×10^{-4}				65
trans-$[Co(en)_2(NH_3)Cl]^{2+}$	1.0	6.6×10^{-5}				65
trans-$[Co(en)_2(NCS)Cl]^{+}$	1.0	1.3×10^{-4}				65
trans-$[Co(en)_2Cl_2]^{+}$	1.0	3.2×10^{-2}				65
trans-$[Co(en)_2BrCl]^{+}$	1.0	3.6×10^{-2}				65
trans-$[Co(en)_2(N_3)Cl]^{+}$	1.0	6.2×10^{-2}				65
trans-$(Co(en)_2(H_2O)Cl]^{2+}$	1.0	2.4×10^{-1}				65

The reaction with $[Co(NH_3)_5SCN]^{2+}$ gives $[Fe(H_2O)_5NCS]^{2+}$ as a transient absorbing maximally at 460 nm, confirming an inner-sphere mechanism [61]. Similar conclusions have been reached with the detection of $[Fe(H_2O)_4(ox)]^{+}$ in the reduction of $[Co(ox)_3]^{3-}$, which is likely to be formed from a doubly bridged transition state. The reactions involving $[Fe(H_2O)_6]^{2+}$ are not in general substitution-controlled and show a strong dependence on non-bridging ligands [65, 66], and this constitutes one of the most important studies of this effect. Although the reduction of *trans*-$[Co(en)_2(OH_2)Cl]^{2+}$ is the only one in which an inner-sphere mechanism has been proved [62], all these reactions are thought to proceed by a chloride bridge with the exception of *trans*-$[Co(en)_2(N_3)Cl]^+$, which prefers an azide bridge. There are trends in the data which are more marked with changing the *trans* ligand than with changing the *cis* ligand. Chelation slows the rate [64], and there is a correlation with ligand field strength (eq. (3.23)), which may reflect the thermodynamic driving force.

$$en < NH_3 < SCN^- < Cl^- < Br^- < H_2O \quad (3.23)$$

The electron exchange reaction for $[Fe(H_2O)_6]^{3+/2+}$ is strongly catalyzed by Cl^- [67]. The dominant pathway for this catalysis involves the reduction of $[Fe(H_2O)_5Cl]^{2+}$ by $[Fe(H_2O)_6]^{2+}$ which has a rate constant of 57.6 M^{-1} s^{-1} at 25.0°C and 3.0 M ionic strength. Comparison of this value with the rate for the $[Fe(H_2O)_6]^{2+}$-catalyzed aquation of $[Fe(H_2O)_5Cl]^{2+}$ (eq. (3.24)), 33.4 M^{-1} s^{-1}, leads to the conclusion that the electron transfer proceeds substantially by an inner-sphere mechanism with transfer of the Cl^--bridge. Further remarks on inner-sphere pathways in the reactions of aqua-metal ion complexes is found in section 3.8.

$$[Fe(H_2O)_5Cl]^{2+} + [Fe(H_2O)_6]^{2+} \rightleftharpoons [Fe(H_2O)_6]^{3+} + [Fe(H_2O)_5Cl]^+ \quad (3.24)$$

In a now classic experiment, advantage was taken of the sluggish electron transfer reactivity and the substitution lability of $[Fe(H_2O)_6]^{2+}$ to design an experiment to detect precursor complex formation [68]. The idea involves stabilization of the precursor by chelation to $[(NH_3)_5CoO_2CCH_2N(CH_2CO_2)]$. The spectra of solutions of the intermediate show absorption characteristics of the reactants only and a structure (3.25) is proposed.

$$(3.25)$$

Electron transfer most likely takes place through the bound carbonyl, but an alternative bridged outer-sphere mechanism is also possible. In the corresponding reduction by $[Cr(H_2O)_6]^{2+}$, the reactions are very rapid and the product is fully chelated consistent with inner-sphere reaction [27].

Studies with the reductant $[Ti(H_2O)_6]^{3+}$ present some experimental difficulties since the ion is slowly oxidized by $[ClO_4]^-$, one of the more important weakly coordinating counter-ions used in kinetic studies so that experimental data are mostly accumulated in Cl^-, $CF_3SO_3^-$, or $CH_3C_6H_4SO_3^-$ media. Reductions by titanium(III) show some unusual features which merit comment. In reactions with $[Co(NH_3)_5Cl]^{2+}$, the rate law indicates a dominant pathway involving an $[H^+]^{-1}$ dependence (eq. (3.26)), for which $k = 0.48$ M^{-1} s^{-1} and $K_h = 4.6 \times 10^{-3}$ M [69]. The latter value is similar to the thermodynamically determined hydrolysis constant for $[Ti(H_2O)_6]^{3+}$, 6.9×10^{-3} M, and the proposed mechanism is given in eqs (3.27)–(3.28).

$$\frac{-d[[Co(NH_3)_5Cl]^{2+}]}{dt} = \frac{k_{OH} K_h [H^+]^{-1}}{1 + K_h[H^+]^{-1}} [[Co(NH_3)_5Cl]^{2+}][[Ti(H_2O)_6]^{3+}]_T \quad (3.26)$$

$$[Ti(H_2O)_6]^{3+} \xrightleftharpoons{K_h} [Ti(H_2O)_5OH]^{2+} + H^+ \quad (3.27)$$

$$[Co(NH_3)_5Cl]^{2+} + [Ti(H_2O)_5OH]^{2+} \xrightarrow{k_{OH}}$$

$$[Co(H_2O)_6]^{2+} + [TiO_{aq}]^{2+} + 5\,NH_4^+ + Cl^- \quad (3.28)$$

A similar rate law is observed in the reduction of $[Co(NH_3)_5N_3]^{2+}$ but in this instance $K_h = 0.075$ M, an order of magnitude larger than the hydrolysis constant [70]. The discrepancy with the thermodynamic value points to the involvement of an inner-sphere steady-state intermediate in the reaction. Two possibilities exist, (3.29)–(3.30) and (3.31)–(3.32).

$$[Co(NH_3)_5N_3]^{2+} + [Ti(H_2O)_6]^{3+} \xrightleftharpoons[k_{-1}]{k_1} [Co(NH_3)_5N_3Ti(H_2O)_4(OH)]^{4+} + H^+ \quad (3.29)$$

$$[Co(NH_3)_5N_3Ti(H_2O)_4(OH)]^{4+} \xrightarrow{k_2} [Co(H_2O)_6]^{2+} + 5\,NH_4^+ + HN_3 + [TiO_{aq}]^{2+} \quad (3.30)$$

$$[Co(NH_3)_5N_3]^{2+} + [Ti(H_2O)_5OH]^{2+} \xrightleftharpoons[k_{-3}]{k_3} [Co(NH_3)_5N_3Ti(H_2O)_4(OH)]^{4+} \quad (3.31)$$

$$[Co(NH_3)_5N_3Ti(H_2O)_4(OH)]^{4+} + H^+ \xrightarrow{k_4}$$

$$[Co(H_2O)_6]^{2+} + 5\,NH_4^+ + HN_3 + [TiO_{aq}]^{2+} + H^+ \quad (3.32)$$

The latter mechanism is unlikely since there is no reason to protonate the intermediate as the final titanium product is an oxo species [47]. Thus the reactions of $[Ti(H_2O)_6]^{3+}$ reveal evidence within the $[H^+]$ dependence of the rate law for the formation of precursor complexes. This fortunate occurrence allows some distinction between inner-sphere and outer-sphere mechanisms. Otherwise free energy relationships must be employed. A listing of rate constants and acidity dependencies is presented in Table 3.10 for a number of reductions by titanium(III).

Table 3.10. Rate constants and mechanistic details of reactions of $[Ti(H_2O)_6]^{3+}$ at 25.0°C

Oxidant	μ (M)	k_{Ti^{3+}/H^+} (s^{-1})	$k_{TiOH^{2+}}$ (M^{-1} s^{-1})	$k_{Ti^{3+}}$ (M^{-1} s^{-1})	K_h	Ref.
$[Co(NH_3)_5Cl]^{2+}$	1.0		0.48		4.6×10^{-3}	69
cis-$[Co(en)_2Cl_2]^+$	1.0		0.75	0.002	4.6×10^{-3}	69
trans-$[Co(en)_2Cl_2]^+$	1.0		2.8	0.009	4.6×10^{-3}	69
$[Co(NH_3)_5N_3]^{2+}$	0.5	47			7.5×10^{-2}	70
$[Co(NH_3)_5OAc]^{2+}$	1.0		1×10^{-2}		3.5×10^{-3}	69
$[Ru(NH_3)_5Cl]^{2+}$	1.0		12		5×10^{-3}	71
$[Ru(NH_3)_5NCS]^{2+}$	2.0			840		72
$[Ru(NH_3)_5OAc]^{2+}$	2.0			700	0.21	73
$[Ru(NH_3)_4(ox)]^+$	2.0			36000	0.13	73
$[Co(ox)_3]^{3-}$	1.0	1.9×10^3				74,75
$[Ru(ox)_3]^{3-}$	1.0	53×10^2				76

Reduction of $[Ru(NH_3)_5Cl]^{2+}$ has been identified [71] as outer-sphere in nature but other ruthenium(III) oxidants including $[Ru(NH_3)_5OAc]^{2+}$ [73], $[Ru(NH_3)_5NCS]^{2+}$ [72], and $[Ru(NH_3)_4(ox)]^+$ [73] show evidence for inner-sphere behavior with inhibition by H^+ (eq. (3.33)). The rates do not show a close correlation with a linear free energy relationship based on comparisons of outer-sphere reactions of $[Ti(H_2O)_5OH]^{2+}$ and $[Ru(NH_3)_6]^{2+}$ [77–79] and are assigned an inner-sphere mechanism. The detailed mechanism (eqs (3.34)–(3.35)) involves protonation of the oxidant which reduces reactivity. Activation parameters for the reaction are substantially different from the normal values for outer-sphere reactions and reflect the fact that the rate of formation of the precursor complex is rate-limiting. Electron transfer is particularly enhanced by bridging oxalate.

$$\frac{-d[[Ru(NH_3)_5OAc]^{2+}]}{dt} = \frac{kK}{K + [H^+]} [[Ru(NH_3)_5OAc]^{2+}][[Ti(H_2O)_6]^{3+}] \quad (3.33)$$

$$[Ru(NH_3)_5OAc]^{2+} + H^+ \underset{}{\overset{K_h}{\rightleftharpoons}} [Ru(NH_3)_5OAcH]^{3+} \quad (3.34)$$

$$[Ru(NH_3)_5OAc]^{2+} + [Ti(H_2O)_6]^{3+} \xrightarrow{k} [Ru(NH_3)_5OAc]^+ + [TiO_{aq}]^{2+} \quad (3.35)$$

Mechanisms for reactions of the three reductants $[Cu_{aq}]^+$, $[Eu_{aq}]^{2+}$ and $[U_{aq}]^{3+}$ are more difficult to decipher because the pattern of outer-sphere behavior is not well established. However, all three reductants have low self-exchange rates, high rates of substitution and behave in a manner which is very similar to that of $[Cr(H_2O)_6]^{2+}$, where inner-sphere reactions are preferred. There is little prospect for detection of inner-sphere products even as transients, and mechanistic assignments are made on the basis of rate comparisons. Rate data for the reagents are presented in Table 3.11. The major differences are that the $[U_{aq}]^{3+}$ reactions are much faster than those for $[Eu_{aq}]^{2+}$ and are affected by the greater charge. There is a much greater range of rates for $[Cu_{aq}]^+$ compared with $[Cr(H_2O)_6]^{2+}$ which reflects the fact that the latter are close to the substitution limit.

Table 3.11. Rate constants for the reduction of selected oxidants by $[Cu_{aq}]^+$, $[Eu_{aq}]^{2+}$ and $[U_{aq}]^{3+}$ at 25.0°C

Oxidant	k^a(Cu(I)) [80] ($M^{-1} s^{-1}$)	k^b(Eu(II)) [10] ($M^{-1} s^{-1}$)	k^a(U(III)) [81] ($M^{-1} s^{-1}$)
$[Co(NH_3)_5F]^{2+}$	1.11	2.6×10^4	5.40×10^5
$[Co(NH_3)_5Cl]^{2+}$	4.88×10^4	3.9×10^2	3.24×10^4
$[Co(NH_3)_5Br]^{2+}$	4.46×10^5	2.5×10^2	1.42×10^4
$[Co(NH_3)_5N_3]^{2+}$	1.50×10^3	1.9×10^2	1.08×10^6
$[Co(NH_3)_5NCS]^{2+}$	≈ 2	≈ 0.7	18.2
$[Co(NH_3)_5OH]^{2+}$	3.8×10^2		—^{c}
$[Co(NH_3)_5OAc]^{2+}$		1.8×10^{-1}	1.50×10^4

^{a}0.20 M ionic strength.
^{b}1.0 M ionic strength.
^{c}Kinetic ambiguity.

3.6 ANIONIC REDUCTANTS

In the preceding studies of reactions of the labile metal aquo-ion reductants, there is good evidence for bridge formation in the inner-sphere mechanism. The reaction energetics are complicated by substitution processes involved in precursor complex formation and successor complex decomposition such that those for the electron transfer process are ill-defined. Measurements on well defined inner-sphere electron transfer reactions are lacking and while there is some kinetic detection of the precursor, stability constants are low and quantitative study has not been possible. Complex formation may be more favored with the use of anionic reductants, and two reagents have been used extensively. The $[Co(CN)_5]^{3-}$ ion is five-coordinate in solution with $K_6 = 10^{-1}$–10^{-4} M^{-1} (eq. (3.36)) and the potential of $[Co(CN)_5(OH_2)]^{2-}$ is strongly reducing. Reactions with complexes such as $[Co(NH_3)_5F]^{2+}$ show competition between inner-sphere pathways involving

[Co(CN)$_5$]$^{3-}$, and outer-sphere pathways involving [Co(CN)$_6$]$^{4-}$, with a rate law of the form of eq. (3.37) [82]. Rate constants are presented in Table 3.12. The inner-sphere nature of the former pathway can be deduced from the detection in the reduction of [Co(NH$_3$)$_5$CN]$^{2+}$ of the N-bonded product, [Co(CN)$_5$(NC)]$^{3-}$, as a reactive transient which rapidly isomerizes to [Co(CN)$_6$]$^{3-}$, the thermodynamic product with a half-life of 0.8 s [83]. Interestingly, where [Co(NH$_3$)$_5$NCS]$^{2+}$ gives the expected product of remote attack, the isomeric species [Co(NH$_3$)$_5$SCN]$^{2+}$ shows no evidence for [Co(CN)$_5$NCS]$^{3-}$, preferring instead adjacent attack to give the thermodynamically favored product [Co(CN)$_5$SCN]$^{3-}$, and contrasting with the corresponding reaction with [Cr(H$_2$O)$_6$]$^{2+}$ [84].

$$[Co(CN)_5]^{3-} + CN^- \overset{K_6}{\rightleftharpoons} [Co(CN)_6]^{4-} \qquad (3.36)$$

$$\frac{-d[[Co(NH_3)_5F]^{2+}]}{dt} = \{k_{IS} + k_{OS}[CN^-]\} [[Co(CN)_5]^{3-}][[Co(NH_3)_5F]^{2+}] \qquad (3.37)$$

$$[Co(NH_3)_5F]^{2+} + [Co(CN)_5]^{3-} \xrightarrow{k_{IS}}$$

$$[Co(CN)_5F]^{3-} + [Co(H_2O)_6]^{2+} + 5\,NH_4^+ \qquad (3.38)$$

$$[Co(NH_3)_5F]^{2+} + [Co(CN)_6]^{4-} \xrightarrow{k_{OS}}$$

$$[Co(CN)_6]^{3-} + [Co(H_2O)_6]^{2+} + 5\,NH_4^+ + F^- \qquad (3.39)$$

Table 3.12. Rate constants for the reduction of cobalt(III) complexes by [Co(CN)$_5$]$^{3-}$

Complex	μ (M)	k_{IS} (M^{-1} s^{-1})	k_{OS} (M^{-2} s^{-1})	Ref.
[Co(NH$_3$)$_5$Cl]$^{2+}$	0.2	$\approx 5 \times 10^7$		82
[Co(NH$_3$)$_5$F]$^{2+}$	0.2	1.8×10^3	1.7×10^4	82
[Co(NH$_3$)$_5$N$_3$]$^{2+}$	0.2	1.6×10^6	$< 8 \times 10^5$	82
[Co(NH$_3$)$_5$NCS]$^{2+}$	0.2	1.1×10^6	$< 5 \times 10^5$	82
[Co(NH$_3$)$_5$OH]$^{2+}$	0.2	9.3×10^4	$< 5 \times 10^4$	82
[Co(NH$_3$)$_5$OAc]$^{2+}$	0.2	$< 1 \times 10^2$	1.1×10^4	82
[Co(NH$_3$)$_5$CN]$^{2+}$	0.2	2.9×10^2	4.5×10^3	83
[Co(NH$_3$)$_5$SCN]$^{2+}$	0.1	$> 10^8$	$< 2 \times 10^9$	84

The reagent [Fe(CN)$_5$OH$_2$]$^{2-}$ is a poor reductant with a reduction potential estimated as 0.54 V [85]. Attempts to use the reagent to study inner-sphere reactions by taking advantage of the high affinity of the reductant for OS(Me)$_2$ (eq. (3.40)) lead to the formation of an inner-sphere complex but decomposition involves substitution and not electron transfer [86].

$$[(NH_3)_5CoOS(CH_3)_2]^{3+} + [Fe(CN)_5(OH_2)]^{3-} \rightleftharpoons$$

$$[(NH_3)_5CoOS(CH_3)_2Fe(CN)_5] \qquad (3.40)$$

A related reaction is the reduction of $[Fe(CN)_6]^{3-}$ by $[Co(edta)]^{2-}$, which takes place rapidly to give a transient, identified by spectroscopy and magnetic susceptibility measurements as a successor complex (eq. (3.41)) [87]. The reductant is labile and the structure proposed for the intermediate involves dissociation of one of the chelate rings to provide a site for substitution by a CN^- bridge from the oxidant. Rate constants are $k_1 = 8 \times 10^4$ M^{-1} s^{-1} and $k_{-1} = 96$ s^{-1} at 25.0°C and 0.66 M ionic strength [88]. Decomposition of the intermediate has a rate of 5.4×10^{-3} s^{-1} [89], giving the substitution inert products, $[Co(edta)]^-$ and $[Fe(CN)_6]^{4-}$ [90]. This observation sheds considerable light on the mechanism of the decomposition reaction. The oxidation of $[Fe(CN)_6]^{4-}$ by $[Co(edta)]^-$ is thermodynamically unfavorable but can be driven to completion by the addition of ascorbate ion. The reaction (eq. (3.42)) has a second-order rate constant $k_{-2} = 0.21$ M^{-1} s^{-1} at 25.0°C and 0.6 M ionic strength, this is too slow to represent substitution of coordinated cyanide into the inner-coordination sphere of $[Co(edta)]^-$, and it is concluded that the reaction is outer-sphere. Consequently the inner-sphere successor complex represents a 'dead-end' complex in the overall outer-sphere reaction and the decomposition of the transient involves dissociation to reactants followed by outer-sphere electron transfer. The rate of the outer-sphere reaction can be calculated from the reduction potentials to be 4.5 M^{-1} s^{-1}, four orders of magnitude slower than the inner-sphere reaction.

$$[Fe(CN)_6]^{3-} + [Co(edta)]^{2-} \underset{k_{-1}}{\overset{k_1}{\rightleftharpoons}} [(edta)Co(NC)Fe(CN)_5]^{5-} \qquad (3.41)$$

$$[Fe(CN)_6]^{3-} + [Co(edta)]^{2-} \underset{k_{-2}}{\overset{k_2}{\rightleftharpoons}} [Fe(CN)_6]^{4-} + [Co(edta)]^- \qquad (3.42)$$

Similar mechanisms are found with other reactions involving $[Co(edta)]^{2-}$, and while it has proved to be an effective inner-sphere reductant [91, 92], outer-sphere mechanisms are often preferred [93].

3.7 STEREOSELECTIVITY IN INNER-SPHERE ELECTRON TRANSFER REACTIONS

Stereoselectivity provides a further probe of the inner-sphere electron transfer mechanism. As a result of the specificity of the inner-sphere transition state, stereoselectivities in inner-sphere reactions might be anticipated to be larger than those for corresponding outer-sphere reactions. However, the mechanisms are more complex. A series of inner-sphere reductants which are both chiral, as a result of stereospecific binding of a chiral pentadentate ligand, and labile, as a result of the available sixth coordination position on iron(II), have been developed [94]. Stereoselectivities are modest and comparable with those in outer-sphere reactions (Table 3.13). The reduction of $[Co(ox)_3]^{3-}$ by $[Co(en)_3]^{2+}$ follows two pathways: an outer-sphere pathway (eq. (3.44)) where $[Co(en)_3]^{2+}$ is the reductant, and an inner-sphere pathway (eq. (3.45)) where $[Co(en)_2]^{2+}$ is the reductant. Stereoselectivities in the latter pathway are very small [95], most likely as a result of the large (approximately 5.5 Å) separation between the chiral centers. The stereoselectivity is sensitive to the position of methyl substituents on the 1,2-diaminoethane ligands, and the sense of the discrimination can be understood by consideration of hydrogen bonding in the assembly of the precursor complex [96]. Much larger stereoselectivities are noted with the more rigid $[Co(bpy)_2]^{2+}$ and especially with the $[Co(phen)_2]^{2+}$ reductants [97].

$$[Co(en)_2]^{2+} + en \rightleftharpoons [Co(en)_3]^{2+} \qquad (3.43)$$

$$[Co(ox)_3]^{3-} + [Co(en)_3]^{2+} \xrightarrow{k_{OS}} [Co(en)_3]^{3+} + [Co(ox)_2]^{2-} + ox^{2-} \qquad (3.44)$$

$$[Co(ox)_3]^{3-} + [Co(en)_2]^{2+} \xrightarrow{k_{IS}} [Co(en)_2(ox)]^{+} + [Co(ox)_2]^{2-} \qquad (3.45)$$

Table 3.13. Inner-sphere electron transfer stereoselectivities at 25.0°C

Oxidant	Reductant	μ (M)	k (M^{-1} s^{-1})	$k_{\Delta\Delta}/k_{\Delta\Lambda}$	Ref.
[Co(bamap)OH]	[Fe(alamp)]	0.10	61	0.55	94
$[Co(ox)_3]^{3-}$	$[Co(en)_2]^{2+}$	0.10	3300	1.03	95
$[Co(ox)_3]^{3-}$	$[Co(N,N-Me_2en)_2]^{2+}$	0.10	—	0.91	96
$[Co(ox)_3]^{3-}$	$[Co(N,N'-Me_2en)_2]^{2+}$	0.10	—	1.17	96
$[Co(ox)_3]^{3-}$	$[Co(bpy)_2]^{2+}$	0.017	160	1.47	97
$[Co(ox)_3]^{3-}$	$[Co(phen)_2]^{2+}$	0.017	94	4.4	97

3.8 THEORETICAL ASPECTS

Inner-sphere reactions show considerable complexity and the development of a useful theoretical framework for discussion of mechanistic details has been much slower than for outer-sphere reactions. On the positive side, there is a detailed picture of the transition state for the reaction which is not available in outer-sphere reactions. To balance this, bond making, bond breaking, precursor complex stability and successor complex stability are all variables which may have a profound effect on the rate besides the dynamics of the electron transfer itself, determined primarily by

Frank–Condon factors as in outer-sphere reactions. In comparing data, therefore, it is essential to establish the rate-determining step in the overall process. The usefulness of linear free energy relationships is much more limited, but where electron transfer is rate limiting and the details of the mechanisms are similar, some comparisons can be made [98–101].

The great facility of the self-exchange process in understanding outer-sphere electron transfer reactions has no exact counterpart in the inner-sphere process because there is an unique bridging atom or group. However, a pseudo-self-exchange process can be defined with $\Delta G° \approx 0$ as shown in eq. (3.46) for the reaction of $[Cr(H_2O)_5OH]^{2+}$ with $[Cr(H_2O)_6]^{2+}$.

$$[Cr(H_2O)_5OH]^{2+} + [^*Cr(H_2O)_6]^{2+} \rightleftharpoons [^*Cr(H_2O)_5OH]^{2+} + [Cr(H_2O)_6]^{2+}$$
(3.46)

This reaction is the $[H^+]^{-1}$ catalyzed pathway in the self-exchange process and it is subject to a proton ambiguity when studied under conditions of pH far from the pK_a values for $[Cr(H_2O)_6]^{3+}$ and $[Cr(H_2O)_6]^{2+}$. The alternative to eq. (3.46) is the reaction of $[Cr(H_2O)_6]^{3+}$ with $[Cr(H_2O)_5OH]^+$ and this is less likely since hydrolysis in $[Cr(H_2O)_6]^{2+}$ is much less favored. Pathways of this sort are not restricted to OH⁻ bridges. In the presence of many halides and pseudo-halides, the reactions show evidence for halide ion catalysis. As indicated in section 1.9, a cautionary note is required in interpreting such dependencies since pathways involving both $[Cr(H_2O)_5Cl]^{2+}$ and $\{[Cr(H_2O)_6]^{3+}, Cl^-\}$ are possible and only the former species gives the appropriate pseudo-exchange reaction. Selected values for pseudo-exchange rates are presented in Table 3.14.

Table 3.14. Pseudo-self-exchange rates for aqua-ion complexes at 25°C

X	μ (M)	$E°$ (V)	k (M⁻¹ s⁻¹)	$\Delta H^\ddagger$ (kJ mol⁻¹)	$\Delta S^\ddagger$ (J K⁻¹ mol⁻¹)	$\Delta V^\ddagger$ (cm³ mol⁻¹)	Ref.
$[Cr(H_2O)_5X]^{2+}/[Cr(H_2O)_6]^{2+}$							
OH	1.0	−0.69	0.69				43
Cl	1.0ᵃ		9				42
F	1.0		2×10^{-2}				42
NCS	1.0		1.2×10^{-4}				42
$[Fe(H_2O)_5X]^{2+}/[Fe(H_2O)_6]^{2+}$							
OH	0.1 (Li)	0.36	1.36×10^3	28	−92	+0.8ᵇ	102,103
Cl	3.0		58				67
$[Co(H_2O)_5X]^{2+}/[Co(H_2O)_6]^{2+}$							
OH	1.0	1.44	7.0×10^2	75	−5		104
$[Fe(H_2O)_5X]^{2+}/[Cr(H_2O)_6]^{2+}$							
OH	1.0 (Na)		3.3×10^6				105
Cl	1.0 (Na)		2×10^7				105
OH	1.0 (Li)		4.4×10^6	19	−54		106
N₃	1.0 (Li)		2.9×10^7				106
NCS	1.0 (Li)		2.8×10^7				106
NCO	1.0 (Li)		2.2×10^5				106
$[Co(H_2O)_5X]^{2+}/[Cr(H_2O)_6]^{2+}$							
OH	3.0 (Li)		3.3×10^6				107

(continues)

Table 3.14. (continued)

X	μ (M)	$E°$ (V)	k (M^{-1} s^{-1})	$\Delta H^{\ddagger}$ (kJ mol^{-1})	$\Delta S^{\ddagger}$ (J K^{-1} mol^{-1})	$\Delta V^{\ddagger}$ (cm^3 mol^{-1})	Ref.
[Co(H$_2$O)$_5$X]$^{2+}$/[Fe(H$_2$O)$_6$]$^{2+}$							
OH	3.0 (Li)		1.4×10^5				108
[Co(H$_2$O)$_5$X]$^{2+}$/[Mn(H$_2$O)$_6$]$^{2+}$							
OH	3.0 (Li)		2.3×10^4				108
[V(H$_2$O)$_5$X]$^{2+}$/[Cr(H$_2$O)$_6$]$^{2+}$							
OH	3.0 (Na)		3.7×10^2				109

^{a}0°C ^{b}2°C, 0.5 M

In principle one can apply a Marcus-type relationship to an inner-sphere electron transfer reaction (eq. (3.47)) by defining the corresponding pseudo-self-exchange processes (eqs (3.48) and (3.49)) [110].

$$A^{ox}\text{-}X + B^{red} \underset{}{\overset{k_{AXB}}{\rightleftharpoons}} A^{red} + B^{ox}\text{-}X \tag{3.47}$$

$$A^{ox}\text{-}X + A'^{red} \underset{}{\overset{k_{AXA}}{\rightleftharpoons}} A^{red} + A'^{ox}\text{-}X \tag{3.48}$$

$$B^{ox}\text{-}X + B'^{red} \underset{}{\overset{k_{BXB}}{\rightleftharpoons}} B^{red} + B'^{ox}\text{-}X \tag{3.49}$$

The relationship with free energy is given by eq. (3.50) in exactly the same manner as in section 2.4 for outer-sphere reactions. However, the major assumption involved, that the structural reorganization required to attain the transition state by a reagent in the cross-reaction is the same as that required for the pseudo-self-exchange, is much more difficult to justify for an inner-sphere reaction than for an outer-sphere reaction. In inner-sphere reactions, the overlap of the metal-centered orbitals involved in the electron transfer is modified by the orbitals on the bridge, and the extent of this interaction will depend on the nature of the reaction partner. Moreover, the inner-sphere precursor stability constant is much more difficult to estimate than the outer-sphere ion association, which is calculable by modifications of the Fuoss equation. There is not an abundance of information to test a relationship such as eq. (3.50) for inner-sphere reactions. A number of [H$^+$]$^{-1}$-catalyzed cross-reactions between metal aqua-ion complexes have been reported (Table 3.14). The reduction of [Co(H$_2$O)$_5$OH]$^{2+}$ by [Fe(H$_2$O)$_6$]$^{2+}$ has a rate constant of 3×10^5 M^{-1} s^{-1} but the rate predicted by eq. (3.50) is 2×10^9 M^{-1} s^{-1} [106]. It is quite clear that the specificity of the inner-sphere interactions precludes any generalized linear free energy relationship.

$$\Delta G^{\ddagger}_{AXB} = \tfrac{1}{2}(\Delta G^{\ddagger}_{AXA} + \Delta G^{\ddagger}_{BXB}) + \tfrac{1}{2}\Delta G^{\circ}_{AXB}\left(1 + \frac{\Delta G^{\circ}_{AXB}}{4\,(\Delta G^{\ddagger}_{AXA} + \Delta G^{\ddagger}_{BXB})}\right) \tag{3.50}$$

Other, more specific aspects of the inner-sphere mechanism have been addressed. One is the role of precursor and successor complex stability which is also related to the free-energy change for the electron transfer process within the reaction assembly. Consider the comparison of the rates in a typical inner-sphere electron transfer reaction, the reductions of $[Co(NH_3)_5F]^{2+}$ and $[Co(NH_3)_5I]^{2+}$ by $[Cr(H_2O)_6]^{2+}$ (eq. (3.51)). This combination follows the 'normal order' with regard to the effectiveness of the halide bridge in promoting electron transfer with $k_I > k_F$, the order also followed in outer-sphere reactions. The second-order rate constant is $k_X = K_{1X}k_{2X}$ where K_{1X} is the precursor association constant and k_{2X} is the electron transfer rate constant.

$$[Co(NH_3)_5F]^{2+} + [Cr(H_2O)_6]^{2+} \underset{k_1}{\overset{k_{-1}}{\rightleftharpoons}} [(NH_3)_5Co^{III}FCr^{II}(H_2O)_5]^{4+} \overset{k_2}{\rightarrow}$$

$$[(NH_3)_5Co^{II}FCr^{III}(H_2O)_5]^{4+} \quad (3.51)$$

$$\frac{k_I}{k_F} = \frac{K_{1I}k_{2I}}{K_{1F}k_{2F}} \quad (3.52)$$

A simple comparison of the rate constants, (eq. (3.52)), may reflect differences in K_{1X} or k_{2X} and is especially difficult to interpret because there are differences in the thermodynamic driving forces for the reactions. Haim [46] has attempted to clarify a parameter to compare the rates by defining a quantity, Q, which represents the relative stabilities of the electron transfer transition states with fluoride and iodide bridges (eq. (3.53)). The assumptions of transition state theory are used to obtain Q from the reaction rates, and the common point for comparison, which takes some account of the differences in driving force for the reactions, is the ion $[Co(NH_3)_5OH_2]^{3+}$, since values for K_{3X} in eq. (3.56) are 0.12, 0.35, 1.11, and 25 M^{-1} respectively for I^-, Br^-, Cl^-, and F^- [111].

$$\{[(NH_3)_5CoFCr(H_2O)_5]^{4+}\}^{\ddagger} + I^- \underset{}{\overset{Q_{IF}}{\rightleftharpoons}} \{[(NH_3)_5CoICr(H_2O)_5]^{4+}\}^{\ddagger} + F^-$$

$$(3.53)$$

$$Q_{IF} = \frac{K_{1I}k_{2I}K_{3I}}{K_{1F}k_{2F}K_{3F}} \quad (3.54)$$

$$[Co(NH_3)_5F]^{2+} + [Cr(H_2O)_6]^{2+} \overset{K_{1F}k_{2F}}{\rightleftharpoons} \{[(NH_3)_5CoFCr(H_2O)_5]^{4+}\}^{\ddagger} \quad (3.55)$$

$$[Co(NH_3)_5OH_2]^{3+} + F^- \overset{K_{3F}}{\rightleftharpoons} [Co(NH_3)_5F]^{2+} + H_2O \quad (3.56)$$

The value for Q_{IF} in this instance is 6.4×10^{-2}. Negative values for $\log Q$ are found where both oxidant and reductant are hard nonpolarizable Lewis acids but positive values are found when at least one of the reagents is a soft polarizable Lewis acid such as $[Co(CN)_5]^-$. It is apparent that at least two factors are required to explain the accumulated data: a term measuring the relative affinity of the halide for the transition state, and a measure of the electron permeability through the bridge.

This latter quantity is emphasized in an alternative approach. It has been found useful to define the inner-sphere advantage for a reaction by comparing the outer-sphere self-exchange process (eq. (3.57)) with an inner-sphere pseudo-self exchange (eq. (3.58)) [112–114]. For both these reactions $\Delta G = 0$ and the inner-sphere advantage is given by $\chi = (k_{is}/k_{os})$.

$$[Cr(H_2O)_6]^{3+} + [*Cr(H_2O)_6]^{2+} \rightleftharpoons [Cr(H_2O)_6]^{2+} + [*Cr(H_2O)_6]^{3+} \quad (3.57)$$

$$[Cr(H_2O)_5Cl]^{2+} + [*Cr(H_2O)_6]^{2+} \rightleftharpoons [Cr(H_2O)_6]^{2+} + [*Cr(H_2O)_5Cl]^{2+} \quad (3.58)$$

Data for chromium give $\chi \geq 10^6$ whereas for the corresponding iron case, $\chi = 10$, revealing that inner-sphere reactivity is greatly dependent on the electronic structure of the donor and acceptor. The amount of data of this sort which is available is not large and rates for cross-reactions such as those in eqs (3.59) and (3.60) can be incorporated by correcting rates for outer-sphere and corresponding inner-sphere reactions [115] to $\Delta G = 0$ with a Marcus-type relationship (eq. (3.61)). Such corrections can be justified at least for small driving force differences since linear free energy relationships of this sort are found between comparable inner-sphere processes.

$$[Co(H_2O)_6]^{3+} + [Ni([14]aneN_4)]^{2+} \rightleftharpoons [Co(H_2O)_6]^{2+} + [Ni([14]aneN_4)]^{3+}$$
$$(3.59)$$

$$[Co(H_2O)_5Cl]^{2+} + [Ni([14]aneN_4)]^{2+} \rightleftharpoons [Co(H_2O)_6]^{2+}$$
$$+ [Ni([14]aneN_4)Cl]^{2+} \quad (3.60)$$

$$k_{AB} \approx \sqrt{(k_{AA} k_{BB} K_{AB})} \quad (3.61)$$

This approximation give $(k_{is}/k_{os})_{\Delta G \to 0}$ which may then be used for comparison purposes. Values of $(k_{is}/k_{os})_{\Delta G \to 0}$ are presented in Table 3.15 for a number of reactions. There is a strong dependence on the nature of the bridging group with $\chi_{N_3^-} > \chi_{Cl^-} > \chi_{CH_3}$ which parallels the homolytic bond dissociation energy [116]. In addition, for a given electronic configuration of the reactants, the inner-sphere advantage falls within fairly narrow limits which reveal the effects of electronic structure on reactivity.

146 The inner-sphere mechanism [Ch. 3

Table 3.15. Inner-sphere advantage for reactions with chloride bridges

Reactants	Orbital symmetry donor/acceptor	$(k_{is}/k_{os})_{\Delta G \to 0}$	Reactants	Orbital symmetry donor/acceptor	$(k_{is}/k_{os})_{\Delta G \to 0}$
V(III)/V(II)	π/π	≈ 10	Co(III)/Ni(II)	σ/σ	10^3–10^6
V(III)/Cr(II)	π/σ	≈ 10^4	Ni(III)/Co(II)	σ/σ	10^3–10^4
Cr(III)/Cr(II)	σ/σ	≈ 10^7	Co(III)/Cu(II)	δ/σ	10
Fe(III)/Fe(II)	π/π	≈ 10	Ni(III)/Ni(II)	σ/σ	10–10^3
Co(III)/Fe(II)	π/σ	≥ 10^3	Cu(III)/Ni(II)	σ/δ	≈ 1
Co(III)/Co(II)	σ/σ	10^5–10^7			

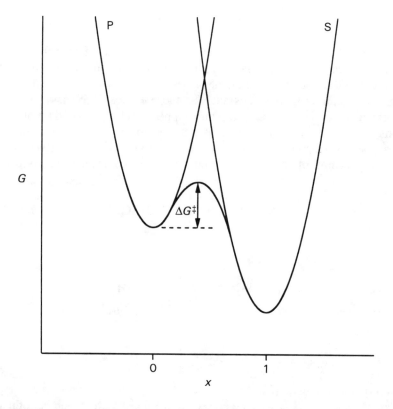

Fig. 3.2. Reaction coordinate diagram for inner-sphere electron transfer. The barrier for activation is lowered relative to the barrier for outer-sphere reaction by formation of the three center inner-sphere bridge which represents a strong interaction between the reactants. The potential energy function is related to the homolytic bond dissociation energy $\{A^{ox}\text{-}X,B^{red}\} \to \{A^{red},X,B^{red}\}$ which represents the activation process at the limit of weak interaction, where the two parabolas intersect.

A simple picture of the reaction is presented in Fig. 3.2. In this case the potential energy function is represented by homolytic bond dissociation in the precursor, P, and the successor, S, complexes since this corresponds to the major structural rearrangement during the course of the electron transfer. The energy of the transition state is determined by consideration of the stability of the bridged intermediate, dominated by a three-center bonding arrangement and involving d_{z^2} orbitals on the two metal centers and an appropriate orbital from the ligand radical fragment. To a first approximation, where the donor and acceptor orbitals on the metal centers do not both have σ symmetry with respect to this three-center bonding arrangement, poor orbital coupling results and there is little inner-sphere advantage [117, 118]. Thus where the donor or acceptor orbitals have π (d_{xy}, d_{xz}, d_{yz}), or δ ($d_{x^2-y^2}$) symmetry, the inner-sphere advantage is small. Where the symmetry is σ, there is a correlation with the electron occupation of the three-center bonding scheme with a decrease in the rate advantage corresponding to filling of the antibonding orbital.

3.9 ELECTRON TRANSFER THROUGH ORGANIC STRUCTURAL UNITS

The realization that electron transfer takes place through bridging units which are composed of several atoms opens a prospect for study of remote inner-sphere electron transfer through various organic bridges. Early studies provided results which implied that these bridges would have interesting properties [98, 119–121], but definitive proof that electron transfer takes place through an organic bridge was first obtained [122] in the reduction of the isonicotinamide complex (eq. (3.62)), by $[Cr(H_2O)_6]^{2+}$ where the identified kinetic product is the carbonyl-bound species (eq. (3.63)). Under conditions of excess $[Cr(H_2O)_6]^{2+}$, the initial product subsequently reacts, also by a remote inner-sphere electron transfer mechanism, to an equilibrium mixture with the pyridine bound complex (eq. (3.64)) prior to complete hydrolysis.

$[(NH_3)_5CoN\bigcirc-CONH_2]^{3+}$ (3.62)

$[HN\bigcirc\overset{O-Cr(H_2O)_5}{\underset{NH_2}{}}]^{4+}$ (3.63)

$[(H_2O)_5CrN\bigcirc-CONH_2]^{3+}$ (3.64)

It is deduced that the carbonyl-bound species (eq. (3.63)) is the initial product of the electron transfer reaction and not a rearrangement product, and that it reflects the transition state for this process (eq. (3.65)).

$[(NH_3)_5CoN\bigcirc\overset{O-Cr(H_2O)_5}{\underset{NH_2}{}}]^{5+\ddagger}$ (3.65)

Interestingly, in the reduction of the amide-bound complex (eq. (3.66)), attack at the adjacent carbonyl oxygen is preferred over remote attack at the pyridine nitrogen, and electron transfer through the organic bridge is not involved. The rate constant, 7.8×10^{-2} M^{-1} s^{-1} (25.0°C, $\mu = 1.0$ M), reflects this mechanistic difference [123].

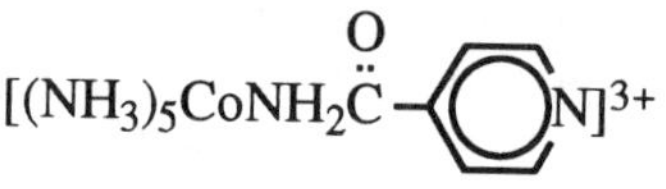

$$[(NH_3)_5CoNH_2\overset{\overset{\displaystyle O}{\|}}{C}-C_5H_4N]^{3+} \tag{3.66}$$

The results of studies of a number of reactions which are thought to involve remote attack through organic units are presented in Table 3.16 [124]. In many instances, the immediate reaction product is subject to hydrolysis and cannot be fully characterized. For example, in the reduction of the species in eq. (3.67), hydrolysis of the immediate product (eq. (3.68)) is rapid (24 s^{-1}) [125]. In such cases, an inner-sphere remote mechanism can be inferred from rate comparisons with competing outer-sphere and adjacent inner-sphere pathways since the latter leads to different products. In some instances with symmetric ligands such as fumarate, (eq. (3.69)), it is not possible to deduce from product analysis whether remote or adjacent attack takes place, and more convoluted arguments must be considered [126].

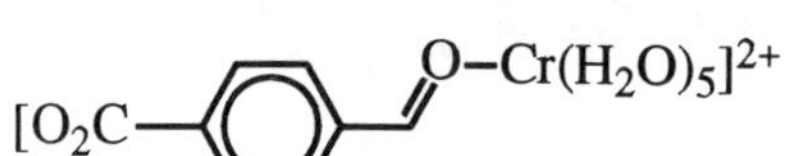

$$[(NH_3)_5CoO_2C-C_6H_4-CHO]^{3+} \tag{3.67}$$

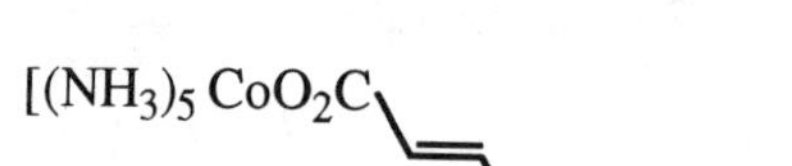

$$[O_2C-C_6H_4-C(O)O-Cr(H_2O)_5]^{2+} \tag{3.68}$$

$$[(NH_3)_5CoO_2C-CH=CH-CO_2]^+ \tag{3.69}$$

Table 3.16. Rate constants and activation parameters for reduction by [Cr(H$_2$O)$_6$]$^{2+}$ through organic structural units at 25.0 °C

Oxidant	μ (M)	k (M^{-1} s^{-1})	$\Delta H^\ddagger$ (kJ mol^{-1})	$\Delta S^\ddagger$ (J K^{-1} mol^{-1})	Ref.
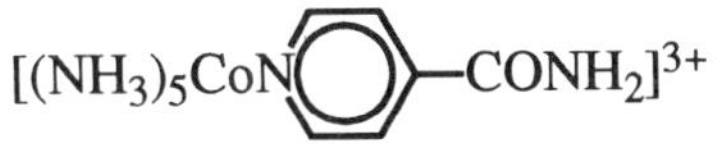 [(NH$_3$)$_5$CoN(C$_6$H$_4$)-CONH$_2$]$^{3+}$ (para)	1.0	17.4	16	-167	122
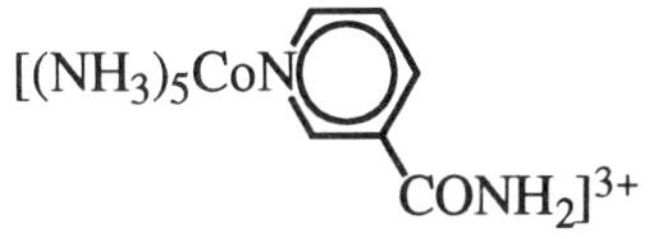 [(NH$_3$)$_5$CoN(C$_6$H$_4$)CONH$_2$]$^{3+}$ (meta)	1.0	0.003	42	-130	122
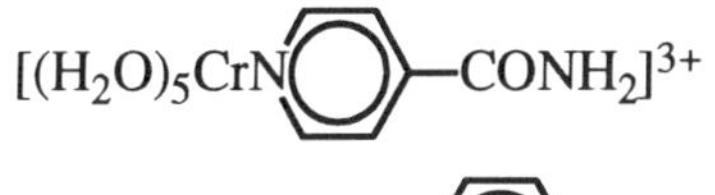 [(H$_2$O)$_5$CrN(C$_6$H$_4$)-CONH$_2$]$^{3+}$	1.0	1.8	24	-159	122
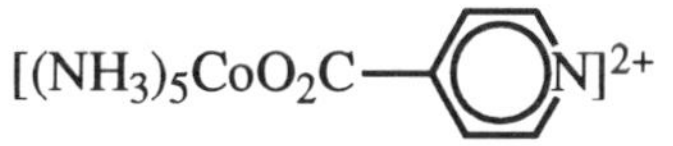 [(NH$_3$)$_5$CoO$_2$C-(C$_6$H$_4$)N]$^{2+}$	3.0	1.5 × 10^3			120
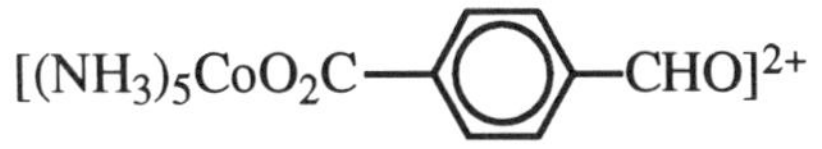 [(NH$_3$)$_5$CoO$_2$C-(C$_6$H$_4$)-CHO]$^{2+}$	1.1	53	11	-176	125
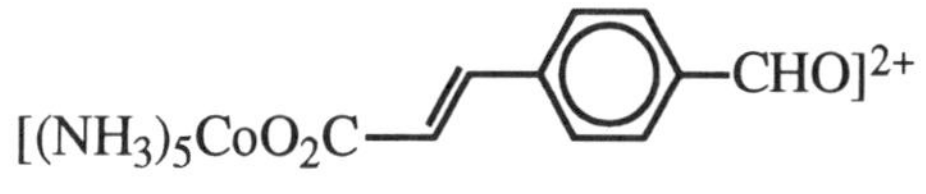 [(NH$_3$)$_5$CoO$_2$C-CH=CH-(C$_6$H$_4$)-CHO]$^{2+}$	1.2	2.6 × 10^2			127
 [(NH$_3$)$_5$CoN(C$_6$H$_4$)-CN]$^{3+}$	1.0	1.24 × 10^2	0.8	-201	124
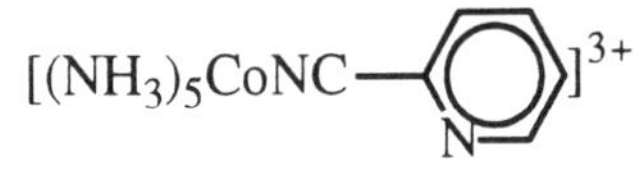 [(NH$_3$)$_5$CoNC-(C$_5$H$_4$N)]$^{3+}$	1.0	≥ 1.6 × 10^4			124
 [(NH$_3$)$_5$RuN(C$_6$H$_4$)-CONH$_2$]$^{3+}$ (para)	0.1	3.92 × 10^5	0	-192	128
[(NH$_3$)$_5$RuN(C$_6$H$_4$)CONH$_2$]$^{3+}$ (meta)	1.0	3.92 × 10^5 2.30 × 10^4	0 4.6	-192 -146	128 128
 [(NC)$_5$CoN(C$_6$H$_4$)-CN]$^{2-}$	1.0	28			129
 [(NC)$_5$CoN(C$_6$H$_4$)N]$^{2-}$	1.0	1.5 × 10^3			129
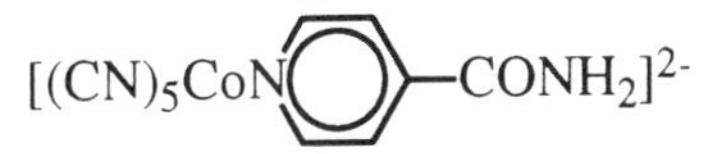 [(CN)$_5$CoN(C$_6$H$_4$)-CONH$_2$]$^{2-}$	1.0	19			129

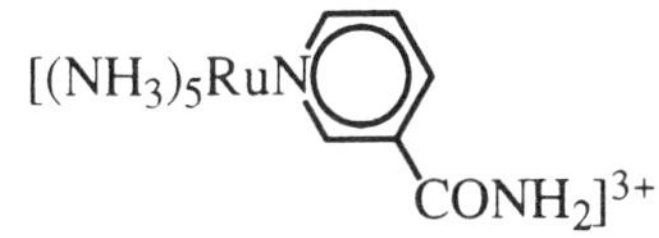

150 The inner-sphere mechanism [Ch. 3]

The remote pathway is detected only where the competing outer-sphere pathways are sufficiently slow. In the reduction of the nicotinamide complex (eq. (3.70)), product analysis reveals that only 71% of the reaction proceeds by the inner-sphere pathway. The remainder of the complex is reduced by outer-sphere reduction with a second-order rate constant 1.4×10^{-2} M^{-1} s^{-1} giving $[Cr(H_2O)_6]^{3+}$ as the initial product.

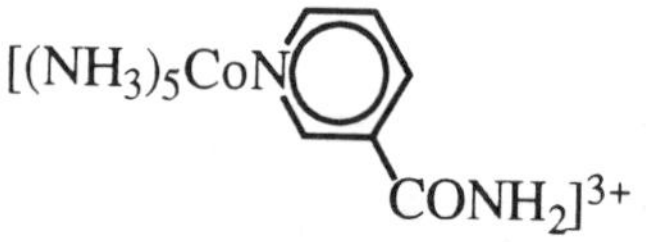

$$[(NH_3)_5CoN\text{–}C_5H_4\text{–}CONH_2]^{3+} \quad (3.70)$$

By contrast, in the isonicotinamide case, the outer-sphere pathway is not competitive. This difference in reactivity lies with the pattern of conjugation in the pyridine ring (eq. (3.71)). The presence of low-lying orbitals capable of accepting an electron is evidenced by the reduction potentials of nicotinamide, $E° = -0.83$ V, and isonicotinamide, $E° = -0.41$ V in 1M HClO$_4$ at 25.0°C [130].

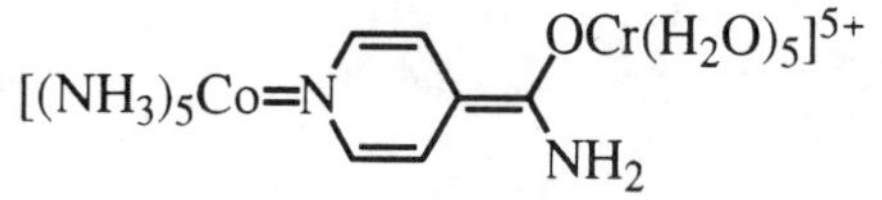

$$[(NH_3)_5Co\text{=}N\text{–}C_5H_4\text{–}C(OCr(H_2O)_5)\text{=}NH_2]^{5+} \quad (3.71)$$

Linear free energy relationships such as those developed by Gould [99, 100, 131] have been important in establishing the importance of the outer-sphere pathway in a number of reactions, particularly those with reductants such as $[V(H_2O)_6]^{2+}$ and $[Eu_{aq}]^{2+}$, but also in reactions with $[Cr(H_2O)_6]^{2+}$ as reductant. For example, in the reduction of the complex in eq. (3.72), the rate calculated by eq. (2.54) for the outer-sphere pathway is 2.7×10^{-2} M^{-1} s^{-1}, much smaller than the observed rate constant, 4.7×10^4 M^{-1} s^{-1}. The ligand shows no groups suitable for an inner sphere adjacent mechanism, and hence a remote inner-sphere mechanism is deduced.

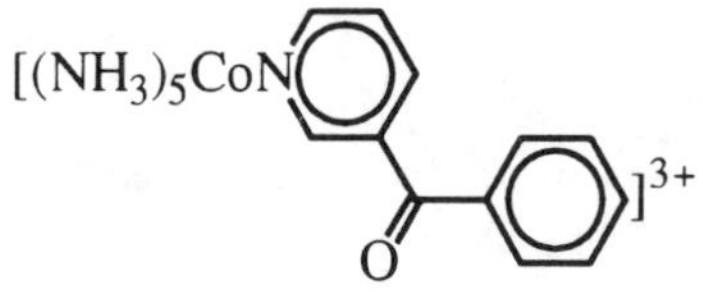

$$[(NH_3)_5CoN\text{–}C_5H_4\text{–}CO\text{–}C_6H_5]^{3+} \quad (3.72)$$

Some of the best examples of this mechanism involve pyridine ligands since the coordinated pyridine has no site available for a competitive adjacent attack. In some cases, however, the ligands are carboxylate-bound and competition from carbonyl attack is possible. Unusual acid catalysis is observed in reactions where the ligand is carboxylate-bound [119, 125]. This is not fully understood but a plausible explanation is that protonation of the carbonyl oxygen atom increases the overlap (conjugation) between the ligand and the metal center (eq. (3.73)).

$$[(NH_3)_5CoO\underset{}{\overset{HO}{>\!\!=\!\!\bigcirc\!\!=\!\!<}}CHO]^{3+} \tag{3.73}$$

The rate constants for the reductions provide an important key to the mechanism of electron transfer in these systems. In the reactions with simple inorganic bridging groups and with some organic groups by adjacent routes, the rate constants show a very strong dependence on the nature of the oxidant metal and its ligand environment. Rate variations from 2.6×10^6 M^{-1} s^{-1} to approximately 90 M^{-1} s^{-1} are found for the $[Cr(H_2O)_6]^{2+}$ reductions of $[Co(NH_3)_5Cl]^{2+}$ and $[Cr(H_2O)_5Cl]^{2+}$ respectively, a factor of approximately 10^5, which is typical of reactions taking place by so-called resonance transfer, outlined in section 3.8. By contrast, in the corresponding reductions by remote attack on the isonicotinamide complexes in eqs (3.62) and (3.64), the rates are very similar (Table 3.16), varying by only a factor of ten [122]. This has led to the proposal that the electron transfer takes place by a different mechanism involving rate-limiting reduction of the coordinated ligand, and generally known as the chemical mechanism. Further evidence that the metal center is less directly involved in the rate-limiting step comes from the isotope effect on the rate [132]. Comparisons of the rates of reduction of $[Co(NH_3)_5Cl]^{2+}$ and its deuterated counterpart $[Co(ND_3)_5Cl]^{2+}$ by $[Cr(H_2O)_6]^{2+}$ reveal an isotope effect $k_H/k_D \approx 1.5$, the result of changes in the N—H bonds on reduction of the cobalt(III) center. The isotope effect in the case of the isonicotinamide complex (eq. (3.62)), and its ammine-deuterated counterpart is much smaller (1.1), since there is less involvement of the cobalt reduction in the rate-limiting step.

Not all complexes which react by a remote outer-sphere mechanism through an organic bridge show evidence for the involvement of a coordinated radical. Reduction of $[Ru(NH_3)_5(isonicotinamide)]^{3+}$ (eq. (3.74)), by $[Cr(H_2O)_6]^{2+}$ yields the chromium-bound successor complex expected for a remote attack [128]. However, the reaction is more rapid than reduction of eq. (3.62) or eq. (3.64), which appears to rule out rate-limiting reduction of the coordinated ligand. In this case electron transfer is direct to the metal center through the ligand by resonance transfer. The difference between the ruthenium and the cobalt reactions can be rationalized by consideration of the symmetry of the ligand orbitals and the acceptor orbitals on the metal center [133]. In the case of $[Co(NH_3)_5(isonicotinamide)]^{3+}$ and $[Cr(OH_2)_5(isonicotinamide)]^{3+}$, the acceptor orbitals on the metal are of σ^* symmetry and result in poor overlap with the π symmetry orbitals on the ligand center. Consequently, the electron is trapped on the ligand orbitals and reduced ligand is an intermediate in the reaction. In the case of $[Ru(NH_3)_5(isonicotinamide)]^{3+}$, the acceptor orbital on the metal is of π symmetry and while the electron transfer may take place through the ligand, reduced ligand is not formed.

$[(NH_3)_5RuN\bigcirc-CONH_2]^{3+}$ (3.74)

The evidence points to two quite distinct mechanisms for remote attack through organic bridging groups. Regardless of the mechanism, all of the bridging groups have a chromophore with an electron pair of sufficient basicity available for coordination to the reductant. In addition, the ligands are unsaturated, and the remote electron pair is in conjugation with the metal center. The symmetry of the acceptor orbital on the metal center dictates whether electron transfer will take place by resonance transfer, smoothly from the donor to the acceptor, or by the chemical mechanism involving the intermediacy of the reduced bridging group. Where the symmetry of the metal acceptor matches the π symmetry of the ligand orbitals, resonance transfer through the organic bridge is permitted. Where there is a mismatch in symmetry, the inner-sphere pathway is disadvantaged and outer-sphere pathways become important. In instances where there is a suitable orbital on the ligand which is of sufficiently low energy that the ligand chromophore is reducible, inner-sphere electron transfer by the chemical mechanism is competitive.

In most instances a chemical mechanism has been implicated by rate comparisons. However, direct evidence for the involvement of ligand radical species in these reactions has been obtained in the reduction by $[Cr(H_2O)_6]^{2+}$ of the pyrazine oxidant (eq. (3.75)) [134].

$[\text{pyrazine-}OCo^{III}(NH_3)_5]^{2+}$ (3.75)

Two intermediates are detected in the reaction. The first of these, formed within the time of mixing of the reactants ($k > 10^6$ M^{-1} s^{-1}) has an absorption maximum at 640 nm and flow EPR studies reveal a signal characteristic of an organic radical with $g = 2.003$ and hyperfine coupling to two nitrogen atoms (Fig. 3.3) [135]. The structure proposed is eq. (3.76). Subsequent transfer of the electron to the metal center with a rate constant of 450 s^{-1} yields the second intermediate, a chelated inner-sphere product (eq. (3.77)).

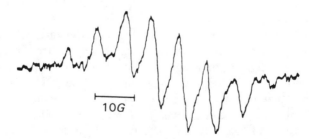

Fig. 3.3. Epr spectrum of the complex in eq. (3.76). The nine-line spectrum indicates coupling to two inequivalent nitrogen atoms, $g = 2.003$ from Spiecker, H., Wieghardt, K. *Inorg. Chem.* 1977, **16**, 1290 with permission.

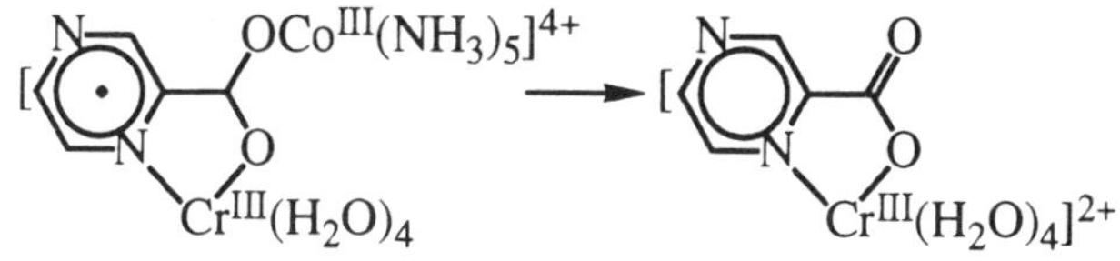

Electron transfer through organic structural units

$$[\text{pyridine-O-Co}^{III}(NH_3)_5]^{4+} \text{ with } Cr^{III}(H_2O)_4 \longrightarrow [\text{product}] \quad Cr^{III}(H_2O)_4]^{2+} \quad (3.76) \text{ and } (3.77)$$

Further light on electron transfer from coordinated radicals to a metal center has been obtained from pulse radiolysis studies on eq. (3.78) which combine the readily reducible nitrobenzene chromophore with the sluggish reactivity of a cobalt(III) center [136].

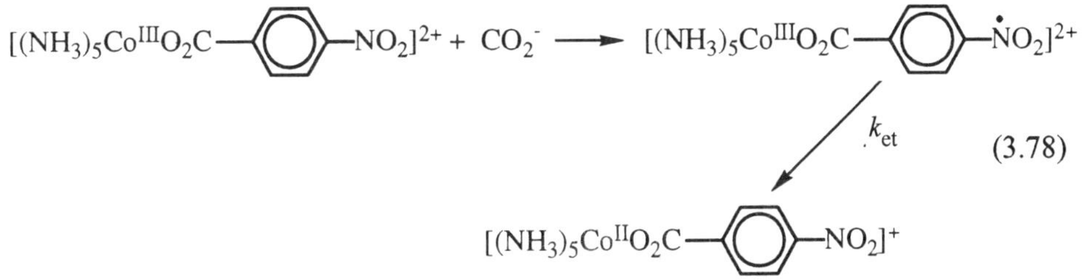

$$[(NH_3)_5Co^{III}O_2C\text{-}C_6H_4\text{-}NO_2]^{2+} + CO_2^- \longrightarrow [(NH_3)_5Co^{III}O_2C\text{-}C_6H_4\text{-}\dot{N}O_2]^{2+}$$

$$\downarrow k_{et}$$

$$[(NH_3)_5Co^{II}O_2C\text{-}C_6H_4\text{-}NO_2]^+ \quad (3.78)$$

The intermediate has an absorption spectrum characteristic of a nitrobenzene radical and the decay by intramolecular electron transfer to cobalt(III) is slow, indicating the poor orbital overlap between the π orbital on the ligand and the acceptor orbital, σ^*, on the metal center (Table 3.17). Examination of the position of the nitro-group on the aromatic ring suggests that the electron transfer rate is determined by the electron spin density adjacent to the bound carboxylate group. A similar rate is found with intramolecular electron transfer in the binuclear complex (eq. (3.79)) [137].

eq. (3.79)

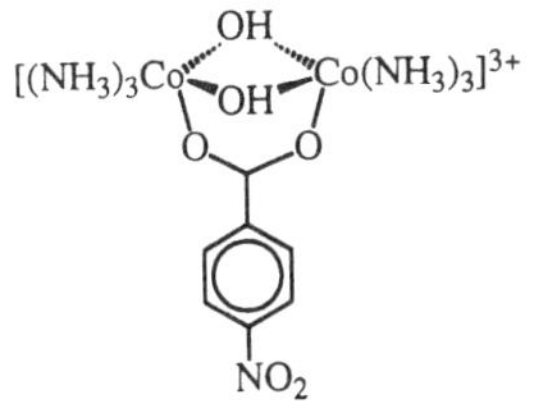

$$[(NH_3)_3Co(OH)_2Co(NH_3)_3(O_2C\text{-}C_6H_4\text{-}NO_2)]^{3+} \quad 3.3 \times 10^3$$

Table 3.17. Intramolecular electron transfer rate constants form coordinated ligand radicals at 25 °C

Complex	k (s^{-1})	$\Delta H^{\ddagger}$ (kJ mol^{-1})
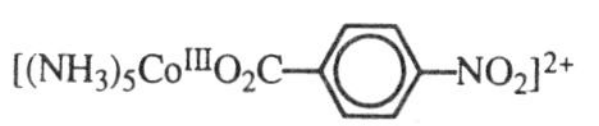 $[(NH_3)_5Co^{III}O_2C\text{-}C_6H_4\text{-}NO_2]^{2+}$ (para)	2.6×10^3	≈ 40
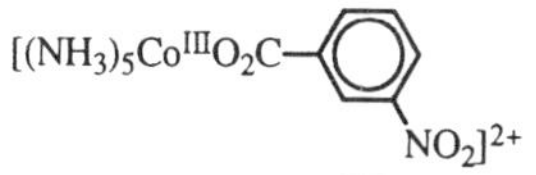 $[(NH_3)_5Co^{III}O_2C\text{-}C_6H_4\text{-}NO_2]^{2+}$ (meta)	1.5×10^2	
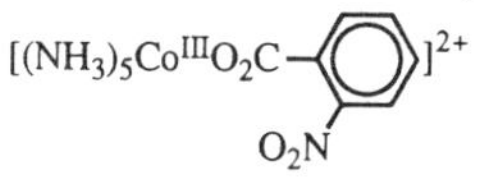 $[(NH_3)_5Co^{III}O_2C\text{-}C_6H_4\text{-}NO_2]^{2+}$ (ortho, O_2N)	4.0×10^5	

The key component of the chemical mechanism is the ready reducibility of the ligands and studies of nicotinic and isonicotinic acid reduction have also been carried out [138, 139]. Although the net reactions involve the transfer of two electrons to the ligands, the mechanisms of reduction of the free ligands by species such as $[Eu_{aq}]^{2+}$, $[Cr(H_2O)_6]^{2+}$ and $[V(H_2O)_6]^{2+}$ have been shown to involve single electron transfer. Interestingly, whereas isonicotinic acid and picolinic acids are readily reduced, the nicotinic acid is much less reactive. Free radicals can also participate in the electron transfer process. Reduction of $[Co(en)_3]^{3+}$ by $[Eu_{aq}]^{2+}$ is slow but has been found to be catalyzed by the addition of isonicotinamide. In fact by appropriate choice of radicals, electron chains can be constructed [140, 141]. Analysis of the catalytic reaction provides a useful method for the determination of rate constants for reduction of the ligand by $[Eu_{aq}]^{2+}$, and selected values are presented in Table 3.18. Note that for the third and fourth entries in the table, the rate of reduction of coordinated isonicotinamide radical is comparable to that of the free isonicotinamide species [142].

Table 3.18. Rate constants for production of radicals and coordinated radicals by $[Eu_{aq}]^{2+}$ at 25.0 °C

Ligand	k (M^{-1} s^{-1})	Ref.
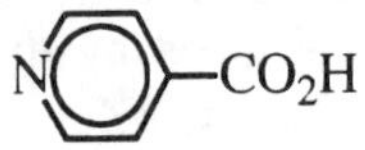 N⌬-CO$_2$H	14	143
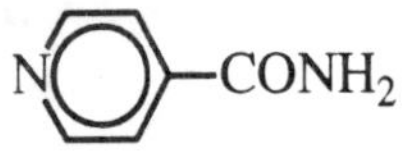 N⌬-CONH$_2$	2.0	143
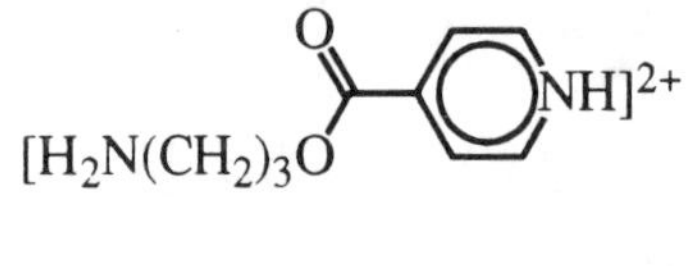 [H$_2$N(CH$_2$)$_3$O-C(O)-⌬-NH]$^{2+}$	3.7	130
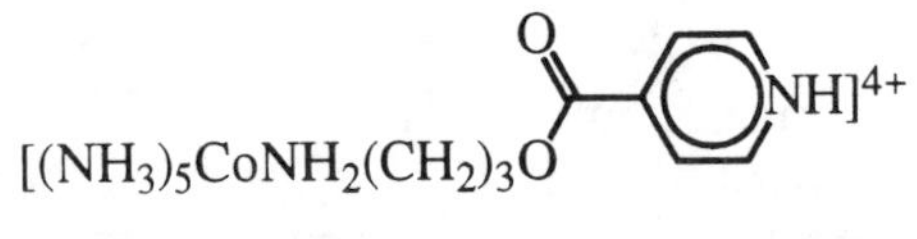 [(NH$_3$)$_5$CoNH$_2$(CH$_2$)$_3$O-C(O)-⌬-NH]$^{4+}$	4.4	130

The rate of reduction of eq. (3.80) by $[Cr(H_2O)_6]^{2+}$ is 6.6×10^2 M^{-1} s^{-1} at 25.0°C and 1.2 M ionic strength. The product is the carboxylate-bound chromium(III) but the rate is much faster than expected for adjacent attack. Electron transfer through the organic bridge is precluded by the methylene group which interrupts the conjugation; however, stabilization of the precursor by charge transfer (eq. (3.81)) is possible.

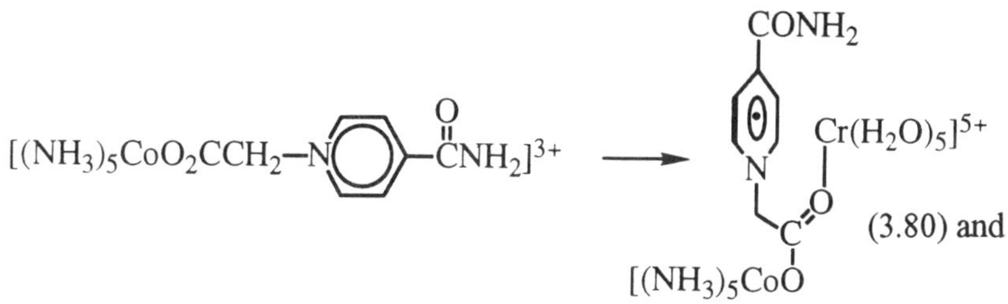

$$[(NH_3)_5CoO_2CCH_2-N\underset{}{\bigcirc}-\overset{O}{\overset{\|}{C}}NH_2]^{3+} \longrightarrow \text{[Cr(H}_2\text{O)}_5\text{]}^{5+} \text{ complex}$$

(3.80) and (3.81)

This work has been extended to include the binuclear cobalt(III) oxidant (eq. (3.82)), where adjacent attack is not possible. Again rate enhancement as a result of outer-sphere intramolecular electron transfer between cobalt(III) and the coordinated radical occurs in the isonicotinamide derivative, $k = 86$ M^{-1} s^{-1} (25.0 °C, $\mu = 1.0$ M), but not in the nicotinamide derivative, $k = 1.03 \times 10^{-1}$ M^{-1} s^{-1}, where ligand reduction is more difficult [144].

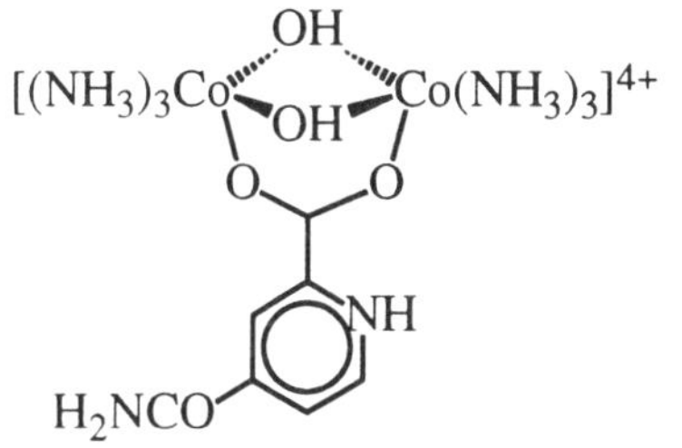

(3.82)

3.10 INTRAMOLECULAR ELECTRON TRANSFER

To this point, evidence for precursor complexes in inner-sphere electron transfer reactions has come indirectly from kinetic and activation parameters. Direct observation of precursor species and the ability to measure rates of intramolecular electron transfer has not been evident and most transients which are detected are either successor complexes or 'dead-end' species. The exception is the reduction of [Co(NH$_3$)$_5$O$_2$CCH$_2$N(CH$_2$CO$_2$)$_2$] by [Fe(H$_2$O)$_6$]$^{2+}$ where the precursor is stabilized by chelation and electron transfer is sluggish. Even in this case the mechanism of electron transfer is not clear and a bridged outer-sphere pathway is possible. Studies designed to optimize the detection of inner-sphere precursor complexes with [Co(CN)$_5$]$^{3-}$ and [Fe(CN)$_5$(H$_2$O)]$^{3-}$ as reductants with simple oxidants have also proved disappointing. However, this latter approach has achieved considerable success when combined with organic bridging groups and intermediates are formed in stoichiometric quantities.

When [Fe(CN)$_5$(H$_2$O)]$^{3-}$ reacts with [Co(NH$_3$)$_4$(4,4'-bpy)]$^{3+}$ (eq. (3.83)), a complex electron transfer reaction takes place [145]. There is spectroscopic evidence for the formation of a precursor complex (eq. (3.84)), with a metal to ligand charge transfer band at 505 nm ($\varepsilon_{505} = 6 \times 10^3$ M^{-1} cm^{-1}), and analysis of the products shows formation of the expected inner-sphere product (eq. (3.85)). The rates of formation,

$k_1 = 5.5 \times 10^3$ M^{-1} s^{-1} and dissociation, $k_{-1} = 4.5 \times 10^{-3}$ s^{-1}, yield a precursor complex formation constant of 1.2×10^6.

$$[(NH_3)_5CoN\underset{}{\bigcirc}\text{—}\underset{}{\bigcirc}N]^{3+} + [Fe(CN)_5(H_2O)]^{3-} \underset{k_{-1}}{\overset{k_1}{\rightleftharpoons}} \quad (3.83)$$

$$[(NH_3)_5CoN\underset{}{\bigcirc}\text{—}\underset{}{\bigcirc}Fe(CN)_5] \quad (3.84)$$

$$[(CN)_5FeN\underset{}{\bigcirc}\text{—}\underset{}{\bigcirc}N]^{2-} \quad (3.85)$$

The electron transfer kinetics are complicated by the dissociation reaction and by interference from the product, [Fe(CN)$_5$(4,4'-bpy)]$^{2-}$, which oxidizes the precursor complex. Addition of ascorbate to reduce this species and pyridine to form [Fe(CN)$_5$py]$^{3-}$ minimizes this problem. Studies of intramolecular electron transfer rates in inner-sphere reactions have now been reported for an extensive series of reagents, and the data are presented in Table 3.19.

The intramolecular electron transfer rates in complexes bridged by the pyrazine and bipyridine derivatives reveal some interesting trends. The rates are relatively slow, leading to the conclusion that they involve transfer through the organic bridge. The donor orbitals have the same π symmetry as the bridge orbitals, but the acceptor orbitals on the cobalt(III) have σ^* symmetry, resulting in the low rate. For those bridging ligands where the conjugated π-system is maintained, there is a reduction in rate as the length of the bridge increases. Further comment on this aspect will be reserved for Chapter 4, where intramolecular electron transfer is discussed in detail.

The effects of conjugation on the rates are of considerable interest. They can be correlated with metal–ligand charge transfer spectra of the intermediates. These spectra represent charge transfer from the [Fe(CN)$_5$]— to pyridine and are shifted to lower energy when the [Co(NH$_3$)$_5$]— group is in conjugation with the iron-bound pyridine ring, but no shift occurs when the pyridine rings are linked by —CH$_2$—, —CH$_2$CH$_2$—, and —CH$_2$CH$_2$CH$_2$—. While a single —CH$_2$— group interrupts the electron transfer, for —CH$_2$CH$_2$— the rate increases. This is most likely the result of a switch in mechanism from a through-bridge inner-sphere mechanism to bridged outer-sphere (eq. (3.86)) where the oppositely charged ends of the molecule are attracted electrostatically. Sufficient bridge flexibility is required for this to take place. In support of this assertion, a comparison can be drawn between the rate of intramolecular electron transfer in the inner-sphere precursor with intramolecular electron transfer in the outer-sphere ion pair (eq. (3.89)) which has a rate constant of 3.12×10^{-3} s^{-1} at 25.0°C and 0.077 M ionic strength [156].

Table 3.19. Rate and activation parameters for intramolecular electron transfer in inner-sphere systems

Complex	μ (M)	k (M^{-1} s^{-1})	$\Delta H^{\ddagger}$ (kJ mol^{-1})	$\Delta S^{\ddagger}$ (J K^{-1} mol^{-1})	$\Delta V^{\ddagger}$ (cm^3 mol^{-1})	Ref.
[(NH$_3$)$_5$CoN◯NFe(CN)$_5$]	0.10	2.6×10^{-3}				145
[(NH$_3$)$_5$CoO$_2$C─◯NFe(CN)$_5$]$^-$	0.10	1.75×10^{-4}				146
[(NH$_3$)$_5$CoO$_2$C─◯NFe(CN)$_5$]$^-$	0.10	$< 3 \times 10^{-5}$				146
[(NH$_3$)$_5$CoN◯NFe(CN)$_5$]	0.15	5.5×10^{-2}	103	75		147
[(NH$_3$)$_5$CoN◯NFe(CN)$_5$]	0.10	4.6×10^{-2}	130	65	38	148
[(NH$_3$)$_5$CoN◯NFe(CN)$_5$]	0.10	4.41×10^{-2}			35	149

(continues)

Table 3.19. (continued)

Complex	μ (M)	k (M^{-1} s^{-1})	$\Delta H^{\ddagger}$ (kJ mol^{-1})	$\Delta S^{\ddagger}$ (J K^{-1} mol^{-1})	$\Delta V^{\ddagger}$ (cm^3 mol^{-1})	Ref.
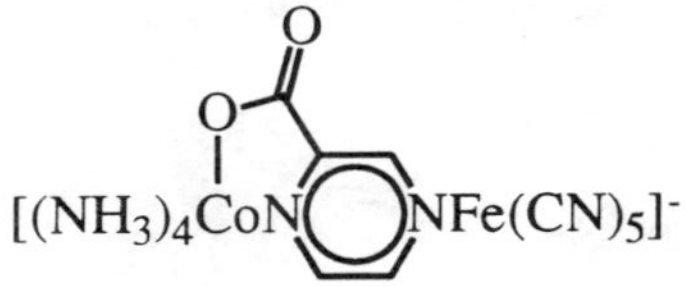 [(NH$_3$)$_4$CoN◯–CO–O NFe(CN)$_5$]$^-$	0.15	1.3×10^{-2}	95	40		147
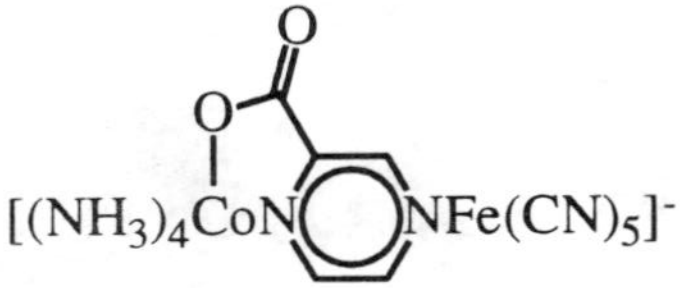 [(NH$_3$)$_4$CoN◯–CO–O NFe(CN)$_5$]$^-$	0.10	1.38×10^{-2}	83	2	27.3	149
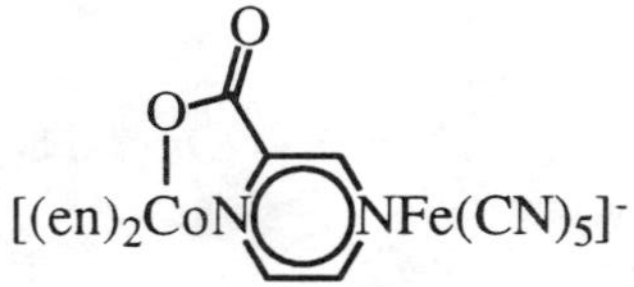 [(en)$_2$CoN◯–CO–O NFe(CN)$_5$]$^-$	—	1.04×10^{-4}			24	150
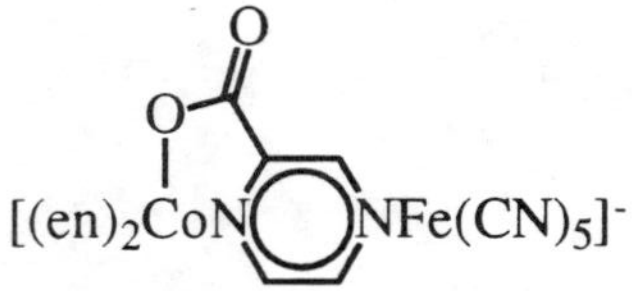 [(en)$_2$CoN◯–CO–O NFe(CN)$_5$]$^-$	0.10	1.27×10^{-3a}			28.4	149
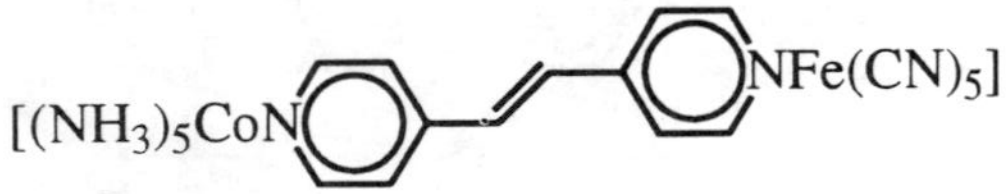 [(NH$_3$)$_5$CoN◯–CH=CH–◯NFe(CN)$_5$]	0.10	1.4×10^{-3}				151

(continues)

Table 3.19. (continued)

Complex	μ (M)	k ($M^{-1} s^{-1}$)	$\Delta H^{\ddagger}$ (kJ mol^{-1})	$\Delta S^{\ddagger}$ (J K^{-1} mol^{-1})	$\Delta V^{\ddagger}$ (cm^3 mol^{-1})	Ref.
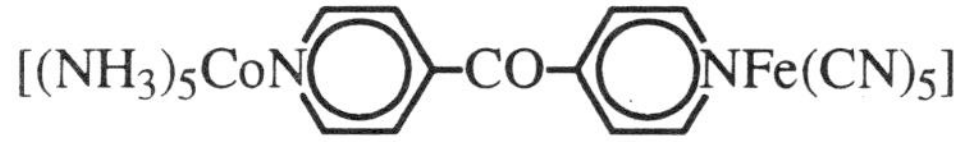 [(NH$_3$)$_5$CoN◯-CO-◯NFe(CN)$_5$]	0.1	$<0.1 \times 10^{-3}$				151
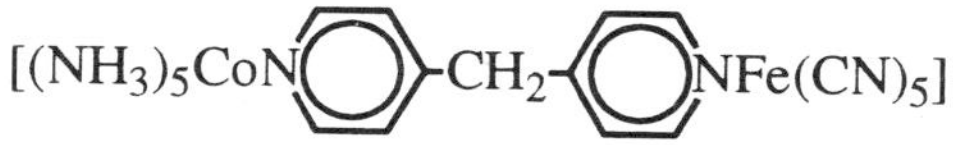 [(NH$_3$)$_5$CoN◯-CH$_2$-◯NFe(CN)$_5$]	0.1	$<0.6 \times 10^{-3}$				151
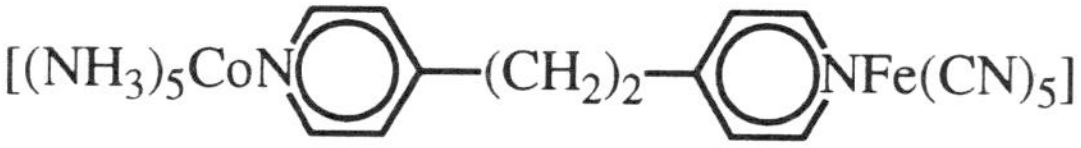 [(NH$_3$)$_5$CoN◯-(CH$_2$)$_2$-◯NFe(CN)$_5$]	0.1	2.0×10^{-3}				151
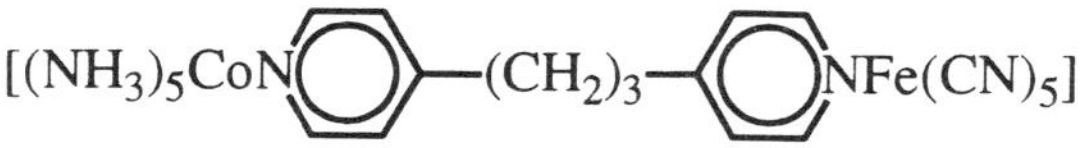 [(NH$_3$)$_5$CoN◯-(CH$_2$)$_3$-◯NFe(CN)$_5$]	0.1	4.8×10^{-3}				151
 [(NH$_3$)$_5$CoN◯NFe(CN)$_5$]$^-$	0.1	1.65×10^{-1}	80	10		152

(continues)

Table 3.19. (continued)

Complex	μ (M)	k (M^{-1} s^{-1})	$\Delta H^\ddagger$ (kJ mol^{-1})	$\Delta S^\ddagger$ (J K^{-1} mol^{-1})	$\Delta V^\ddagger$ (cm^3 mol^{-1})	Ref.
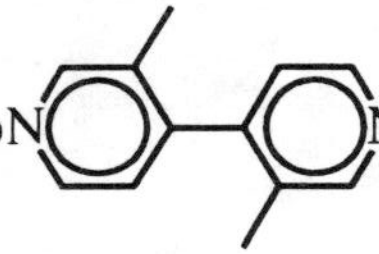 [(NH$_3$)$_5$CoN⟨⟩–⟨⟩NFe(CN)$_5$] (dimethylbiphenyl)	0.1	2.3×10^{-3}				153
 [(NH$_3$)$_5$CoN⟨⟩–C≡C–⟨⟩NFe(CN)$_5$]	0.1	1.7×10^{-3}				153
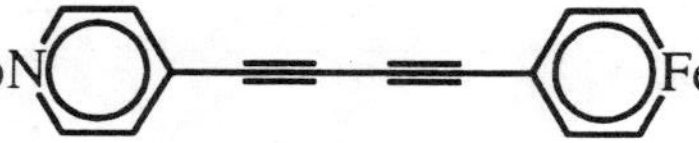 [(NH$_3$)$_5$CoN⟨⟩–C≡C–C≡C–⟨⟩NFe(CN)$_5$]	0.1	0.69×10^{-3}				153
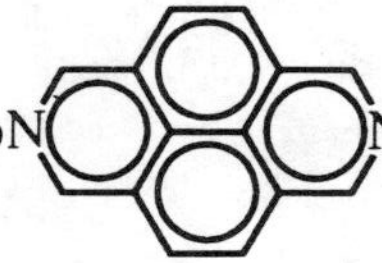 [(NH$_3$)$_5$CoN(pyrene)NFe(CN)$_5$]	0.1	4.2×10^{-3}				153
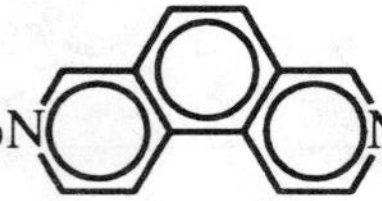 [(NH$_3$)$_5$CoN(phenanthrene)NFe(CN)$_5$]	0.1	9.3×10^{-3}				153

(continues)

Table 3.19. (continued)

Complex	μ (M)	k (M^{-1} s^{-1})	$\Delta H^\ddagger$ (kJ mol^{-1})	$\Delta S^\ddagger$ (J K^{-1} mol^{-1})	$\Delta V^\ddagger$ (cm^3 mol^{-1})	Ref.
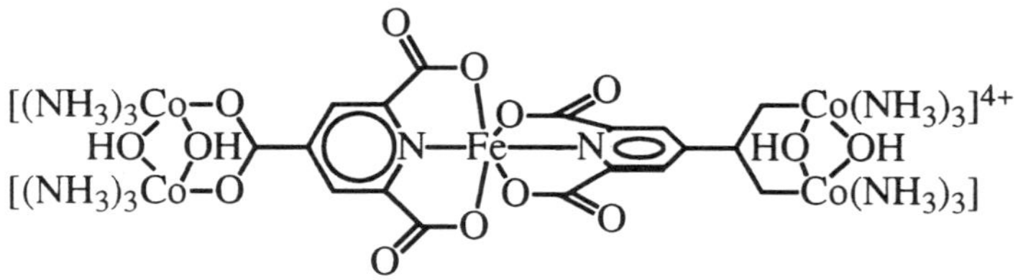 [(NH$_3$)$_3$Co(OH)(O)Fe(O)(OH)Co(NH$_3$)$_3$]$^{4+}$ bis-pyridine-dicarboxylate bridged	0.04	3.7×10^{-3}	100	46		154
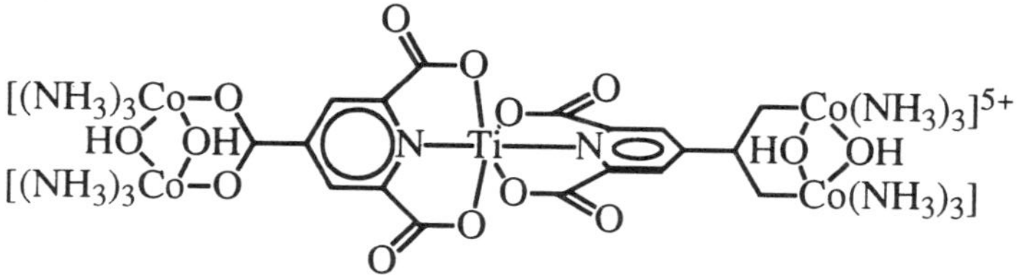 [(NH$_3$)$_3$Co(OH)(O)Tl(O)(OH)Co(NH$_3$)$_3$]$^{5+}$ bis-pyridine-dicarboxylate bridged	0.05	32	59	−25		155

[a] 45°C.

$[(NH_3)_5CoN\langle pyrazine \rangle$
$[(CN)_5FeN\langle pyrazine \rangle]$ (3.86)

$[(NH_3)_5CoN\langle\bigcirc\rangle\langle\bigcirc\rangle]^{3+}$, $[(CN)_5FeN\langle\bigcirc\rangle\langle\bigcirc\rangle N]^{3-}$ (3.87)

$\begin{matrix} [(NH_3)_3Co-O \\ HO\ OH \\ [(NH_3)_3Co-O \end{matrix}\rangle\langle N-Fe-N\rangle\langle\begin{matrix} -Co(NH_3)_3]^{4+} \\ HO\ OH \\ -Co(NH_3)_3] \end{matrix}$ (3.88)

$[(NH_3)_3Co\langle\begin{matrix}OH\\OH\end{matrix}\rangle Co(NH_3)_3]^{3+}$, $[Fe(dipic)_2]^{2-}$ (3.89)

Similar conclusions are drawn with the complex (3.88), where the rate can be compared with the rate of intramolecular electron transfer in the reaction in eq. (3.89), 3.71×10^{-3} s−1 at 25.0°C and 0.10 M ionic strength [155]. This recurring theme that outer-sphere electron transfer rates are comparable with the corresponding inner-sphere electron transfer rates leads to an assertion that both processes can be treated similarly from one point of view to theory except where there is strong coupling in the inner-sphere case. As with outer-sphere reactions, activation parameters, particularly $\Delta V\ddagger$, are consistent with a decrease in the amount of electrostricted water as the complexes rearrange prior to electron transfer [149].

The reduction of the pyrazine derivatives is photocatalyzed. Irradiation of the metal–ligand charge transfer band transfers the electron from the [Fe(CN)$_5$]— to the pyrazine ligand (eq. (3.90)), and interestingly $\Delta V\ddagger$ for the photocatalyzed electron transfer is much smaller, indicating that transfer of the electron to the cobalt(III) involves little change in electrostricted water.

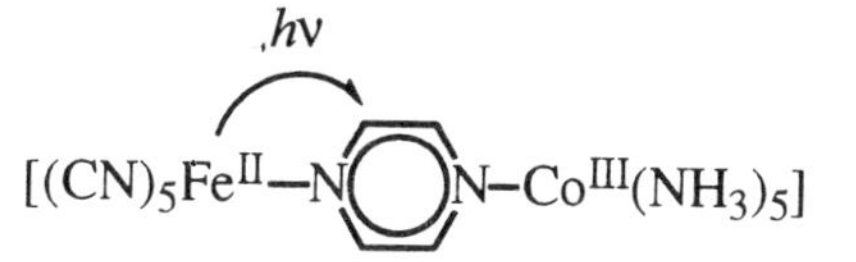

$$[(CN)_5Fe^{II}-N\underset{h\nu}{\frown}N-Co^{III}(NH_3)_5] \qquad (3.90)$$

In these studies of inner-sphere intramolecular electron transfer, there has been some ability to examine the effect of changing the electronic structure of the metal ions, but the scope of this is limited. Attempts to examine similar reactions with other metal ion reductants such as $[Ru(NH_3)_5OH_2]^{2+}$ fail because outer-sphere electron transfer is competitive with the rate of complex formation [157]. Fortunately there have been many clever methods devised to obtain data of this sort, and these, together with a more detailed look at the electron transfer process, form the subject of Chapter 4.

QUESTIONS

3.1 The rate law for the reduction of $[Co(NH_3)_5Cl]^{2+}$ by excess $[Fe_{solv}]^{2+}$ in DMF solution and 25.0 °C is of the form:

$$-\frac{d[[Co(NH_3)_5Cl]^{2+}]}{dt} = \frac{k[[Co(NH_3)_5Cl]^{2+}][[Fe_{solv}]^{2+}]}{1 + K[[Fe_{solv}]^{2+}]}$$

where $k = 1.87 \times 10^{-2}$ M^{-1} s^{-1} and $K = 16.1$ M^{-1}. Suggest a mechanism for the reaction. In other solvents, DMSO and H_2O, second-order behavior is found, and rate and activation parameters for all three solvents are presented in the table. Provide an explanation for these data.

Solvent	k (M^{-1} s^{-1})	$\Delta H^{\ddagger}$ (kJ mol^{-1})	$\Delta S^{\ddagger}$ (J K^{-1} mol^{-1})
DMF	1.87×10^{-2}	87	13
DMSO	9.7×10^{-3}	90	16
H_2O	1.6×10^{-3}	63	−96

(Beckham, K. R.; Watts, D. W. *Aust. J. Chem.* 1979, **32**, 1425–1431.)

3.2 The rate law for reduction of the iron(III) siderophore complexes, ferrichrome (Fc), by $[V(H_2O)_6]^{2+}$ is:

$$-\frac{d[Fc]}{dt} = k[[V(H_2O)_6]^{2+}][Fc]$$

where $k = 25$ M^{-1} s^{-1} at 25 °C and 1.0 M ionic strength. Activation parameters are $\Delta H^{\ddagger} = 16$ kJ mol^{-1}, $\Delta S^{\ddagger} = -167$ J K^{-1} mol^{-1}. Propose a detailed mechanism for this reaction, outlining the criteria which are used to assess inner-sphere and outer-sphere processes for $[V(H_2O)_6]^{2+}$.

The corresponding reaction with $[Cr(H_2O)_6]^{2+}$ has a more complex rate law:

$$-\frac{d[Fc]}{dt} = \frac{k_a[H^+] + k_b}{1 + k_c[[Cr(H_2O)_6]^{2+}]} [[Cr(H_2O)_6]^{2+}][Fc]$$

where $k_a = 5.5 \times 10^4$ M^{-2} s^{-1}, $k_b = 20.3$ M^{-1} s^{-1}, and $k_c = 188$ M^{-1}. Again propose a mechanism for the reaction and compare this with the reaction for $[V(H_2O)_6]^{2+}$. (Kazmi, S. A.; Shorter, A. L.; McArdle, J. V. *Inorg. Chem.* 1984, **23**, 4332–4341.)

3.3 The rate laws for reduction of $[Co(ox)_3]^{3-}$ and $[Co(pic)_3]$ by $[Fe(pic)_3]^-$ and $[Fe(pic)_2]$ are of the form:

$$-\frac{d[[Co(pic)_3]]}{dt} = \{k_1[[Fe(pic)_3]^-] + k_2[[Fe(pic)_2]]\}\,[[Co(pic)_3]]$$

Rate constants are presented in the table.

Reaction	k_1 (M^{-1} s^{-1})	k_2 (M^{-1} s^{-1})
$[Co(ox)_3]^{3-} + [Fe(pic)_3]^-$	3.4	
$[Co(pic)_3] + [Fe(pic)_3]^-$	3.5	
$[Co(ox)_3]^{3-} + [Fe(pic)_2]$		160
$[Co(pic)_3] + [Fe(pic)_2]$		200

Use these data and the known properties of the oxidants and reductants (Table 2.9 and Question 1.5) to argue on behalf of inner-sphere or outer-sphere mechanisms for the reactions.
(Lannon, A. M.; Lappin, A. G.; Segal, M. G. *J. Chem. Soc., Dalton Trans.* 1986, 619–624.)

3.4 The following data refer to heterogeneous rate constants for the reductive dissolution of Fe_2O_3 by a number of well-characterized metal ion complexes under a variety of conditions. Use the information to classify the preferred mechanism for reduction of Fe_2O_3 in terms of inner-sphere and outer-sphere pathways.

Reductant	Rel. Rate
$[Cr(H_2O)_6]^{2+}/ClO_4^-$	$< 10^{-4}$
$[Cr(H_2O)_6]^{2+}/HCl$	1
$[Cr(bpy)_3]^{2+}$	25
$[V(H_2O)_6]^{2+}/ClO_4^-$	50
$[V(pic)_3]^-$	> 100

(Segal, M. G.; Sellers, R. M. *J. Chem. Soc., Chem. Commun.* 1980, 991–993.)

3.5 Explain the following observations:
   (a) Rate constants for the $[Cr(H_2O)_6]^{2+}$ reductions of $[(NH_2)_5CoNCS]^{2+}$ and $[(NH_3)_5CoSCN]^{2+}$ are 19 M^{-1} s^{-1} and 8×10^4 M^{-1} s^{-1} respectively at 25°C.
   (b) The rate law for the reduction of $[(NH_3)_5CoO_2CCO_2H]^{2+}$ by $[Cr(H_2O)_6]^{2+}$ is of the form:

$$-\frac{d[[(NH_3)_5CoO_2CCO_2H]^{2+}]}{dt} = \{a + b\,[H^+]^{-1}\}[[Cr(H_2O)_6]^{2+}][[(NH_3)_5CoO_2CCO_2H]^{2+}]$$

3.6 The rate law for the partial aquation of $[Cr(histamine)(H_2O)_2(ox)]^+$ catalyzed by $[Cr(H_2O)_6]^{2+}$ gives $[Cr(H_2O)_4(ox)]^+$ as product and follows the rate law shown:

$$-\frac{d[[Cr(histamine)(H_2O)_2(ox)]^+]}{dt} = \frac{a\,[[Cr(H_2O)_6]^{2+}][[Cr(histamine)(H_2O)_2(ox)]^+]}{1 + b\,[[Cr(H_2O)_6]^{2+}]}$$

where $a = 1.93 \times 10^{-2}$ M^{-1} s^{-1} and $b = 3.2$ M^{-1} at 25°C and 1.0 M ionic strength with $\Delta H^{\ddagger} = 37$ kJ mol^{-1}, $\Delta S^{\ddagger} = -154$ J K^{-1} mol^{-1}, and $\Delta H = 24$ kJ mol^{-1}, $\Delta S = 88$ J K^{-1} mol^{-1}, respectively. Provide an explanation (Hussain, I.; Chatlas, J.; Kita, P. *Polish J. Chem.* 1991, **65**, 1577—1583.)

3.7 Cite two experimental observations which provide proof for remote attack in inner-sphere electron transfer reactions, outline any difficulties which arise in interpretation, and provide a summary of the criteria which are required for such mechanistic pathways.

REFERENCES

[1] Candlin, J. P.; Halpern, J. *Inorg. Chem.* 1965, **4**, 766–767.
[2] Taube, H.; Myers, H.; Rich, R. L. *J. Am. Chem. Soc.* 1953, **75**, 4118–4119.
[3] Taube, H.; Meyers, H. *J. Am. Chem. Soc.* 1954, **76**, 2103–2111.
[4] Taube, H.; King, E. L. *J. Am. Chem. Soc.* 1954, **76**, 4053–4054
[5] Halpern, J.; Orgel, L. E. *Discuss. Farad. Soc.* 1960, **29**, 32–41.
[6] Green, M.; Schug, K.; Taube, H. *Inorg. Chem.* 1965, **4**, 1184–1186.
[7] Toppen, D. L.; Link, R. G. *Inorg. Chem.* 1971, **10**, 2635–2636.
[8] Birk J. P.; Espenson, J. H. *J. Am. Chem Soc.* 1968, **90**, 1153–1162
[9] Shea, C.; Haim, A. *J. Am. Chem. Soc.* 1971, **93**, 3055–3056.
[10] Candlin, J. P.; Halpern, J.; Trimm, D. L. *J. Am. Chem. Soc.* 1964, **86**, 1019–1022.
[11] Barrett, M. B.; Swinehart, J. H.; Taube, H. *Inorg. Chem.* 1972, **11**, 1965–1967.
[12] Holwerda, R.; Deutsch, E.; Taube, H. *Inorg. Chem.* 1971, **10**, 1983–1989.
[13] Srinivasan, V. S.; Singh, A. N.; Radlowski, C. A.; Gould, E. S. *Inorg. Chem.* 1982, **21**, 1240–1246.

[14] Balahura, R. J.; Jordan, R. B. *J. Am. Chem. Soc.* 1970, **92**, 1533–1539.
[15] Balahura, R. J.; Jordan, R. B. *J. Am. Chem. Soc.* 1971, **93**, 625–631.
[16] Balahura, R. J.; Johnson, M.; Black, T. *Inorg. Chem.* 1989, **28**, 3933–3937.
[17] Splinter, R. C.; Harris, S. J.; Tobias, R. S. *Inorg. Chem.* 1968, **7**, 897–902.
[18] Zwickel, A.; Taube, H. *J. Am. Chem. Soc.* 1959, **81**, 1288–1291.
[19] Kruse, W.; Taube, H. *J. Am. Chem. Soc.* 1960, **82**, 526–528.
[20] Diebler, H.; Dodel, P. H.; Taube, H. *Inorg. Chem.* 1966, **5**, 1688.
[21] Espenson, J. H.; Birk, J. P. *J. Am. Chem. Soc.* 1965, **87**, 3280–3281.
[22] Scott, K. L.; Sykes, A. G. *J. Chem. Soc., Dalton Trans.* 1972, 2364–2369.
[23] Wieghardt, K.; Sykes, A. G. *J. Chem. Soc., Dalton Trans.* 1974, 651–654.
[24] Glennon, C. S.; Edwards, J. D.; Sykes, A. G. *Inorg. Chem.* 1978, **17**, 1654–
[25] Butler, R. D.; Taube, H. *J. Am. Chem. Soc.* 1965, **87**, 5597–5602.
[26] Price, H. J.; Taube, H. *Inorg. Chem.* 1968, **7**, 1–9.
[27] Cannon, R. D.; Gardiner, J. *J. Chem. Soc., Dalton Trans.* 1976, 622–626.
[28] Ogino, H.; Tsukahara, K.; Tanaka, N. *Inorg. Chem.* 1980, **19**, 255–259
[29] Kupferschmidt, W. C.; Jordan, R. B. *Inorg. Chem.* 1981, **20**, 3469–3473.
[30] Moore, M. C.; Keller, R. N. *Inorg. Chem.* 1971, **10**, 747–753.
[31] Patel, R. C.; Ball, R. E.; Endicott, J. F.; Hughes, R.G. *Inorg. Chem.* 1970, **9**, 23–29.
[31] Benjarvongkulchai, S.; Cannon, R. D. *Polyhedron* 1992, **11**, 517–521.
[32] Haim, A.; *J. Am. Chem. Soc.* 1966, **88**, 2324–2325.
[33] Ward, J. R.; Haim, A. *J. Am. Chem. Soc.* 1970, **92**, 475–482.
[34] Hwang, C.; Haim, A. *Inorg. Chem.* 1970, **9**, 500–505.
[35] Wood, P. B.; Higginson, W. C. E. *J. Chem. Soc., A* 1966, 1645–1652.
[36] Lane, R. H.; Sedor, F. A.; Gilroy, M. J.; Eisenhardt, P. F.; Bennett, J. P.; Ewall, R. X.; Bennett, L. E. *Inorg. Chem.* 1977, **16**, 93–101.
[37] Williams, R. D.; Pennington, D. E.; Smith, W. R. *Inorg. Chim. Acta* 1982, **58**, 45–51.
[38] Lane, R. H.; Bennett, L.E. *J. Am. Chem. Soc.* 1970, **92**, 1089–1090.
[39] Liteplo, M. P.; Endicott, J.F. *J. Am. Chem. Soc.* 1969, **91**, 3982–3983.
[40] Ogard, A. E.; Taube, H. *J. Am. Chem. Soc.* 1958, **80**, 1084–1089
[41] Williams, T. J.; Garner, C. S. *Inorg. Chem.* 1970, **9**, 2058–2064.
[42] Ball, D. L.; King, E. L. *J. Am. Chem. Soc.* 1958, **80**, 1091–1094.
[43] Anderson, A.; Bonner, N. A. *J. Am. Chem. Soc.* 1954, **76**, 3826–3830.
[44] Snellgrove, R.; King, E. L. *J. Am. Chem. Soc.* 1962, **84**, 4609–4610.
[45] Movius, W. G.; Linck, R. G. *J. Am. Chem. Soc.* 1970, **92**, 2677–2683.
[46] Haim, A. *Inorg. Chem.* 1968, **7**, 1475–1478.
[47] Haim, A. *Prog. Inorg. Chem.* 1983, **30**, 273–357.
[48] Stritar, J. A.; Taube, H. *Inorg. Chem.* 1969, **11**, 2281–2292.
[49] Sykes, A. G.; Thornley, R. N. F. *J. Chem. Soc. A* 1970, 232–238.
[50] Melvin, W. S.; Haim, A. *Inorg. Chem.* 1977, **16**, 2016–2020.
[51] Sutin, N. *Acc. Chem. Res.* 1968, **1**, 225–231.
[52] Guenther, P. R.; Linck, R. G. *J. Am. Chem. Soc.* 1969, **91**, 3769–3773.
[53] Espenson, J. H. *J. Am. Chem. Soc.* 1967, **89**, 1276–1278.
[54] Grossman, B.; Haim, A. *J. Am. Chem. Soc.* 1971, **93**, 6490–6494.

[55] Orhanović, M.; Po, H. N.; Sutin, N. *J. Am. Chem. Soc.* 1971, **90**, 7224–7229.
[56] Espenson, J. H. *Inorg. Chem.* 1965, **4**, 121–123.
[57] Diebler, H.; Taube, H. *Inorg. Chem.* 1965, **4**, 1029–1032.
[58] Candlin, J. P.; Halpern, J. *Inorg. Chem.* 1965, **4**, 1029–1032.
[59] van Eldik, R. *Inorg. Chem.* 1982, **21**, 2501–2502.
[60] Haim, A. *J. Am. Chem. Soc.* 1963, **85**, 1016–1017.
[61] Fay, D. P.; Sutin, N. *Inorg. Chem.* 1970, **9**, 1291–1293.
[62] Haim, A.; Sutin, N. *Inorg. Chem. Soc.* 1966, **88**, 5343–5344.
[63] Ohashi, K. *Bull. Chem. Soc. Jpn.* 1972, **45**, 3093–3095.
[64] Kurimura, Y.; Ohashi, K. *Bull Chem. Soc. Jpn.* 1971, **44**, 1797–1800.
[65] Benson, P.; Haim, A. *J. Am. Chem. Soc.* 1965, **87**, 3826–3835.
[66] Bifano, C.; Linck, R. G. *J. Am. Chem. Soc.* 1967, **89**, 3945–3947.
[67] Campion, R. J.; Conocchioli, T. J.; Sutin, N. *J. Am. Chem. Soc.* 1964, **86**, 4591–4594.
[68] Cannon, R. D.; Gardiner, J. *J. Am. Chem. Soc.* 1970, **92**, 3800–3801.
[69] Orhanović, M.; Earley, J. E. *Inorg. Chem.* 1975, **14**, 1478–1481.
[70] Birk, J. P. *Inorg. Chem.* 1975, **14**, 1724–1726.
[71] Chalilpoyil, P.; Davies, K. M.; Earley, J. E. *Inorg. Chem.* 1977, **16**, 3344–3346.
[72] Lee, R. A.; Earley, J. E. *Inorg. Chem.* 1981, **20**, 1739–1742.
[73] Adegite, A.; Earley, J. E.; Ojo, J. F. *Inorg. Chem.* 1979, **18**, 1535–1537.
[74] Akinyugha, N.; Ige, J.; Ojo, J. F.; Olubuyide, O.; Simoyi, R. *Inorg. Chem.* 1978, **17**, 218–221.
[75] Oytunji, A. O.; Olubuyide, O.; Ojo, J. F.; Earley, J. E. *Polyhedron*, 1991, **10**, 829–835.
[76] Olubuyide, O.; Lu, K.; Oyetunji, A. O.; Earley, J. E. *Inorg. Chem.* 1986, **25**, 4798–4799.
[77] Thompson, G. A. K.; Sykes, A. G. *Inorg. Chem.* 1976, **15**, 638–642.
[78] Bakac, A.; Marcec, R.; Orhanović, M. *Inorg. Chem.* 1977, **16**, 3133–3135.
[79] Davies, K. M.; Earley, J. E. *Inorg. Chem.* 1978, **17**, 3350–3354.
[80] Parker, O. J.; Espenson, J. H. *J. Am. Chem. Soc.* 1969, **91**, 1968–1974.
[81] Wang, R. T.; Espenson, J. H. *J. Am. Chem. Soc.* 1971, **93**, 380–386.
[82] Candlin, J. P.; Halpern, J.; Nakamura, S. *J. Am. Chem. Soc.* 1963, **85**, 2517–2518.
[83] Halpern, J.; Nakamura, S. *J. Am. Chem. Soc.* 1965, **87**, 3002–3003.
[84] Shea, C. J.; Haim, A. *Inorg. Chem.* 1973, **12**, 3013–3015.
[85] Stasiw, R.; Wilkins, R. G. *Inorg. Chem.* 1969, **8**, 156–157.
[86] Toma, H. *Inorg. Chim. Acta* 1977, **22**, 269–275.
[87] Adamson, A. W.; Gonick, E. *Inorg. Chem.* 1963, **2**, 129–132.
[88] Huchital, D. H.; Wilkins, R. G. *Inorg. Chem.* 1967, **6**, 1022–1027.
[89] Huchital, D. H.; Hodges, R. J. *Inorg. Chem.* 1973, **12**, 998–1000.
[90] Rosenhein, L.; Speiser, D.; Haim, A. *Inorg. Chem.* 1974, **13**, 1571–1575.
[91] Huchital, D. H.; Lepore, J. *Inorg. Chem.* 1978, **17**, 1134–1138.
[92] Phillips, J.; Haim, A. *Inorg. Chem.* 1980, **19**, 1616–1619.
[93] Phillips, J.; Haim, A. *Inorg. Chem.* 1980, **19**, 76–79.

[94] Bernauer, K.; Pousaz, P.; Porret, J.; Jeanguenat, A. *Helv. Chim. Acta* 1988, **71**, 1339–1348.
[95] Marusak, R. A.; Osvath, P.; Kemper, M.; Lappin, A. G. *Inorg. Chem.* 1989, **28**, 1542–1548.
[96] Marusak, R. A.; Ivanca, M. A.; Haller, K. J.; Lappin, A. G. *Inorg. Chem.* 1991, **30**, 618–623.
[97] Warren, R. M. L.; Lappin, A. G.; Tatehata, A. *Inorg. Chem.* 1992, **31**, 1566–1547.
[98] Gould, E. S. *J. Am. Chem. Soc.* 1966, **88**, 2983–2994.
[99] Chen, J. C.; Gould, E. S. *J. Am. Chem. Soc.* 1973, **95**, 5539–5544.
[100] Fan, F.-R. F.; Gould, E. S. *Inorg. Chem.* 1974, **13**, 2639–2646.
[101] Loar, M. K.; Sends, M. A.; Loar, G. W.; Gould, E. S. *Inorg. Chem.* 1978, **17**, 330–333.
[102] Brunschwig, B. S.; Creutz, C.; Macartney, D. H.; Sham, T.-K.; Sutin, N. *Faraday Discuss. Chem. Soc.* 1982, **74**, 113–127.
[103] Jolley, W. H.; Stranks, D. R.; Swaddle, T. W. *Inorg. Chem.* 1990, **29**, 1948–1951.
[104] Habib, H. S.; Hunt, J. P. *J. Am. Chem. Soc.* 1966, **88**, 1668–1671.
[105] Dulz, G.; Sutin, N. *J. Am. Chem. Soc.* 1964, **86**, 829–832.
[106] Carlyle D. W.; Espenson, J. H. *J. Am. Chem. Soc.* 1969, **91**, 599–606.
[107] Hyde, M. R.; Davies, R.; Sykes, A. G. *J. Chem. Soc., Dalton Trans.* 1972, 1838–1843.
[108] Davies, G. *Inorg. Chem.* 1971, **10**, 1155–1159.
[109] Rotzinger, F. P. *Inorg. Chem.* 1986, **25**, 4570–4572.
[110] Alberry, W. J. *Ann. Rev. Phys. Chem.* 1980, **31**, 227–263.
[111] Langord, C. H. *Inorg. Chem.* 1965, **4**, 265–266.
[112] Rotzinger, F. P.; Kumar, K.; Endicott, J. F. *Inorg. Chem.* 1982, **21**, 4111–4112.
[113] Kumar, K.; Rotzinger, F. P.; Endicott, J. F. *J. Am. Chem. Soc.* 1983, **105**, 7064–7074.
[114] Endicott, J. F.; Kumar, K.; Ramasami, T.; Rotzinger, F. P. *Prog. Inorg. Chem.* 1983, **30**, 141–187.
[115] Durham, B.; Endicott, J. F.; Wong, C.-L.; Rillema, D. P. *J. Am. Chem. Soc.* 1979, **101**, 847–857.
[116] Endicott, J. F.; Wong, C.-L.; Ciskowski, J. M.; Balakrishnan, K. P. *J. Am. Chem. Soc.* 1979, **102**, 2100–2103.
[117] Halpern, J.; Orgel, L. E. *Discuss. Farad. Soc.* 1960, **29**, 32–41.
[118] Burdett, J. K. *Inorg. Chem.* 1978, **17**, 2537–2551.
[119] Sebera, D. K.; Taube, H. *J. Am. Chem. Soc.* 1961, **83**, 1785–1791.
[120] Gould, E. S.; Taube, H. *J. Am. Chem. Soc.* 1964, **86**, 1318–1328.
[121] Gould, E. S. *J. Am. Chem. Soc.* 1967, **89**, 5792–5796.
[122] Nordmeyer, F.; Taube, H. *J. Am Chem. Soc.* 1968, **90**, 1162–1173.
[123] Balahura, R. J. *Inorg. Chem.* 1974, **13**, 1350–1354.
[124] Balahura, R. J. *J. Am. Chem. Soc.* 1976, **98**, 1487–1492.
[125] Zanella, A.; Taube, H. *J. Am. Chem. Soc.* 1972, **94**, 6403–6440.
[126] Hurst, J. K.; Taube, H. *J. Am. Chem. Soc.* 1968, **90**, 1178–1186.
[127] Gould, E. S. *J. Am. Chem. Soc.* 1974, **96**, 2372–2377.

[128] Gaunder, R. G.; Taube, H. *Inorg. Chem.* 1970, **9**, 2627–2639.
[129] Balahura, R. J.; Johnson, M. D. *Inorg. Chem.* 1989, **28**, 4548–4551.
[130] Fanchiang, Y.-T.; Thomas, J. C.; Heh, J. C. K.; Gould, E. S. *Inorg. Chem.* 1977 **16**, 1942–1945.
[131] Fan, F.-R. F.; Gould, E. S. *Inorg. Chem.* 1974, **13**, 2647–2651.
[132] Itzkowitz, M. M.; Nordmeyer, F. R. *Inorg. Chem.* 1975, **14**, 2124–2129.
[133] Endicott, J. F. *Acc. Chem. Res.* 1988, **21**, 59–66.
[134] Gould, E. S. *J. Am. Chem. Soc.* 1972, **94**, 4360–4361.
[135] Spiecker, H.; Wieghardt, K. *Inorg. Chem.* 1977, **16**, 1290–1294.
[136] Simic, M. G.; Hoffman, M. Z.; Brezniak, N. V. *J. Am. Chem. Soc.* 1977, **99**, 2166–2172.
[137] Wieghardt, K.; Cohen, H.; Meyerstein, D. *Ber. Bunsenges. Phys. Chem.* 1978, **82**, 388–392.
[138] Vrachnou-Astra, E.; Katakis, D. *J. Am. Chem. Soc.* 1973, **95**, 3814–3815.
[139] Vrachnou-Astra, E.; Katakis, D. *J. Am. Chem. Soc.* 1975, **97**, 5357–5363.
[140] Fanchiang, Y.-T.; Gould, E. S. *J. Am. Chem. Soc.* 1977, **99**, 5226–5227.
[141] Fanchiang, Y.-T.; Gould, E. S. *Inorg. Chem.* 1978, **17**, 1138–1141.
[142] Goli, U. B.; Gould, E. S. *Inorg. Chem.* 1984, **23**, 221–227.
[143] Fanchiang, Y.-T.; Gould, E. S. *Inorg. Chem.* 1977, **16**, 2516–2519.
[144] Srinivasan, V. S.; Rajasekar, N.; Singh, A. N.; Radlowski, C. A.; Heh, J. C.-K.; Gould, E. S. *Inorg. Chem.* 1982, **21**, 2824–2828.
[145] Gaswick, D.; Haim, A. *J. Am. Chem. Soc.* 1974, **96**, 7845–7846.
[146] Jwo, J.-J.; Haim, A. *J. Am. Chem. Soc.* 1976, **98**, 1172–1176.
[147] Malin, J. M.; Ryan, D. A.; O'Halloran, T. V. *J. Am. Chem. Soc.* 1978, **100**, 2097–2102.
[148] Sasaki, Y.; Ninomiya, T.; Nagasawa, A.; Endo, K.; Saito, K. *Inorg. Chem.* 1987, **26**, 2164–2168.
[149] Guardado, P.; van Eldik, R. *Inorg. Chem.* 1990, **29**, 3473–3477.
[150] Ali, R. B.; Blandamer, M. J.; Burgess, J.; Guardado, P.; Sanchez, F. *Inorg. Chim. Acta* 1987, **131**, 59–63.
[151] Jwo, J.-J.; Gaus, P. L.; Haim, A. *J. Am. Chem. Soc.* 1979, **101**, 6189–6197.
[152] Szecsy, A. P.; Haim, A. *J. Am. Chem. Soc.* 1981, **103**, 1679–1683.
[153] Lee, G.-H.; Cianna, L. D.; Haim, A. *J. Am. Chem. Soc.* 1989, **111**, 2535–2541.
[154] Bertram, H.; Wieghardt, K. *Angew. Chem. Int. Ed. Engl.* 1978, **17**, 205–206.
[155] Bertram, H.; Wieghardt, K. *Inorg. Chem.* 1979, **18**, 1799–1807.
[156] Gaus, P. L.; Villanueva, J. L. *J. Am. Chem. Soc.* 1980, **102**, 1934–1938.
[157] Isied, S. S.; Taube, H. *J. Am. Chem. Soc.* 1973, **95**, 8198–8199.

4

Intramolecular electron transfer

4.1 INTRODUCTION

In Chapters 2 and 3 it was seen that careful choice of the reagents for bimolecular electron transfer by both outer-sphere and inner-sphere mechanisms allows direct observation of the rate of intramolecular electron transfer within the precursor assembly. This is important because the rate can be related directly to the predictions of theory without the complications which arise from the energetics of bringing the two reactants together. Further, the basic theory outlined in Chapter 2 is, for the most part, adequate for describing the intramolecular process in both outer-sphere and weakly coupled inner-sphere reactions. The barriers to electron transfer are basically the same: inner-sphere and solvent reorganization to attain the activated complex geometry. However, instances where intramolecular electron transfer can be observed unambiguously in these bimolecular systems are not common since the systems are rather inflexible and impose limits on the structural trends which can be examined. Fortunately, the problem of the direct measurement of intramolecular electron transfer rates can be approached from other directions, most notably by assembling, prior to the induction of electron transfer, the required binuclear structure with both metal centers either fully oxidized or fully reduced. This rather obvious strategy has some serious constraints in that initial reduction or oxidation of the binuclear species must produce a kinetic rather that a thermodynamic product if a subsequent intramolecular process is to be detected. The equivalent of the self-exchange process is the situation where both components of the binuclear species are identical and hence no subsequent electron transfer can be detected by conventional methods. However, the products of the initial oxidation or reduction are mixed-valence species for which theory is sufficiently advanced to allow intramolecular electron transfer information to be gleaned from spectroscopic measurements. These mixed-valence species are not restricted to systems where both ends of the molecule are identical and so they also provide information on intramolecular processes where the free energy change is not zero.

4.2 OPTICAL ELECTRON TRANSFER

Mixed-valence complexes have been classified according to the strength of the interaction between the metal centers [1]. In Class I mixed-valence compounds, the interaction is very weak and the properties of the mixed-valence species are those of the discrete, unperturbed metal centers. This can be the result of an insulating bridge which prevents communication between the centers of asymmetry in the environment. An example of this is the ion $[(NC)_5FeCNFe(CN)_5]^{5-}$ where the ambidentate cyanide bridge results in stabilization of iron(II) at the C-bonded end and iron(III) at the N-bonded end [2]. In Class III mixed-valence compounds, the interaction is very strong and the metal centers are identical. In between these two extremes are Class II systems where the metal centers are almost identical and the unpaired electron is localized but weakly coupled. These latter systems correspond most closely to the model which has evolved in Chapter 2 for electron transfer.

Fig. 4.1 shows the reaction coordinate for electron transfer in a symmetric, weakly coupled, localized system, identical to the intramolecular self-exchange process illustrated in Fig. 2.8. The vertical or intervalence transfer transition between the reactant and product surfaces has energy E_{op} and since the upper energy function takes a value λ when $x = 0$, detection of this transition in the absorption spectrum of the complex gives a direct measure of λ and hence $\Delta G^{\ddagger}$ for the thermal electron transfer process (eq. (4.1)). This value of λ should reflect the reorganization of the donor and acceptor and the reaction medium required for electron transfer [3].

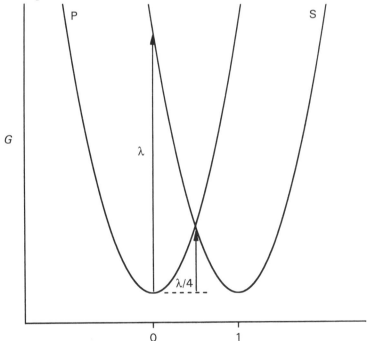

Fig. 4.1. Reaction coordinate diagram for optical electron transfer in a symmetric, weakly coupled, localized system. The nergy of the optical transition from the precursor to the successor is λ.

$$E_{op} = \lambda = 4\,\Delta G^{\ddagger} \tag{4.1}$$

Further detailed information about the nature of the interaction between the two metal centers can be obtained from the spectroscopic parameters. The bandwidth at half-intensity, $\Delta\bar{\upsilon}_{1/2}$, is related to the band maximum, $\bar{\upsilon}_{max}$, by eq. (4.2) which aids in the correct identification of the absorption band. In addition, the oscillator strength, f, determined from the molar extinction coefficient for the band, ε_{max}, is related to the distance, r, between the metal centers and the degree of delocalization of the electron in the ground state, represented by α^2 in eq. (4.3). The value of α^2 can be calculated by recasting this relationship as eq. (4.4) and is directly related to the electronic coupling between the metal centers, H_{AB}, eq. (4.5) [4].

$$\Delta\bar{\upsilon}_{1/2} = (2310\,\bar{\upsilon}_{max})^{1/2} \tag{4.2}$$

$$f = 4.6 \times 10^{-9}\,\Delta\bar{\upsilon}_{1/2}\,\varepsilon_{max} = 1.085 \times 10^{-5}\,\bar{\upsilon}_{max}\,\alpha^2\,r^2 \tag{4.3}$$

$$\alpha^2 = 4.24 \times 10^{-4}\,\frac{\varepsilon_{max}\,\Delta\bar{\upsilon}_{1/2}}{\bar{\upsilon}_{max}\,r^2} \tag{4.4}$$

$$H_{AB} = \bar{\upsilon}_{max}\,\alpha \tag{4.5}$$

One of the most widely celebrated mixed-valence compounds is the Creutz–Taube ion, prepared by the reaction of two equivalents of $[(NH_3)_5Ru(H_2O)]^{2+}$ with pyrazine (eq. (4.6)) [5]. The complex can be oxidized in two distinct one-electron steps with reduction potentials 0.98 V and 0.59 V (vs n.h.e.) in aqueous solution at 25.0°C (eq. (4.7)).

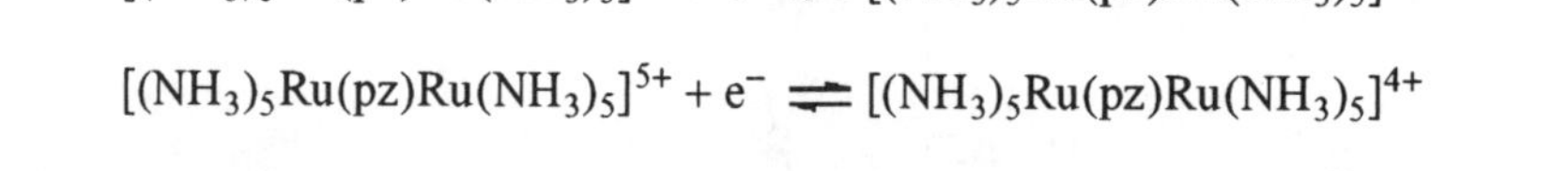

$$2\,[(NH_3)_5Ru(H_2O)]^{2+} + N\!\!\bigcirc\!\!N \rightleftharpoons [(NH_3)_5RuN\!\!\bigcirc\!\!NRu(NH_3)_5]^{4+} \tag{4.6}$$

$$[(NH_3)_5Ru(pz)Ru(NH_3)_5]^{6+} + e^- \rightleftharpoons [(NH_3)_5Ru(pz)Ru(NH_3)_5]^{5+}$$

$$[(NH_3)_5Ru(pz)Ru(NH_3)_5]^{5+} + e^- \rightleftharpoons [(NH_3)_5Ru(pz)Ru(NH_3)_5]^{4+} \tag{4.7}$$

$$[(NH_3)_5Ru(pz)Ru(NH_3)_5]^{6+} + [(NH_3)_5Ru(pz)Ru(NH_3)_5]^{4+} \overset{K_{com}}{\rightleftharpoons}$$

$$2[(NH_3)_5Ru(pz)Ru(NH_3)_5]^{5+} \tag{4.8}$$

The equilibrium constant for comproportionation, K_{com} (eq. (4.8)), of the mixed-valence species is 4×10^6 which suggests that there is a significant interaction between the two centers. A weak interaction, required to obtain information on the electron transfer process from the spectroscopic parameters, would be expected to have a comproportion equilibrium constant much closer to the statistical value of 4. The mixed-valence complex $[(NH_3)_5Ru(pz)Ru(NH_3)_5]^{5+}$ does possess a unique asymmetric spectroscopic feature at 1570 nm with an extinction coefficient of 5.5×10^3 M^{-1} cm^{-1} which was initially assigned as the intervalence transfer transition of a Class II compound. It is absent in the spectra of the fully oxidized and fully reduced complexes. However, $\Delta\bar{\upsilon}_{1/2}$ is very much smaller than predicted by eq. (4.2) for a Class II compound and the absorption maximum of this band is solvent-independent [6] which also suggests that the Creutz–Taube ion is better considered as an example of a Class III mixed-valence compound. However, related complexes (Table 4.1) do show characteristics of Class II mixed-valence species, and these will now be discussed.

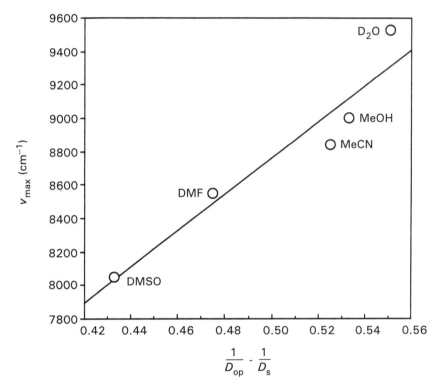

Fig. 4.2. Plot υ_{max} against $(1/D_{op} - 1/D_s)$ showing the effect of solvent on the intervalence transfer band for the complex $[(NH_3)_5Ru(4,4'-bpy)Ru(NH_3)_5]^{5+}$, from Ref. [7].

The complex $[(NH_3)_5Ru(4,4'-bpy)Ru(NH_3)_5]^{5+}$, where the metal centers are significantly farther apart than in the Creutz–Taube ion, shows an intervalence transfer band at 1030 nm in D_2O which is affected by changing solvent as predicted by eq. (2.77) (Fig. 4.2). The reduction potentials of the two metal centers are 0.41 V and 0.33 V, much closer together than with the Creutz–Taube ion, indicating that communication between the centers is diminished. As a consequence, comproportionation is much more important with $K_{com} = 23$. As the distance between the metal centers

Table 4.1. Spectroscopic information relating to intramolecular electron transfer rate data in Type II mixed-valence complexes

	K_{com} (M^{-1})	$\bar{\nu}_{max}$ (cm^{-1})	ε (M^{-1} cm^{-1})	r (Å)	H_{AB}	k (s^{-1})	Ref.
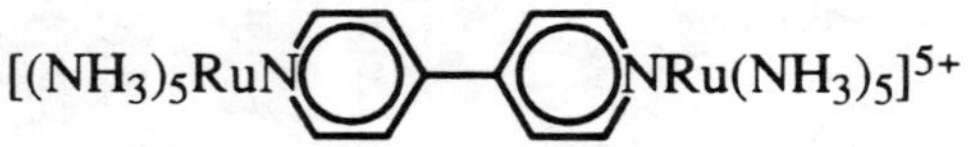 [(NH$_3$)$_5$RuN⟨⟩-⟨⟩NRu(NH$_3$)$_5$]$^{5+}$	20	9.7	920	11.3	390	5×10^8	7
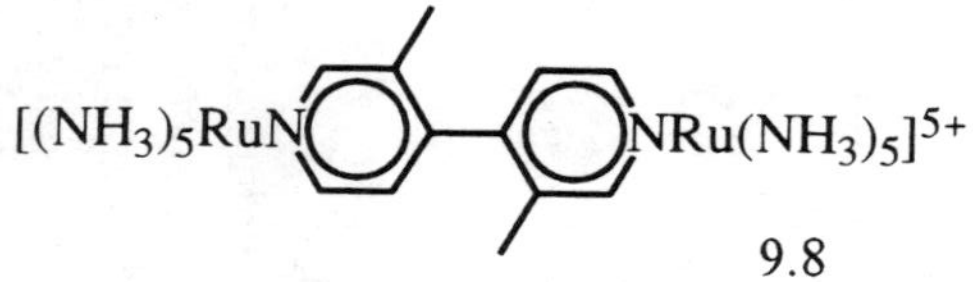 [(NH$_3$)$_5$RuN⟨⟩-⟨⟩NRu(NH$_3$)$_5$]$^{5+}$	9.8	11.2	165	11.3	195	8×10^6	7
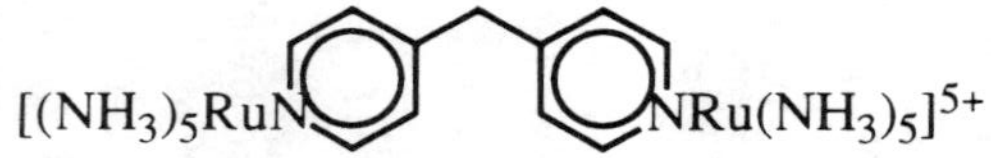 [(NH$_3$)$_5$RuN⟨⟩-CH$_2$-⟨⟩NRu(NH$_3$)$_5$]$^{5+}$	6.7	12.3	30	10.5	100	1×10^5	7
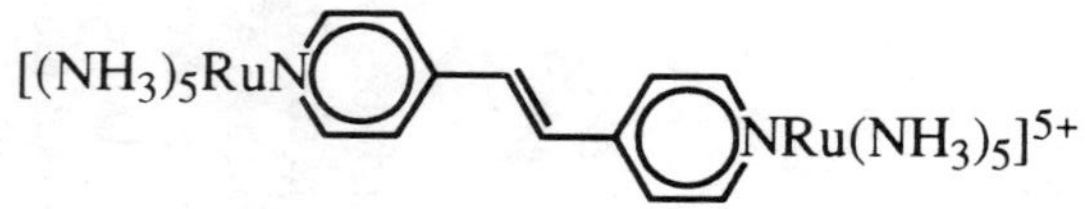 [(NH$_3$)$_5$RuN⟨⟩-CH=CH-⟨⟩NRu(NH$_3$)$_5$]$^{5+}$	14	10.4	760	13.8	305	5×10^8	7
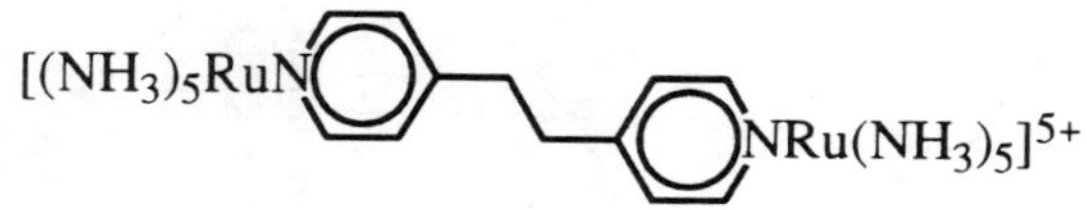 [(NH$_3$)$_5$RuN⟨⟩-CH$_2$-CH$_2$-⟨⟩NRu(NH$_3$)$_5$]$^{5+}$	—		<10				7

(*continued*)

Table 4.1 (continued)

	K_{com} (M^{-1})	$\bar{\nu}_{max}$ (cm^{-1})	ε (M^{-1} cm^{-1})	r (Å)	H_{AB}	k (s^{-1})	Ref.
 [(bpy)$_2$ClRuN⬡NRuCl(bpy)$_2$]$^{3+}$	100	7.69	450	6.8	400	3×10^9	8
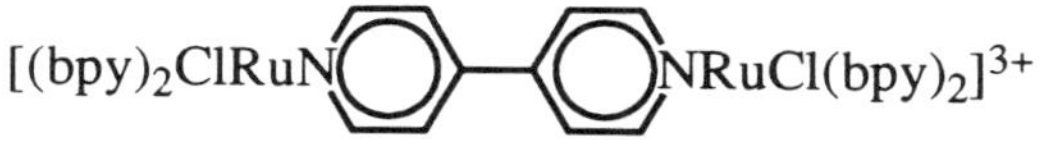 [(bpy)$_2$ClRuN⬡⬡NRuCl(bpy)$_2$]$^{3+}$	≈ 4	10.2	100	11.3	150	1×10^8	8
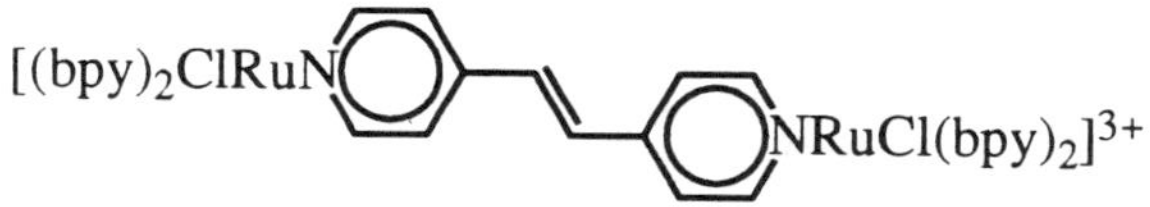 [(bpy)$_2$ClRuN⬡=⬡NRuCl(bpy)$_2$]$^{3+}$	≈ 4	10.8	200	13.8	180	2×10^7	8
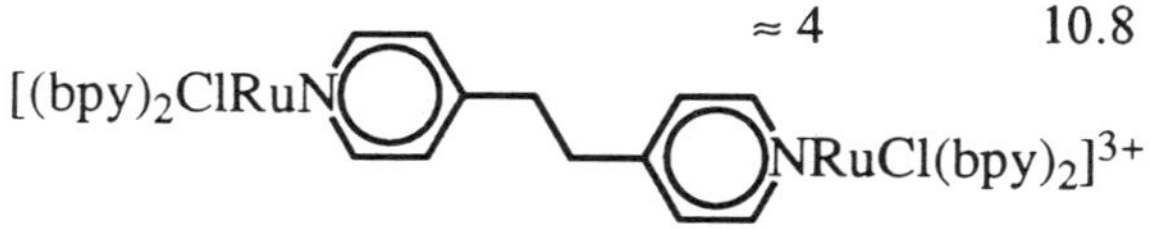 [(bpy)$_2$ClRuN⬡−⬡NRuCl(bpy)$_2$]$^{3+}$	≈ 4	—	—	—	—	—	8
 [(NC)$_5$FeN⬡NFe(CN)$_5$]$^{5-}$	50	8.3	2200	6.8	900		9
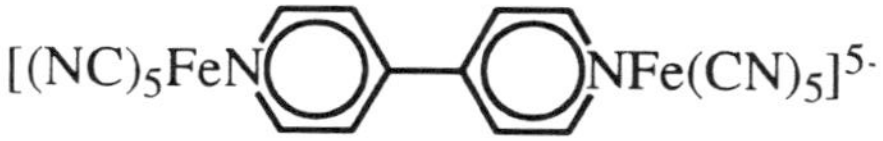 [(NC)$_5$FeN⬡⬡NFe(CN)$_5$]$^{5-}$	4	8.3	1100	11.3	390	2×10^9	9

(continues)

Table 4.1. (continued)

	K_{com} (M^{-1})	$\bar{\nu}_{max}$ (cm^{-1})	ε (M^{-1} cm^{-1})	r (Å)	H_{AB}	k (s^{-1})	Ref.
 [(NH$_3$)$_5$RuS◇◇SRu(NH$_3$)$_5$]$^{5+}$	<10	11.0	43	11.3	140	8.0×10^7	11
 [(NH$_3$)$_5$RuS◇◇◇SRu(NH$_3$)$_5$]$^{5+}$		12.4	9	14.4	55	4.9×10^6	12
 [(NH$_3$)$_5$RuS◇◇◇◇SRu(NH$_3$)$_5$]$^{5+}$		14.5	2.3	17.6	25	2.5×10^4	12

is increased, the energy of the transition increases, though the coupling H_{AB} is less sensitive. This is behavior expected where the outer-sphere reorganization λ_o is important, and while the model used in eq. (2.77) does not adequately describe the behavior in a quantitative fashion [13], the qualitative picture is consistent with adiabatic electron transfer. Similar observations are true for $[(bpy)_2ClRu(pz)RuCl(bpy)_2]^{3+}$ and related systems reported in Table 4.2 [9].

With the value for $\Delta G^{\ddagger}$, the rate for thermal electron transfer can be estimated from eq. (2.18) [3, 14]. In the case of the complex, eq. (4.9), where the separation between the centers is around 11 Å, a comparison can be made with the pseudo-self-exchange rate for related binuclear reaction in eq. (4.10). The reaction has $\Delta G° = 0$ and $k = 4.9 \times 10^7$ M^{-1} s^{-1} in dilute acetonitrile solution, at 25.0°C and $\mu = 0$.

$$[(bpy)_2ClRuN\text{–}\langle\text{py}\rangle\text{–}\langle\text{py}\rangle\text{–}NRuCl(bpy)_2]^{3+} \quad (4.9)$$

$$[Ru(phen)_2pyCl]^{2+} + [Ru(bpy)_2pyCl]^{+} \rightleftharpoons [Ru(phen)_2pyCl]^{+} + [Ru(bpy)_2pyCl]^{2+} \quad (4.10)$$

The mechanism of the bimolecular reaction involves formation of a precursor assembly where the metal–metal distance is calculated to be around 13 Å (eqs (4.11)–(4.12)), and when account is taken of the stability of the electron transfer precursor species with K_o estimated as 0.6 M^{-1}, $k_{et} = 8 \times 10^7$ s^{-1} in good agreement with the value determined for the mixed-valence complex.

$$[Ru(phen)_2pyCl]^{2+} + [Ru(bpy)_2pyCl]^{+} \rightleftharpoons \{[Ru(phen)_2pyCl]^{2+},[Ru(bpy)_2pyCl]^{+}\} \quad (4.11)$$

$$\{[Ru(phen)_2pyCl]^{2+},[Ru(bpy)_2pyCl]^{+}\} \rightleftharpoons \{[Ru(phen)_2pyCl]^{+},[Ru(bpy)_2pyCl]^{2+}\} \quad (4.12)$$

$$\{[Ru(phen)_2pyCl]^{+},[Ru(bpy)_2pyCl]^{2+}\} \rightleftharpoons [Ru(phen)_2pyCl]^{+} + [Ru(bpy)_2pyCl]^{2+} \quad (4.13)$$

Coupling between the metal centers is interrupted by the presence of saturated alkyl linkages between metal-bound pyridine in the bridges, suggesting that a conjugated system is essential for efficient electron transfer. The resulting low extinction coefficients make detection of the intervalence transfer band very difficult. The series of $[(NH_3)_5Ru(SC_nS)Ru(NH_3)_5]^{5+}$ complexes ($n = 1$–3) has been investigated and the small extinction coefficients suggest that electronic coupling is very weak so that these reactions fall into the realm of non-adiabatic electron transfer with $\kappa_{el} \ll 1$. The rate constants are calculated from the spectroscopic parameters with the use of an electronic tunneling model, representing a change in mechanism from the adiabatic reactions considered to this point [15]. Further consideration of electronic tunneling is presented in section 4.4. There is an exponential decrease of the rate with increasing separation between the metal centers. Additional experimental complications are observed in this very weakly coupled system. The mixed-valence complexes are generated by the addition of a strong oxidant to the fully reduced complex. Not only is there a dependence of the E_{op} on the reduction potential of the oxidant [16], but there is a strong ionic strength dependence [17].

Intervalence transfer spectra are also detected in non-symmetric mixed-valence complexes where there is a thermodynamic preference for localization of the electron on one center. This situation is illustrated in Fig. 4.3, where the energy of the vertical transition is given by eq. (4.14), and E_{ex} is a correction required if the product is in an electronically excited state.

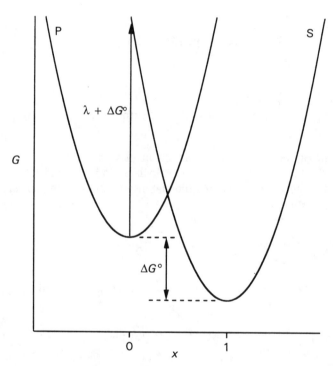

Fig. 4.3. Intervalence electron transfer for a non-symmetric mixed valence complex. The energy of the optical transition from the precursor to the successor surface is $\lambda + \Delta G°$.

$$E_{op} = \lambda + \Delta G° \ (+ E_{ex}) \tag{4.14}$$

A number of experimental parameters are presented in Table 4.2. Values for $\Delta G°$ are difficult to determine directly and are generally estimated from models for the individual components. For example, the complex $[(NC)_5Fe^{III}(pz)Ru^{III}(NH_3)_5]^+$ shows two reduction processes at 0.72 V and 0.49 V corresponding to formation of [(NC)₅Fe^{II}(pz)Ru^{III}(NH₃)₅] and [(NC)₅Fe^{II}(pz)Ru^{II}(NH₃)₅]⁻ respectively. The driving force for the reduction to the thermodynamically unstable mixed-valence form [(NC)₅Fe^{III}(pz)Ru^{II}(NH₃)₅] can be estimated from the reduction potential of [(NC)₅Co^{III}(pz)Ru^{III}(NH₃)₅]⁺, which is 0.64 V. In other systems, potentials are estimated from monomeric analogues.

Table 4.2. Spectroscopic parameters for symmetrical mixed-valence complexes

	$\Delta E°$ (V)	$\bar{\upsilon}_{max}$ (cm⁻¹)	ε (M⁻¹ cm⁻¹)	Ref.
 [(bpy)₂ClRu^{II}N⟨pyrazine⟩NRu^{III}(NH₃)₅]⁴⁺	0.42	1040	5.3 × 10⁻²	18,19
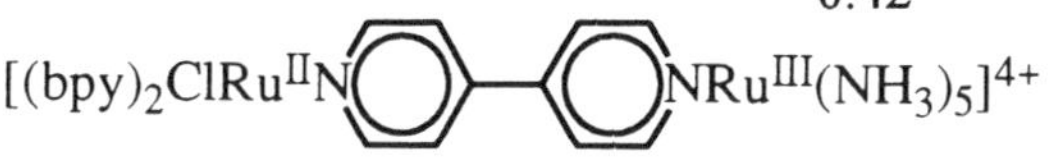 [(bpy)₂ClRu^{II}N⟨biphenyl⟩NRu^{III}(NH₃)₅]⁴⁺	0.40	1440	>3 × 10⁻²	18,19
[(NC)₅Fe^{II}(pz)Ru^{III}(NH₃)₅]	0.24	730	—	20,21
[(NC)₅Ru^{II}(pz)Ru^{III}(NH₃)₅]⁻	—	1464	2.8 × 10³	22
[(NC)₅Fe^{II}(pz)Ru^{III}(NH₃)₅]⁻	0.52	1020	3.0 × 10³	23

Recently the rates of electron transfer, k_{et}, after photoexcitation of the metal–metal charge transfer, have been directly measured for [(NC)₅Fe^{II}CNRu^{III}(NH₃)₅]⁻ and [(NC)₅Ru^{II}CNRu^{III}(NH₃)₅]⁻ (eq. (4.15)) [24, 25]. The back electron transfer occurs on a very fast timescale, less than 0.5 ps, before thermal equilibration of the excited state can occur and opens up a new and revealing area for research.

$$[(NC)_5Fe^{II}CNRu^{III}(NH_3)_5]^- \underset{k_{et}}{\overset{h\upsilon}{\rightleftharpoons}} [(NC)_5Fe^{III}CNRu^{II}(NH_3)_5]^- \qquad (4.15)$$

Intervalence transfer bands are also found in outer-sphere ion pairs such as {[Fe(CN)₆]³⁻,[Ru(NH₃)₅py]²⁺} [26], where the ions are of opposite charge, or under conditions of high concentration, {[Fe(CN)₆]³⁻,[Fe(CN)₆]⁴⁻} [27]. It is found that the electronic coupling, H_{AB}, for the ion pairs is comparable with bridged analogues and in the case of {[Fe(CN)₆]³⁻,[Ru(NH₃)₅py]²⁺}, the relative insensitivity of the interaction to changing the metal in the cyano complex to rutheium and osmium suggests that ligand orbitals mediate the interaction. These comparisons with the corresponding thermal processes give a more complete picture of the interactions between the ions.

4.3 CHEMICALLY INDUCED INTRAMOLECULAR ELECTRON TRANSFER

The preparation of the mixed-valence binuclear complex (eq. (4.16)) by direct reaction of $[(NH_3)_5Co(isonic)]^{2+}$ with $[(NH_3)_5Ru(OH_2)]^{2+}$ is not possible because outer-sphere electron transfer exceeds the rate of substitution. However, an alternative strategy to the design of binuclear compounds of this type where thermally activated intramolecular electron transfer can be detected was developed by Isied and Taube [28]. The lability of the ruthenium complex is enhanced in $trans$-$[Ru(NH_3)_4SO_2Cl]^+$, which reacts with $[(NH_3)_5Co(isonic)]^{2+}$ to give the fully oxidized sulfato-complex after treatment with HCl and H_2O_2 (eq. (417)).

$$[(NH_3)_5CoO_2C\text{—}\langle\text{C}_6H_4\rangle\text{—}NRu(NH_3)_5]^{4+} \qquad (4.16)$$

$$[(NH_3)_5CoO_2C\text{—}\langle\text{C}_6H_4\rangle\text{—}N]^{2+} + trans\text{-}[Ru(NH_3)_4SO_2Cl]^+ \underset{(2)\ HCl\ H_2O_2}{\overset{(1)\ \text{complexation}}{\rightleftharpoons}}$$

$$[(NH_3)_5CoO_2C\text{—}\langle\text{C}_6H_4\rangle\text{—}NRu(NH_3)_4(SO_4)]^{3+} \qquad (4.17)$$

In both acidic and neutral solutions, the thermodynamic product of one-electron reduction results in the release of the labile cobalt(II) and advantage can be taken of the greater kinetic reactivity of the $[(NH_3)_5RuN\text{-}]$ end to outer-sphere electron transfer compared with $[\text{-}OCo(NH_3)_5]$ to prepare the kinetic product (eq. (4.18)). Reaction with $[Ru(NH_3)_6]^{2+}$ is outer-sphere and results in the desired intermediate. It should be noted that the coordinated sulfato group on ruthenium(II) is relatively labile ($k \approx 0.3\ s^{-1}$) and in the reactions is lost prior to electron transfer to give the aqua ion, (eq. (4.19)). Decay of this aqua intermediate to the thermodynamic products is shown to be an intramolecular process (eq. 4.20)), and the rate and activation parameters for this and a number of related reactions are shown in Table 4.3.

$$[(NH_3)_5Co^{III}O_2C\text{—}\langle\text{C}_6H_4\rangle\text{—}NRu^{III}(NH_3)_4(SO_4)]^{3+} + [Ru(NH_3)_6]^{2+}$$

$$\longrightarrow [(NH_3)_5Co^{III}O_2C\text{—}\langle\text{C}_6H_4\rangle\text{—}NRu^{II}(NH_3)_4(SO_4)]^{2+} + [Ru(NH_3)_6]^{3+} \qquad (4.18)$$

$$[(NH_3)_5Co^{III}O_2C-\langle\bigcirc\rangle-NRu^{II}(NH_3)_4(SO_4)]^{2+} \longrightarrow$$

$$[(NH_3)_5Co^{III}O_2C-\langle\bigcirc\rangle-NRu^{II}(NH_3)_4(H_2O)]^{4+} \quad (4.19)$$

$$[(NH_3)_5Co^{III}O_2C-\langle\bigcirc\rangle-NRu^{II}(NH_3)_4(H_2O)]^{4+}$$

$$\longrightarrow [(NH_3)_5Co^{II}O_2C-\langle\bigcirc\rangle-NRu^{III}(NH_3)_4(H_2O)]^{4+} \xrightarrow{k_2} \text{products}$$

$$(4.20)$$

There is an added complication in these reactions because the ruthenium(III) product released in the intramolecular electron transfer serves as an oxidant for the intermediate. Experimentally this can be overcome by the use of initial rates for the reaction or by scavenging the ruthenium(III) with a large excess of ascorbate ion for reduction to ruthenium(II) and a ligand such as nicotinamide which will bind rapidly to the reduced form [29].

There are a number of conclusions which can be drawn. Although the rates show a dependence on bridging groups which largely parallels the trends found with limiting inner-sphere bimolecular rate constants and intervalence transfer data (sections 3.10 and 4.2) the dependence is not large. The presence of a saturated linkage in the bridge decreases the rate of intramolecular electron transfer and may indicate non-adiabatic behavior; however the effect is much less than the reduction in coupling found for intervalence transfer. This raises a question involving the nature of the rate-limiting step, which may be the intramolecular rate constant, k_1, or may involve a contribution from dissociation of the immediate product K_1k_2. The strength of the evidence points to the former interpretation [31]. Changes in the reduction potential of the ruthenium center, effected by a change in the ligand *trans* to the bridge, should be fully reflected in the product K_1k_2 but the actual dependence on the rate is much smaller and closer to the square root dependence predicted by the Marcus relationship for electron transfer.

An increase in the distance between the metal centers results in a rate decrease as found with the other investigations, and there has been considerable discussion of the source of this dependence. A plot of $\Delta G^{\ddagger}$ for the intramolecular electron transfer rate constants in the species [Co^{III}-Ru^{II}] against λ derived from the intervalence transfer bands is shown in Fig. 4.4 [33]. The correlation is very good and processes may be considered adiabatic with changes in λ_{out}, the solvational term, eq. (4.21), representing the major contributor to the dependence on r, the distance between the donor and acceptor metal complex chromophores. However, it must be

Table 4.3. Rate and activation parameters for intramolecular electron transfer in binuclear comlexes at 25.0°C

Complex	μ (M)	k (M^{-1} cm^{-1})	$\Delta H^{\ddagger}$ (kJ mol^{-1})	$\Delta S^{\ddagger}$ (J K^{-1} mol^{-1})	Ref.
[(NH$_3$)$_5$Co^{III}O$_2$C—⌬—NRu^{II}(NH$_3$)$_4$(H$_2$O)]$^{4+}$	0.1	1.24×10^{-2}			28,30
[(NH$_3$)$_5$Co^{III}O$_2$C—⌬—NRu^{II}(NH$_3$)$_4$(H$_2$O)]$^{4+}$	0.1	1.6×10^{-3}			28
[(NH$_3$)$_5$Co^{III}N—⌬—CH$_2$—⌬—NRu^{II}(NH$_3$)$_4$(H$_2$O)]$^{5+}$	0.4	2.1×10^{-3}	77.0	38	31
[(NH$_3$)$_5$Co^{III}N—⌬—⌬—NRu^{II}(NH$_3$)$_4$(H$_2$O)]$^{5+}$	0.4	4.4×10^{-2}	84.1	11	32
[(NH$_3$)$_5$Co^{III}N—⌬—CH=CH—⌬—NRu^{II}(NH$_3$)$_4$(H$_2$O)]$^{5+}$	0.4	1.87×10^{-2}	84.5	5	32

(continues)

Table 4.3 (continued)

Complex	μ (M)	k ($M^{-1}\,cm^{-1}$)	$\Delta H^{\ddagger}$ ($kJ\,mol^{-1}$)	$\Delta S^{\ddagger}$ ($J\,K^{-1}\,mol^{-1}$)	Ref.
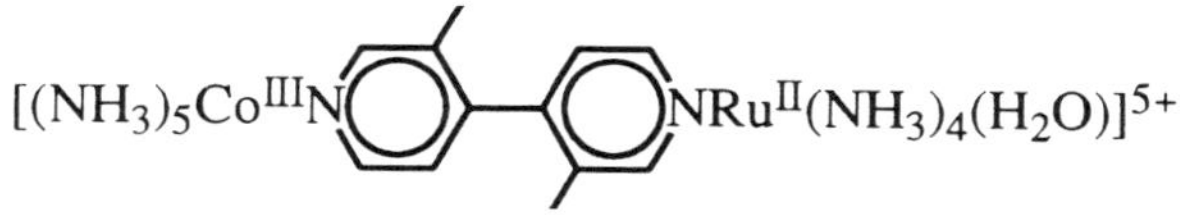 $[(NH_3)_5Co^{III}N\text{−}\langle\text{tolyl-tolyl}\rangle\text{−}NRu^{II}(NH_3)_4(H_2O)]^{5+}$	0.4	5.5×10^{-3}	84.5	−5	32
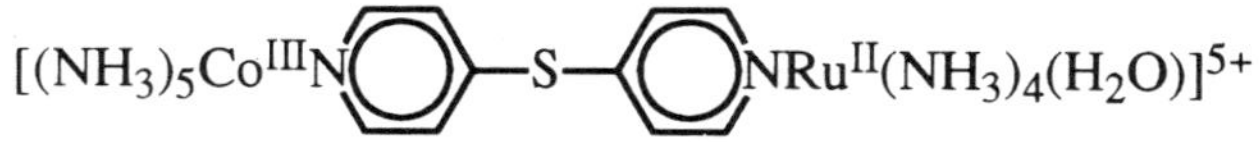 $[(NH_3)_5Co^{III}N\text{−}\langle\text{Ph-S-Ph}\rangle\text{−}NRu^{II}(NH_3)_4(H_2O)]^{5+}$	0.4	4.9×10^{-3}	84.5	−8	32
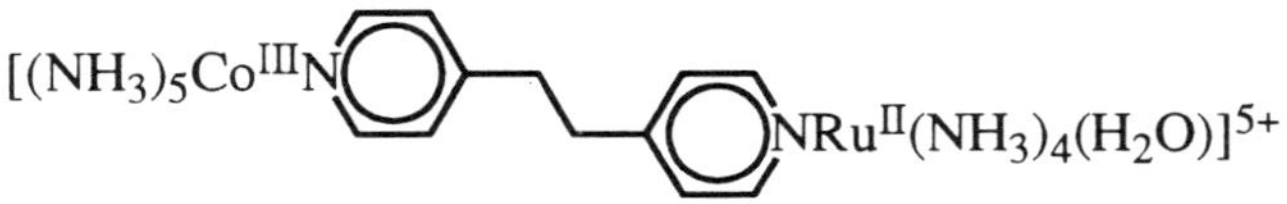 $[(NH_3)_5Co^{III}N\text{−}\langle\text{Ph-CH}_2\text{CH}_2\text{-Ph}\rangle\text{−}NRu^{II}(NH_3)_4(H_2O)]^{5+}$	0.4	1.0×10^{-3}	81.6	−27	32

remarked that the changes in rate are predominantly reflected in $\Delta S^{\ddagger}$ and a significant non-adiabatic contribution cannot be ruled out.

$$\lambda_{out} = (\Delta e)^2 \left(\frac{a_A}{2} + \frac{a_B}{2} - \frac{1}{r} \right) \left(\frac{1}{D_{op}} - \frac{1}{D_s} \right) \quad (4.21)$$

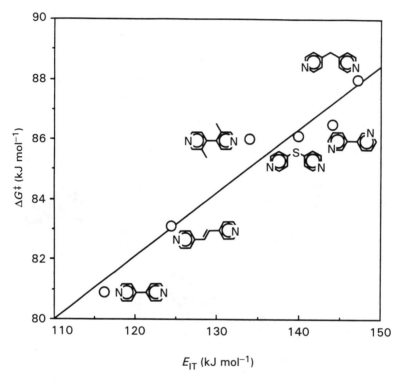

Fig. 4.4. Plot of activation $\Delta G^{\ddagger}$ for intramolecular electron transfer in $[(H_2O)(NH_3)_4Ru^{II}$-L-$Co^{III}(NH_3)_5]^{5+}$ against the energy, E_{IT}, of the intervalence-transfer band in $[(NH_3)_5Ru^{II}$-L-$Ru^{III}(NH_3)_5]^{5+}$ for six bridging ligands, from Ref. [33].

4.4 LONG-RANGE ELECTRON TRANSFER

In an extension of the approach to study of intramolecular electron transfer outlined in section 4.3, Isied and coworkers have used a variety of peptide spacers to increase the distance between the reaction centers [34, 35]. The intramolecular electron transfer rates are readily detected (Table 4.4) and decrease as the distance between the reaction centers increases. However, only the oligoprolyl residues have significant rigidity to prevent competition from a bridged outer-sphere electron transfer mechanism. The preferred conformation in solution of the oligoprolyl chain is a *trans*-helical structure and *trans/cis* isomerization is relatively slow with $t_{1/2} \approx 100$ s.

The electron transfer rate constant decreases significantly as the number of prolyl residues increases from one to two. A further increase in the number of prolines might be expected to provide a further reduction in rate. However, the increased flexibility on the longer peptide chain allows the bridged outer-sphere mechanism to compete with the very slow intramolecular electron transfer, and the measured rates reverse the trend. This change in mechanism is reflected in the activation parameters.

The driving force in these reactions is estimated to be +0.5 V, and the electron transfer is driven only by the subsequent aquation of the cobalt(II) product. Larger rates as a consequence of a more favorable driving force would allow less competition from the bridged outer-sphere mechanism, the rate of which is governed by the rate of *trans/cis* isomerization of the oligoproline spacer. A number of other electron donor and acceptor complexes have been used with considerable success and for the reactions with the largest driving forces, electron transfer over as many as nine proline residues has been observed [36]. Pulse radiolysis is used to initiate these faster reaction processes. The results of some of these studies are presented in Table 4.5.

With the $[Co^{III}\text{-}Os^{II}]$ complexes, interference from the different conformational isomers is again observed when the number of proline residues exceeds two. With the $[Ru^{III}\text{-}Os^{III}]$ reactions, initial pulse radiolytic reduction by CO_2^- or $(CH_3)_2COH^-$ is distributed almost equally between the thermodynamic product and the kinetic product (Scheme 4.1).

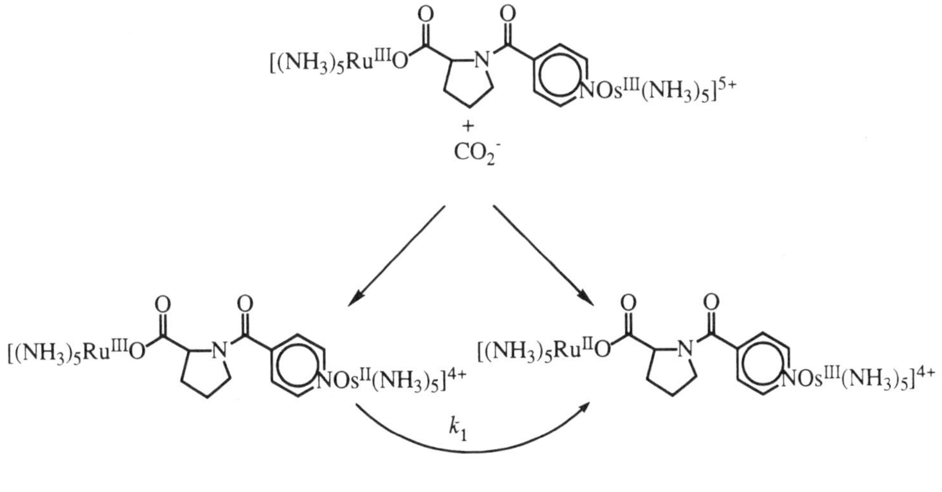

Scheme 4.1.

Table 4.4. Rate and activation parameters for electron transfer through oligopeptide spacers for [Co^{III}–Ru^{II}] at 25.0°C

Complex	k (s^{-1})	$\Delta H^{\ddagger}$ (kJ mol^{-1})	$\Delta S^{\ddagger}$ (J K^{-1} mol^{-1})	Ref.
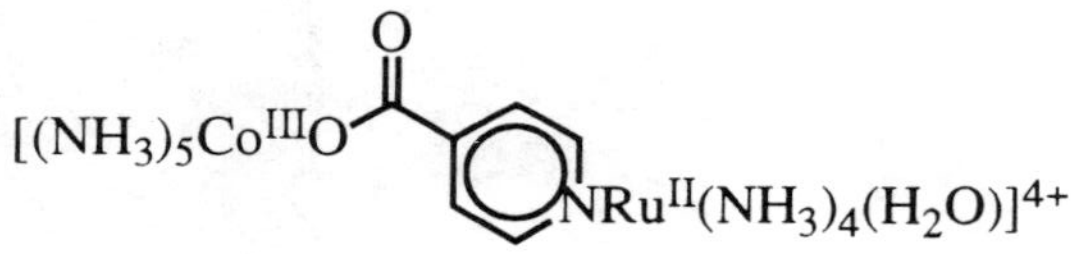 [(NH$_3$)$_5$Co^{III}–OOC–C$_6$H$_4$–NRu^{II}(NH$_3$)$_4$(H$_2$O)]$^{4+}$	1.2×10^{-2}	82	–5	30
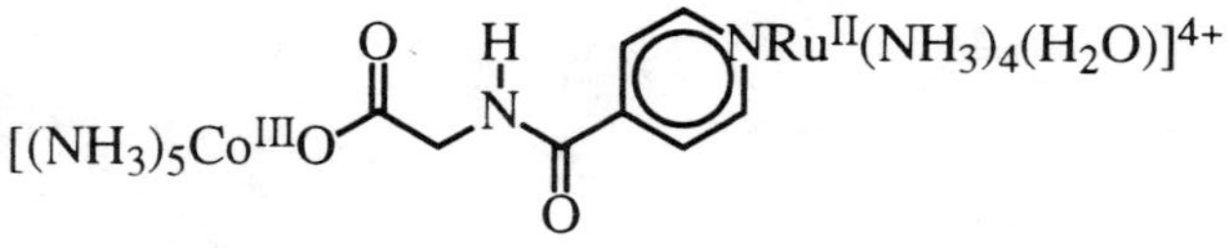 [(NH$_3$)$_5$Co^{III}–OOC–CH$_2$–NH–CO–C$_6$H$_4$–NRu^{II}(NH$_3$)$_4$(H$_2$O)]$^{4+}$	3.8×10^{-5}	83	–50	34
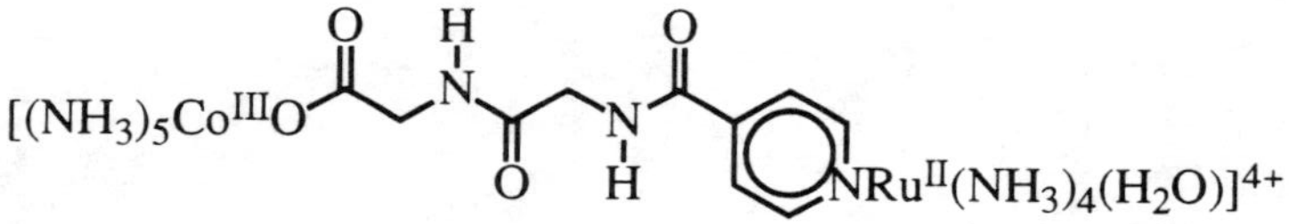 [(NH$_3$)$_5$Co^{III}–OOC–CH$_2$–NH–CO–CH$_2$–NH–CO–C$_6$H$_4$–NRu^{II}(NH$_3$)$_4$(H$_2$O)]$^{4+}$	9.9×10^{-6}	56	–155	34

(continues)

Table 4.4 (continued)

Complex	k (s^{-1})	$\Delta H^{\ddagger}$ (kJ mol^{-1})	$\Delta S^{\ddagger}$ (J K^{-1} mol^{-1})	Ref.
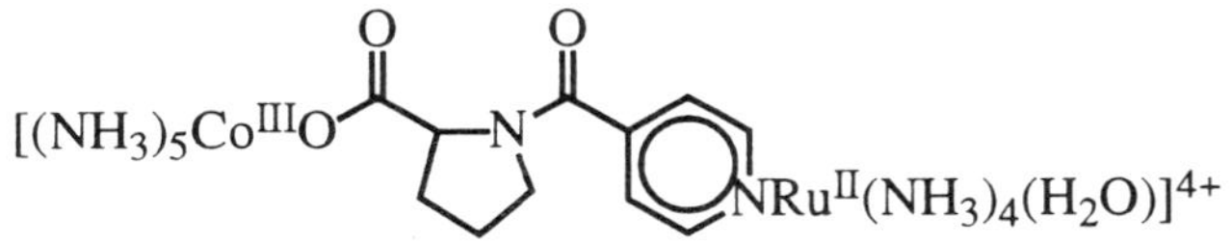 [(NH$_3$)$_5$Co^{III}O—...—NRu^{II}(NH$_3$)$_4$(H$_2$O)]$^{4+}$	1.0×10^{-4}	75	−67	35
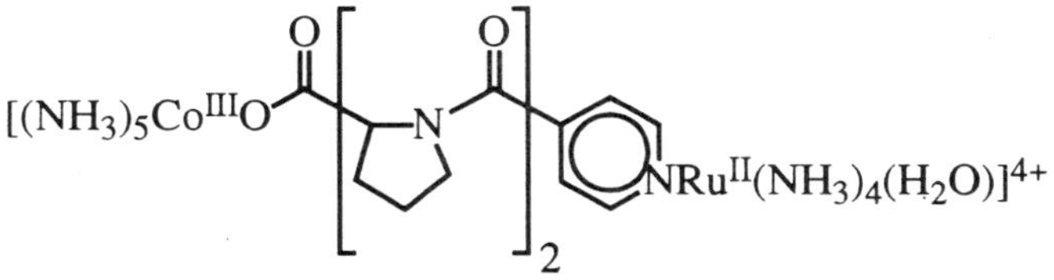 [(NH$_3$)$_5$Co^{III}O—...—NRu^{II}(NH$_3$)$_4$(H$_2$O)]$^{4+}$ (2)	6.4×10^{-6}	78	−84	35
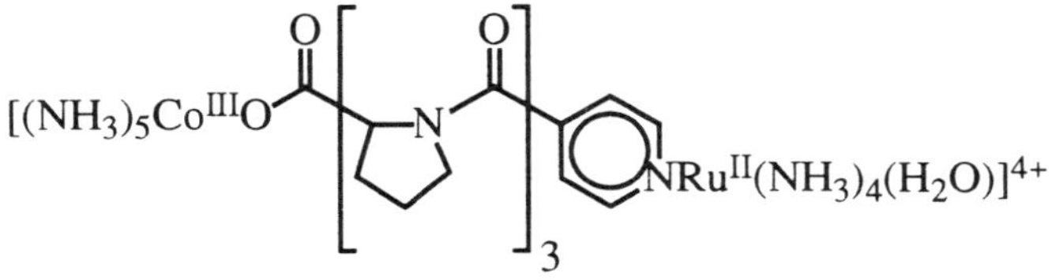 [(NH$_3$)$_5$Co^{III}O—...—NRu^{II}(NH$_3$)$_4$(H$_2$O)]$^{4+}$ (3)	5.6×10^{-5}	61	−121	35

(continues)

Table 4.4 (continued)

Complex	k (s^{-1})	$\Delta H^{\ddagger}$ (kJ mol^{-1})	$\Delta S^{\ddagger}$ (J K^{-1} mol^{-1})	Ref.
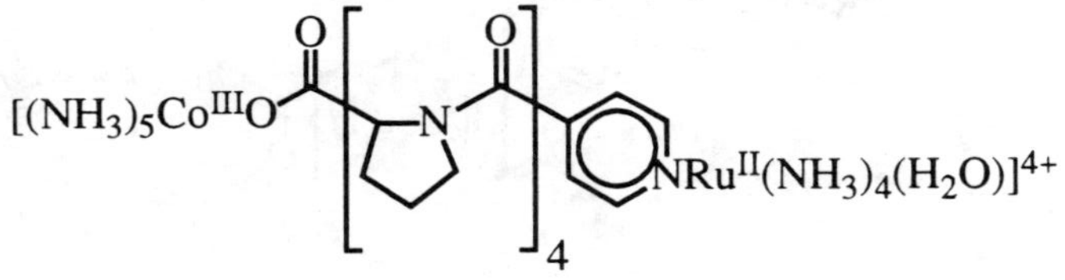 [(NH$_3$)$_5$Co^{III}O–(pyrrolidinyl-C(O)-C$_6$H$_4$)–NRu^{II}(NH$_3$)$_4$(H$_2$O)]$^{4+}$	1.4×10^{-4}	42	−180	35

Table 4.5. Rate and activation parameters for intramolecular electron transfer for [Co^{III}-Os^{II}] and [Ru^{III}-Os^{II}] complexes at 25.0°C

Complex	$E_{1/2}$ (V)	[Ru^{III}-Os^{II}] Dist(+)	k (s^{-1})	$\Delta H^{\ddagger}$ (kJ mol^{-1})	$\Delta S^{\ddagger}$ (J K^{-1} mol^{-1})	Ref.
[Co^{III}-Os^{II}]						
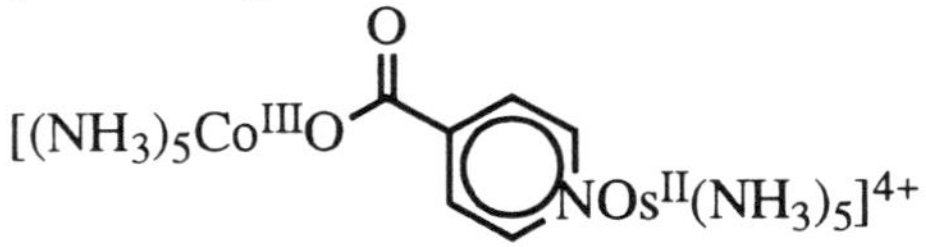 [(NH$_3$)$_5$Co^{III}O–(benzoate)–NOs^{II}(NH$_3$)$_5$]$^{4+}$	−0.23	9.0	1.9×10^5	43	0	37
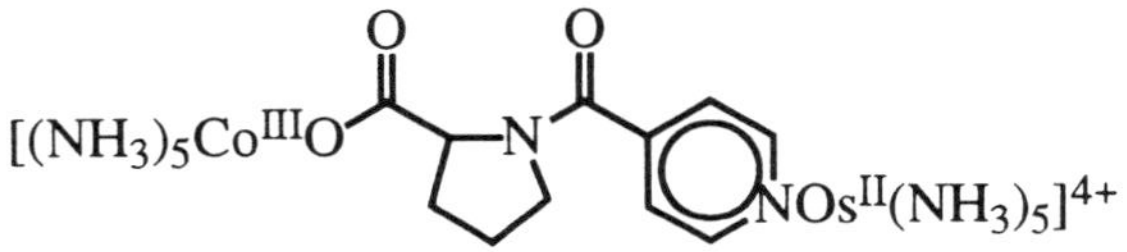 [(NH$_3$)$_5$Co^{III}O–(prolyl-benzamide)–NOs^{II}(NH$_3$)$_5$]$^{4+}$	−0.27	12.2	2.7×10^2	49	−33	37
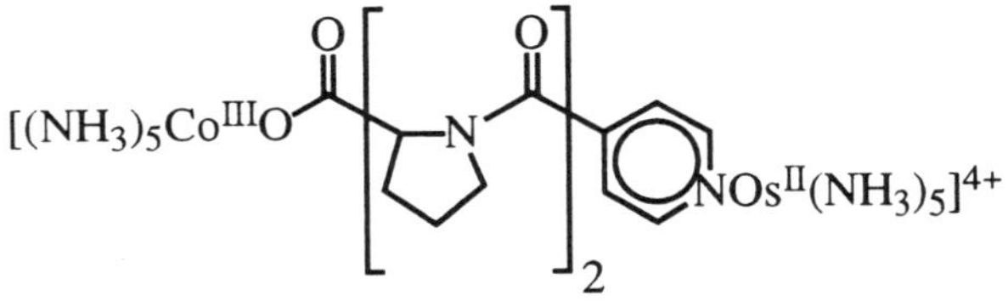 [(NH$_3$)$_5$Co^{III}O–(bis-prolyl-benzamide)–NOs^{II}(NH$_3$)$_5$]$^{4+}$	−0.27	14.8	0.74	53	−67	37

(continues)

Table 4.5. *(continued)*

Complex	$E_{1/2}$ (V)	[Ru^{III}-Os^{II}] Dist(+)	k (s^{-1})	$\Delta H^{\ddagger}$ (kJ mol^{-1})	$\Delta S^{\ddagger}$ (J K^{-1} mol^{-1})	Ref.
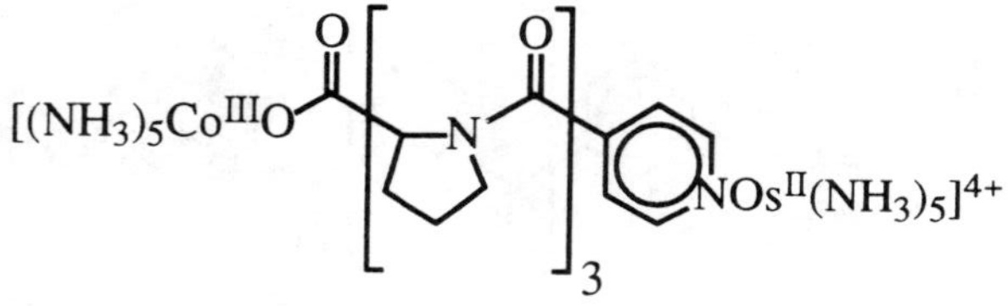 [(NH$_3$)$_5$Co^{III}O−⟨pyrrolidine-C(O)−C$_6$H$_4$⟩$_3$−NOs^{II}(NH$_3$)$_5$]$^{4+}$	−0.25	18.1	<0.9×10^{-1}	52	−92	37
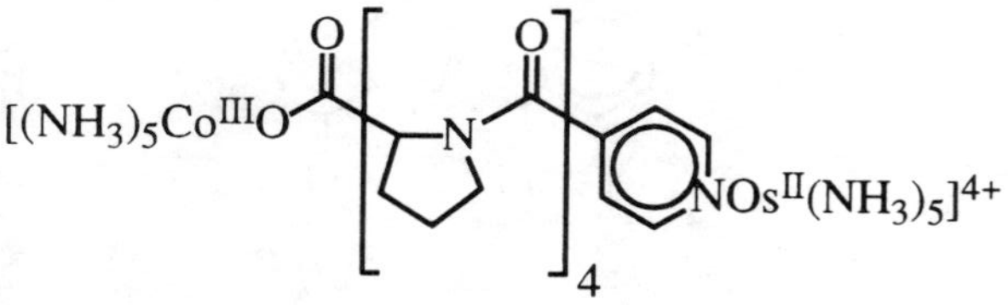 [(NH$_3$)$_5$Co^{III}O−⟨pyrrolidine-C(O)−C$_6$H$_4$⟩$_4$−NOs^{II}(NH$_3$)$_5$]$^{4+}$	−0.26	21.3	<0.9×10^{-1}	48	−105	37
[Ru^{III}-Os^{II}]						
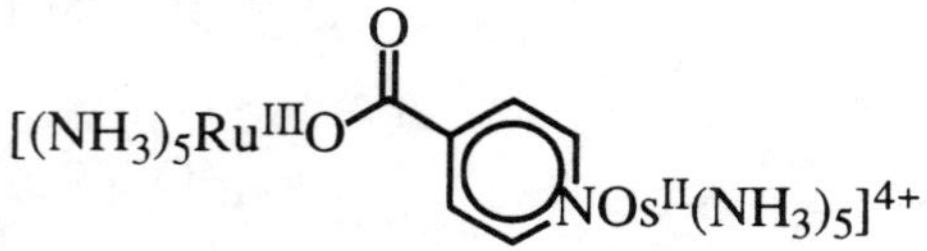 [(NH$_3$)$_5$Ru^{III}O−C(O)−C$_5$H$_4$N−Os^{II}(NH$_3$)$_5$]$^{4+}$	−0.24	9.0	>5×10^9	—	—	38

(continues)

Table 4.5. (continued)

Complex	$E_{1/2}$ (V)	$[\text{Ru}^{\text{III}}\text{-Os}^{\text{II}}]$ Dist(+)	k (s^{-1})	$\Delta H^{\ddagger}$ (kJ mol^{-1})	$\Delta S^{\ddagger}$ (J K^{-1} mol^{-1})	Ref.
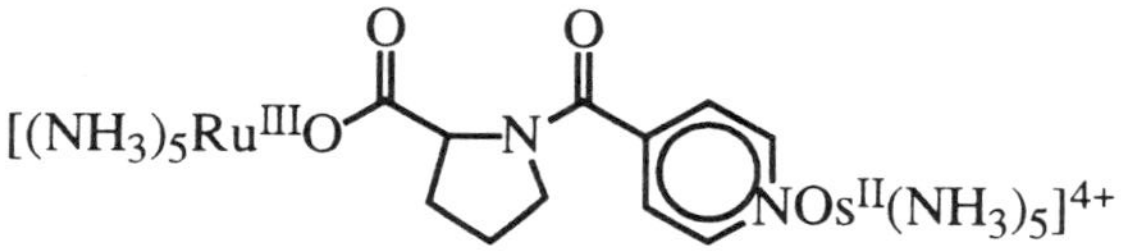 [(NH$_3$)$_5$Ru^{III}O–...–NOs^{II}(NH$_3$)$_5$]$^{4+}$	−0.30	12.3	3.1×10^6	18	−63	38
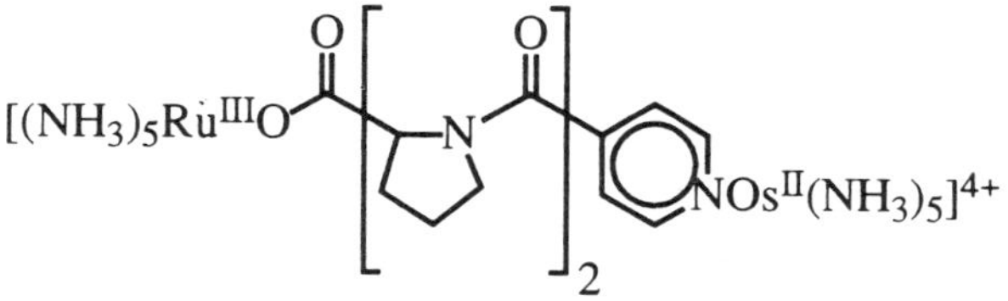 [(NH$_3$)$_5$Ru^{III}O–...–NOs^{II}(NH$_3$)$_5$]$^{4+}$ (2)	−0.30	14.9	3.7×10^4	25	−79	38
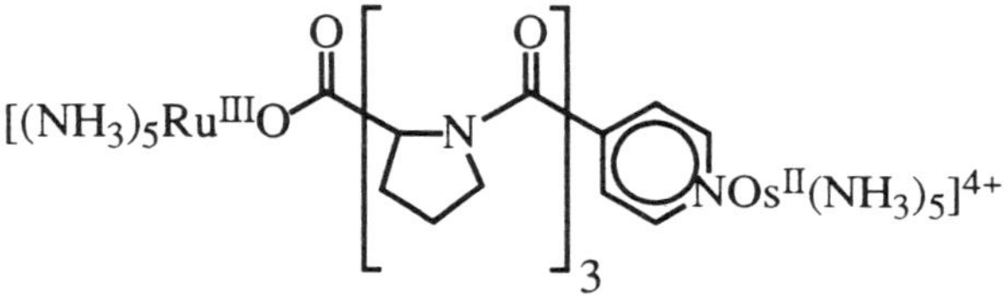 [(NH$_3$)$_5$Ru^{III}O–...–NOs^{II}(NH$_3$)$_5$]$^{4+}$ (3)	−0.30	18.1	3.2×10^2	31	−96	38

(continues)

Table 4.5. (continued)

Complex	$E_{1/2}$ (V)	[Ru^{III}-Os^{II}] Dist(+)	k (s^{-1})	$\Delta H^{\ddagger}$ (kJ mol^{-1})	$\Delta S^{\ddagger}$ (J K^{-1} mol^{-1})	Ref.
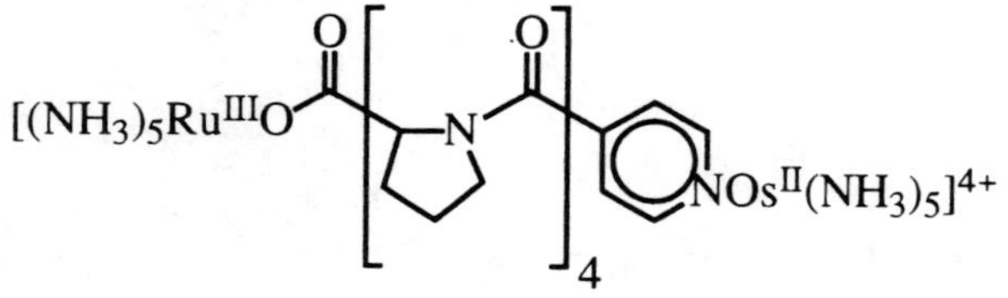 [(NH$_3$)$_5$Ru^{III}O–...–Os^{II}(NH$_3$)$_5$]$^{4+}$	−0.30	21.3	≈50	—	—	38

Intramolecular electron transfer to the thermodynamic product, k_1, can be detected in this latter species for up to four proline residues. The rates decrease by over eight orders of magnitude as the distance between the metal centers increases from 9Å to 21Å (Fig. 4.5). Interpretation of this distance dependence must be viewed with care. In section 4.3 electron transfer over distances up to approximately 12Å were adequately described by an adiabatic model where the distance dependence of the rates results from solvent rearrangement. However, in the present case, weaker electronic coupling is also a factor and separating these two components is not a trivial problem. It forms the basis for the next section.

4.5 NON-ADIABATIC ELECTRON TRANSFER

In Chapter 2, the expression derived for the intramolecular rate constant indicated a dependence on the the electronic transmission coefficient, κ_{el}, an effective frequency, υ_{eff}, and the Frank–Condon activation barrier, $\Delta G^{\ddagger} = (\lambda + \Delta G^{\circ})^2/4\lambda$, as shown in eq. (4.22). In the adiabatic regime, $\kappa_{el} = 1$ and the effective frequency is a nuclear vibration which carries the reactants to the products. However, in the non-adiabatic regime where electronic coupling between the donor and acceptor is very weak, the electronic transmission coefficient assumes greater importance.

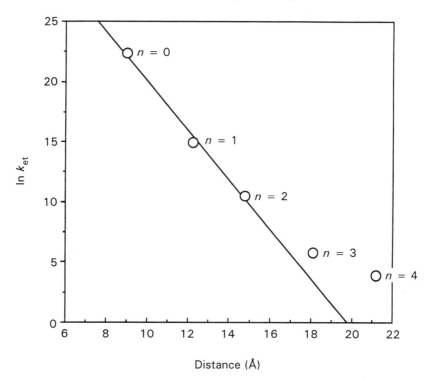

Fig. 4.5. Plot of $\ln k_{et}$ against metal–metal distance for intramolecular electron transfer in $[(NH_3)_5Os^{III}LRu^{III}(NH_3)_5]^{4+}$, L = iso(pro)$_n$, n = 0 to 4, from Ref. [38].

$$k_{et} = \upsilon_{eff} \kappa_{el} \exp(-\Delta G^{\ddagger}/RT) \qquad (4.22)$$

An expression for the electronic transmission coefficient, κ_{el}, is given in eq. (4.23), and is a function of the ratio $\upsilon_{el}/2\upsilon_{eff}$ where υ_{eff} is the frequency of the nuclear vibration which carries reactants to products and υ_{el} is the electronic frequency determined by the rate at which the electron hops between electronic surfaces. In the adiabatic regime, $\upsilon_{el} \gg \upsilon_{eff}$ such that $\kappa_{el} = 1$ but when the electronic coupling, H_{AB}, is small, as found for example from the intervalence transfer spectroscopic measurements for compounds of the type, $[(NH_3)_5RuSC_nSRu(NH_3)_5]^{5+}$, $\upsilon_{el} \ll \upsilon_{eff}$ so that $\kappa_{el} \ll 1$ and the reactions are non-adiabatic (eq. (4.24)). The rates are determined by the rate of electronic hopping and not by the nuclear motion.

$$\kappa_{el} = \frac{2(1 - \exp(-\upsilon_{el}/2\upsilon_{eff}))}{(2 - \exp(-\upsilon_{el}/2\upsilon_{eff}))} \qquad (4.23)$$

$$\upsilon_{el} = \frac{2 H_{AB}^2}{h} \left\{ \frac{\pi^3}{(\lambda_{out} + \lambda_{in})RT} \right\}^{1/2} \qquad (4.24)$$

The electronic coupling shows a strong dependence on distance (eq. (4.25)), where r is the distance between the metal centers and r_o is some minimal distance where the reaction is adiabatic [39]. Thus the factor $\kappa_{el}\upsilon_{eff}$ will decrease in a similar manner with distance from the value expected for an adiabatic process and can be approximated by $10^{13}\exp[-\beta(r-r_o)]$. The rate expression is then eq. (4.26).

$$H_{AB} = H^{\circ}_{AB} \exp[-\beta(r-r_o)] \qquad (4.25)$$

$$k_{et} = 10^{13} \exp[-\beta(r-r_o)] \exp(-\Delta G^{\ddagger}/RT) \qquad (4.26)$$

An exponential dependence of the rate on the distance between the metal centers can be expected, and for the compounds $[(NH_3)_5RuSC_nSRu(NH_3)_5]^{5+}$, such a plot is shown in Fig. 4.6.

The complicating feature is the dependence of the component of the Frank–Condon barrier, λ_{out}, on distance (eq. (4.21)), and separation of this dependence from the effects of decreased electronic coupling presents a considerable problem. Some enlightenment can be found by the realization that λ will contribute only to the distance dependence of $\Delta H^{\ddagger}$ if the static and dynamic dielectric constants, D_s and D_{op}, are independent of temperature and $\Delta S^{\circ} = 0$ for the reaction, resulting in the approximate relationships in eqs (4.27) and (4.28) [40].

$$\Delta H^{\ddagger} \approx -RT \ln \kappa_n \qquad (4.27)$$

$$\Delta S^{\ddagger} \approx -R \ln \kappa_{el} \qquad (4.28)$$

The distance dependence of $\Delta H^{\ddagger}$ and $\Delta S^{\ddagger}$ can be examined for the complexes in Table 4.5 with oligoproline spacer groups (Fig. 4.5). The distance dependence of the solvent rearrangement as measured by $\Delta H^{\ddagger}$ is significantly larger than that for the decrease in electronic coupling, measured by $\Delta S^{\ddagger}$. A value of β obtained from these studies is 0.65 Å^{-1}. It should be noted that in electron transfer in weakly coupled photoinduced electron transfer in systems with saturated organic spacers, a value of $\beta = 0.9$Å is obtained, and it is observed that the coupling is more dependent on the number of bonds separating the donor form the acceptor than a direct measure of distance [41]. Theoretical calculation of the electronic coupling in the oligoproline complexes gives comparable results when the whole bridge [42] is compared with the product of couplings evaluated for individual subunits [43]. It can be concluded that although the reduction in electronic coupling contributes to the distance dependence of the electron transfer rate, Frank–Condon factors associated with rearrangement of the reaction medium are dominant even at distances up to 20 Å.

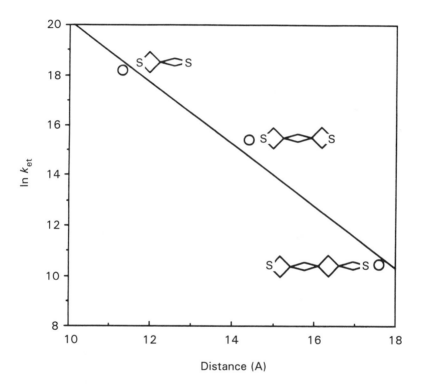

Fig. 4.6. Plot of lnk_{et} calculated from the energies of intervalence electron transfer specta against metal–metal distance for intramolecular electron transfer in [(NH$_3$)$_5$RuSC$_n$SRu(NH$_3$)$_5$]$^{5+}$, n = 1 to 3 from Ref. [12].

4.6 REACTIONS OF METALLOPROTEINS

The most extensively studied examples of electron transfer reactions in weakly coupled systems are those involving small metalloproteins. A variety of small metalloproteins are involved in electron transfer processes in biological systems and

196   **Intramolecular electron transfer**   [Ch. 4

these are the focus of this attention. Much has been learned about the mechanisms of electron transfer of these systems from studies of reactions with small metal ion complexes of the type discussed to this point. The field is extensive and delves deeply into the realms of biochemistry; however, some aspects are very relevant to the subject of this chapter. In this section, the characterization and bimolecular reaction chemistry of two protein types is considered, in order to lay the groundwork for section 4.7, devoted to an examination of intramolecular processes.

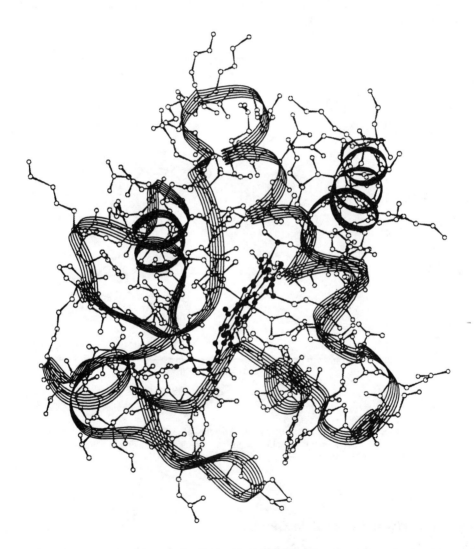

Fig. 4.7. Ribbon diagram of the polypeptide chain of oxidized cytochrome c showing the site of the exposed heme edge, the heme atoms are detailed in bloack, from Ref. [46].

The cytochromes are an important class of electron transfer proteins found in both plant and animal sources [44, 45]. The most common example is cytochrome c, with a molecular weight of 11 700; the reduction potential of the protein from horse heart is 0.26 V. Structural data for the proteins are available (Fig. 4.7) and reveal some differences in structure at the tertiary level between the oxidized and reduced forms [46]. At pH 7.0, the oxidized protein has a net +7 charge, owing to the presence of a large number of lysine residues on the surface. The protein contains the heme group (Fig. 4.8), with an imidazole and a methionine ligand coordinated axially, and the iron center shuttles between the low spin +3 and +2 states. About 4% of the heme ligand of this prosthetic group is exposed at the protein surface. NMR line broadening methods have been used to determine a self-exchange rate for the protein of 10^3–10^4 M^{-1} s^{-1} [47]. Comparison of this value with data for the low spin $[Fe(phen)_3]^{3+/2+}$ self-exchange, which is 10^7 M^{-1} s^{-1}, leads to proposal that the difference may be attributed to a steric factor. Electron transfer takes place only at exposed heme edge which accounts for approximately 0.06% of the total surface area of the protein.

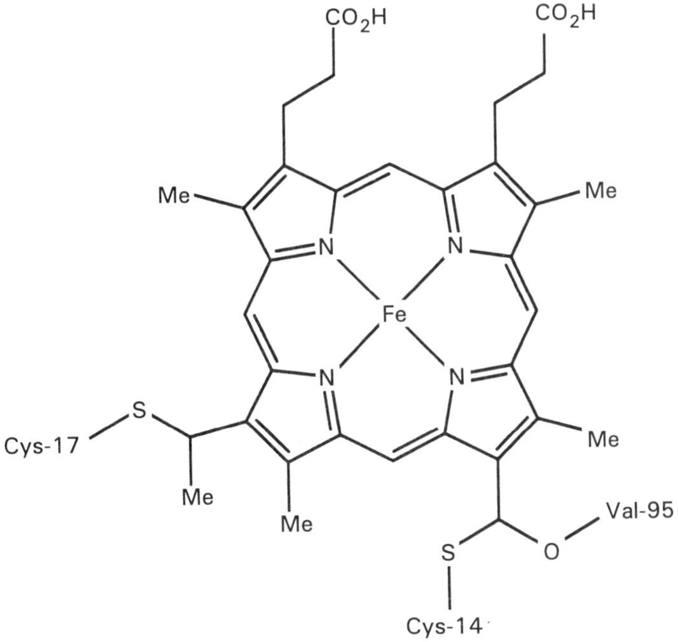

Fig. 4.8. Schematic view of the heme group from cytochrome c.

Electron transfer reactions of cytochrome c generally show simple kinetic behavior, first-order in both reagents, and second-order rate constants are presented in Table 4.6. The reactions show medium effects caused by the specific association of anions, particularly phosphate, to the surface of the protein, and the best studies avoid this by using cacodylate as buffer [48]. Ionic strength dependencies have been investigated extensively. The Debye–Hückel expression (eq. (1.59)), does not provide an adequate description of the behavior and more complex models are required to take account of the fact that the protein is significantly larger than the metal complexes with which it interacts and that the charge distribution is inhomogeneous [49, 50]. Attempts to use Debye–Hückel-type expressions which incorporate localized charge distributions have also met with limited success [51]. Both analyses suggest that electron transfer involves metal ion complexes associated with particular localized sites on the protein.

Table 4.6. Second-order rate constants for reactions of cytochrome c, pH 7.0, 25.0°C

Oxidant	Reductant	μ (M)	k ($M^{-1} s^{-1}$)	Ref.
$[Fe(CN)_6]^{3-}$	cyt c(II)	0.10	9.3×10^6	52
$[Co(ox)_3]^{3-}$	cyt c(II)	0.5	5.5	53
$[Co(phen)_3]^{3+}$	cyt c(II)	0.1	1.5×10^3	54
$[Ru(NH_3)_5py]^{3+}$	cyt c(II)	0.1	6×10^3	55
$[Fe(cp)]^+$	cyt c(II)	0.1	6.2×10^6	56
cyt c(III)	$[Ru(NH_3)_6]^{2+}$	0.1	3.8×10^4	57
cyt c(III)	$[Fe(edta)]^{2-}$	0.1	2.6×10^4	58

Localized sites have been pinpointed by elegant studies in which the effect on the electron transfer rate constants of modifying individual lysine residues on the surface of the protein as a 4-carboxy-2,6-dinitrophenyl derivative (eq. (4.29)) [52]. These substituents reduce the positive charge and provide a steric interaction. Results of studies in the oxidations of cyt c by $[Fe(CN)_6]^{3-}$ and $[Co(phen)_3]^{3+}$ are presented in Table 4.7. It is clear that modifications at residues 72 and 13 which are close to the exposed heme edge (Fig. 4.9) have the most dramatic effect.

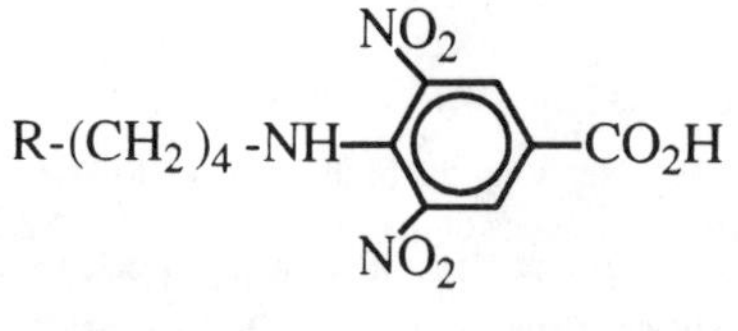

$$R\text{-}(CH_2)_4\text{-}NH\text{-}\underset{NO_2}{\overset{NO_2}{\bigcirc}}\text{-}CO_2H \qquad (4.29)$$

Table 4.7. Rate constants for reactions of modified cyt c

		10^{-3} rate for modified residue			
Oxidant	Native	lys-60	lys-87	lys-72	lys-13
$[Fe(CN)_6]^{3-}$	9300	8100	7800	3900	3300
$[Co(phen)_3]^{3+}$	1.5	1.7	1.9	3.6	4.4

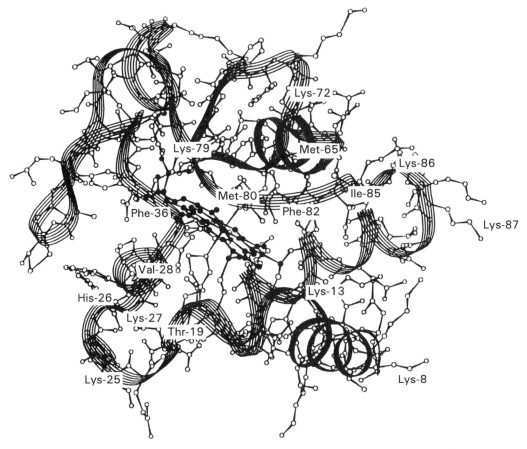

Fig. 4.9. The exposed heme edge of cytochrome c, showing the binding sites of $[Cr(CN)_6]^{3-}$. There are two binding sites close to the heme crevice, one encompassing Met-80, Phe-82, Ile-85 and heme methyl-3, Site 3, and one encompassing Thr-19, His-26 and Val-28, Site 2. A third, less well defined binding site at the rear of the molecule is close to Phe-36 and Met-65, Site 1. Lysine residues thought to be involved in these binding sites are also shown. From Ref. [59].

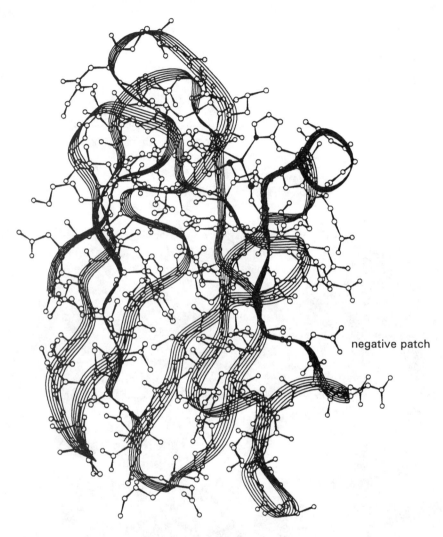

Fig. 4.10. Ribbon diagram showing the polypeptide backbone of oxidized plastocyanin from Ref. [65]. The copper binding site is delineated in black and the negative patch comprising Asp-42, Glu-43, Asp-44, Glu-59 and Glu-60 is n oted.

Further information on the sites of association of the metal complexes on the surface of the protein have been obtained with ^{1}H NMR studies in which paramagnetic analogues are used to broaden signals from specific amino-acid residues close to the point of binding. For $[Cr(CN)_6]^{3-}$ three areas are mapped, corresponding to sites 1, 2, and 3 (Fig. 4.9) [59]. Site 3 shows the strongest binding and encompasses lysines 72 and 13. Other reagents associate with a different distribution of sites [60]. This difference in reaction sites is evident in other ways. While reactions of cyt c with $[Fe(CN)_6]^{3-}$ are pH independent, reactions with cationic oxidants such as

Sec. 4.6] Reactions of metalloproteins 201

[Co(phen)$_3$]$^{3+}$ show a 30% rate reduction over the pH range 8 to 5 as a result of protonation of histidine 33. When the imidazole nitrogen of the residue is conjugated with diethyl pyrocarbonate, the pH effect disappears [61].

Although the detailed mechanisms of electron transfer to and from cytochrome c are complex, application of Marcus Theory with electrostatic corrections gives reasonably good agreement between observed and calculated values [44]. This may be the result of a single site for electron transfer, the exposed heme edge, which means that all the reactions have similar steric and electrostatic requirements. Not all reactions of metalloproteins are as well behaved. Reactions of the blue copper protein plastocyanin in particular show discrepancies of many orders of magnitude in Marcus correlations [62] and some consideration of the reasons for this are outlined next.

The blue copper protein plastocyanin is found in both plant and bacterial sources where it functions as an electron transport agent [63, 64]. The protein has a molecular weight of 10 500 and contains copper(II) in the intensely blue or oxidized form; this can be reduced to a colorless copper(I) form with a reduction potential of 0.37 V at pH 7.0 and 0.10 M ionic strength. Crystal structure data are known for both oxidation states (Fig. 4.10), and for the reduced form as a function of pH [65, 66]. The protein is in the form of a barrel of seven β-pleated sheets with the copper at one end, shielded from the solvent by histidine-87, which functions as a ligand. The other ligands of the prosthetic group are histidine-37, cysteine-84 and methionine-92 in a distorted tetrahedral arrangement (Fig. 4.11). In the reduced protein, histidine-87 dissociates on protonation with pK_a 5.7 [67], leading to an increase in the reduction potential. The oxidized protein has a net charge of −6 at neutral pH, the result of a number of aspartate and glutamate residues on the protein surface. Distribution of

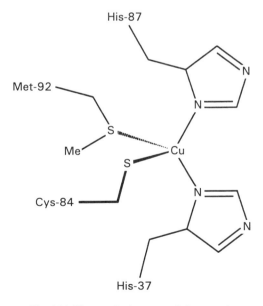

Fig. 4.11. The prosthetic group of plastocyanin.

the charges is very inhomogeneous and there is a significant negative patch on the protein connected to the copper center by a channel of hydrophobic residues.

Outer-sphere electron transfer reactions of plastocyanins have been extensively investigated and show a number of complications. With positively charged oxidants in excess the reaction rate shows evidence for limiting first-order behavior, Fig. 4.12 [68]. The preferred interpretation requires two pathways: a dominant pathway involving formation of a precursor complex between the protein and the oxidant (eqs (4.30)–(4.31)), and a minor parallel pathway which is independent of this complex (eq. (4.32)) [69].

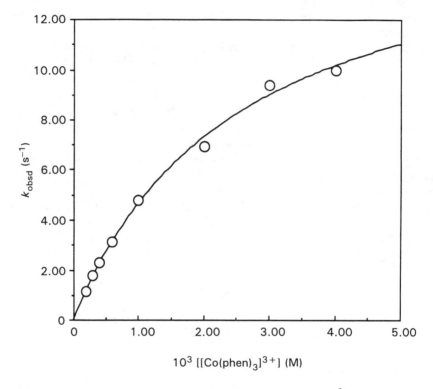

Fig. 4.12. Plot of pseudo-first-order rate constant k_{obsd} against $[[Co(phen)_3]^{3+}]$ for the reaction of spinach plastocyanin, PCu(I), with $[Co(phen)_3]^{3+}$, at pH 7.5, 25°C, $\mu = 0.10$ M (NaCl), from Ref. [68].

$$PCu(I) + [Co(phen)_3]^{3+} \underset{}{\overset{K_o}{\rightleftharpoons}} \{PCu(I),[Co(phen)_3]^{3+}\} \quad (4.30)$$

$$\{PCu(I),[Co(phen)_3]^{3+}\} \overset{k_{et}}{\rightarrow} PCu(II) + [Co(phen)_3]^{2+} \quad (4.31)$$

$$\text{PCu(I)} + [\text{Co(phen)}_3]^{3+} \xrightarrow{k_2} \text{PCu(II)} + [\text{Co(phen)}_3]^{2+} \qquad (4.32)$$

$$\frac{d[\text{PCu(II)}]}{dt} = \left\{ \frac{K_o k_{et} [[\text{Co(phen)}_3]^{3+}]}{1 + K_o [[\text{Co(phen)}_3]^{3+}]} + k_2 [[\text{Co(phen)}_3]^{3+}] \right\} [\text{PCu(II)}] \quad (4.33)$$

The rate law (eq. (4.33)) is consistent with the dependence in Fig. 4.12, and for $[\text{Co(phen)}_3]^{3+}$, $K_o = 340$ M^{-1}, $k_{et} = 5.7$ s^{-1}, and $k_2 = 5.1 \times 10^2$ M^{-1} s^{-1} at 25°C and 0.10 M ionic strength. At low $[[\text{Co(phen)}_3]^{3+}]$, the reaction is competitively inhibited by $[\text{Cr(phen)}_3]^{3+}$ (eq. (4.34)), where $K_I = 360$ M^{-1} is the binding constant for the inhibitor (eq. (4.35)), and a number of other highly charged redox inert complexes such as $[\text{Pt(NH}_3)_6]^{4+}$, (Fig. 4.13). Interestingly the ratio $(K_o k_{et})/k_2$ is the same for the different inhibitors but does vary with the nature of the oxidant (Table 4.8).

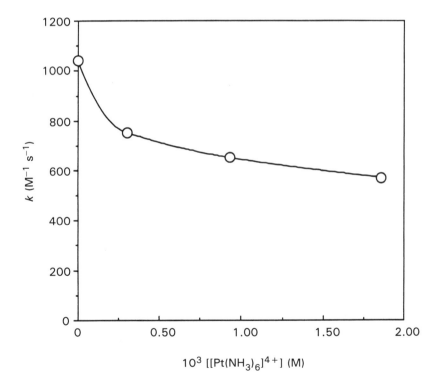

Fig. 4.13. Effect of $[\text{Pt(NH}_3)_6]^{4+}$ on the second order rate constant for the oxidation of PCu(I) parsley plastocyanin by $[\text{Co(phen)}_3]^{3+}$ at pH 5.8, 25°C, $\mu = 0.10$ M (NaCl), from Ref. [69].

$$k_{obsd} = \frac{K_o k_{et} [[\text{Co(phen)}_3]^{3+}]}{1 + K_I [[\text{Cr(phen)}_3]^{3+}]} + k_2 [[\text{Co(phen)}_3]^{3+}] \qquad (4.34)$$

$$\text{PCu(I)} + [\text{Cr(phen)}_3]^{3+} \underset{}{\overset{K_I}{\rightleftharpoons}} \{\text{PCu(I)},[\text{Cr(phen)}_3]^{3+}\} \qquad (4.35)$$

Table 4.8. Percent reaction by $K_o k_{et}$ and k_2 pathways for reductions of plastocyanins

Oxidant	%$K_o k_{et}$	%k_2	Ref.
$[\text{Co(phen)}_3]^{3+}$	75	25	69
$[\text{Fe(CN)}_6]^{3-}$	0	100	69
$[\text{Fe(cp)}_2]^+$	28	72	70

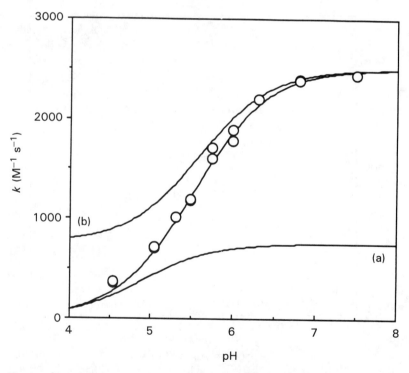

Fig. 4.14. Variation of second-order rate constant with pH at 25°C for the oxidation of PCu(I) spinach plastocyanin by $[\text{Co(phen)}_3]^{3+}$, $\mu = 0.10$ M (NaCl). Simulated curves indicate the relative influence of protonation of His-87 at the active site (a) and protonation of the negative patch (b). From Ref. [72].

The inhibitor, [Cr(phen)$_3$]$^{3+}$ is paramagnetic and induces line broadening in tyrosine-83 which is close to the negative patch, thereby identifying the probable site of reaction for the inhibitable pathway [71]. When [Fe(CN)$_6$]$^{3-}$ is used as the oxidant, no inhibition is observed and NMR studies with [Cr(CN)$_6$]$^{3-}$ reveal a site of interaction close to histidines 37 and 87, which are active site ligands. It is suggested that electron transfer takes place at two sites on the protein surface, a direct pathway at the ligands adjacent to the binding site and an inhibitable pathway through the remote binding site at the negative patch. The relative contribution from each site is dependent on the charge on the oxidant. Protons also inhibit the oxidation of plastocyanin. While this may be expected on the basis of the protonation of the copper(I) site with pK_a = 5.5, the kinetic data for the reduction of [Co(phen)$_3$]$^{3+}$ show inhibition at much higher pH (Fig. 4.14). This may be treated as the addition of two protons, one at the negative patch which destroys the cation binding site (pK_a = 5.8), and leads to approximately 60% inhibition, and the other at the active site which renders the reduction thermodynamically unfavorable [72]. The reverse reaction shows a cation binding site (pK_a = 5.1) consistent with this explanation [73].

The mechanistic picture which emerges is complex and it is little wonder that the application of Marcus Theory to reactions of plastocyanins is limited [53]. That there are two independent pathways for electron transfer is by no means unambiguous. An alternative explanation in which binding by the cationic reagent, whether it be [Co(phen)$_3$]$^{3+}$, [Cr(phen)$_3$]$^{3+}$, or H$^+$ at the remote site, merely reduces the reactivity of the protein is also consistent with the facts, and this 'dead-end' complex mechanism has been considered by other workers [74]. Both pathways have the same thermodynamic driving force and there appears to be no significant kinetic advantage associated with electron transfer through the remote site [75]. In spite of these ambiguities, the studies provide a source of limiting intramolecular electron transfer rate constants through complex media, behavior noted with a number of other proteins [76]. Overwhelmingly $\Delta S^\ddagger$ for the intramolecular electron transfer is strongly negative, contrasting with studies in small molecule systems.

4.7 LONG-RANGE ELECTRON TRANSFER INVOLVING METALLOPROTEINS

The intramolecular rates obtained from the bimolecular kinetic studies are subject to randomness in that the spatial relationship between the donor and acceptor is not well defined since it is the result of non-covalently bonded electrostatic outer-sphere interactions. A much better defined system results if the redox agent is bound covalently to the protein surface. Electron transfer over well-defined distances can be determined. The first example of this approach took advantage of the substitution lability of the aqua ligand in [Ru(NH$_3$)$_5$OH$_2$]$^{2+}$ and the affinity of the chromophore nitrogen donors to attach [Ru(NH$_3$)$_5$–]$^{2+}$ to histidine-33 of horse cytochrome c. It has been shown that addition of the ruthenium center to the protein does not result in a significant change in the protein conformation, and the reduction potentials of both centers are close to those found in the free species, 0.26 V for Fe$^{III/II}$ and 0.13

206 Intramolecular electron transfer [Ch. 4

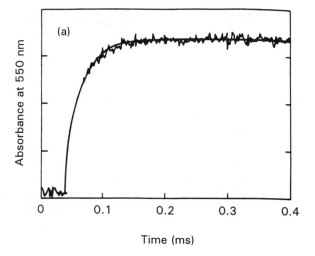

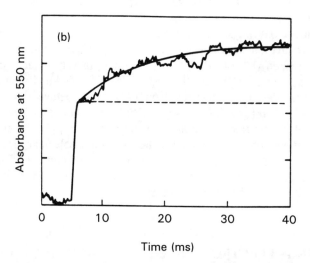

Fig. 4.15. Plot of absorbance against time monitored at 550 nm for the reduction of (a) horse-heart cytochrome c by CO_2^-, (b) $(NH_3)_5Ru$-His-33-modified horse-heart cytochrome c by CO_2^-, studied by pulse radiolysis. The latter experiment shows a biphasic process indicated by the dashed line. The fast step corresponds to direct reduction of the heme center while the slower process involves initial reduction of covalently bound ruthenium and reduction of the heme center by intramolecular electron transfer, from Isied, S. S.; Worosila, G.; Atherton, S. J. *J. Am. Chem. Soc.* 1982, **104**, 7659, with permission.

V for $Ru^{III/II}$. Reaction of the covalently bound fully oxidized system [$(NH_3)_5Ru^{III}$his-33-cyt c(Fe^{III})] with the strong reductants [*$Ru(bpy)_3$]$^{2+}$ [77, 78] and CO_2^- [79] results in a reaction trace (Fig. 4.14) which is biphasic with a fast decay consistent with direct intramolecular reduction of the heme center to give the thermodynamic product, [$(NH_3)_5Ru^{III}$his-33-cyt c(Fe^{II})], directly and the slower intramolecular reaction from the kinetic product, $[Ru^{II}(NH_3)_5his-33-cyt\ c(Fe^{III})]$, where ruthenium(III) is initially reduced. It should be noted that the intramolecular reaction rate is independent of the concentration of modified cyt c present, indicating that competition from intermolecular reactions of the type in eq. (4.39) can be ignored. There is some dependence of the intramolecular rate constant on the nature of the reducing species but a value of 53 s^{-1} for the reaction in which CO_2^- is the reductant at 25°C, pH 7.0 in 0.1 M phosphate buffer and 0.1 M hydrogen carbonate has been most recently reported [80]. Activation parameters are $\Delta H^{\ddagger} = 15$ kJ mol^{-1}, $\Delta S^{\ddagger} = -163$ J K^{-1} mol^{-1}, and $\Delta V^{\ddagger} = -17.7$ cm^3 mol^{-1}, comparable to values obtained for other weakly coupled systems [80, 81].

$$[(NH_3)_5Ru^{III}his\text{-}33\text{-}cyt\ c(Fe^{III})] + CO_2^- \rightarrow$$
$$[(NH_3)_5Ru^{III}his\text{-}33\text{-}cyt\ c(Fe^{II})] + CO_2 \quad (4.36)$$

$$[(NH_3)_5Ru^{III}his\text{-}33\text{-}cyt\ c(Fe^{III})] + CO_2^- \rightarrow$$
$$[(NH_3)_5Ru^{II}his\text{-}33\text{-}cyt\ c(Fe^{III})] + CO_2 \quad (4.37)$$

$$[(NH_3)_5Ru^{II}his\text{-}33\text{-}cyt\ c(Fe^{III})] \rightarrow [(NH_3)_5Ru^{III}his\text{-}33\text{-}cyt\ c(Fe^{II})] \quad (4.38)$$

$$[(NH_3)_5[Ru^{III}his\text{-}33\text{-}cyt\ c(Fe^{III})] + [(NH_3)_5Ru^{II}his\text{-}33\text{-}cyt\ c(Fe^{III})] \rightarrow$$
$$[(NH_3)_5[Ru^{III}his\text{-}33\text{-}cyt\ c(Fe^{II})] + [(NH_3)_5Ru^{III}his\text{-}33\text{-}cyt\ c(Fe^{III})] \quad (4.39)$$

Interpretation of this result is not at all straightforward. It must be demonstrated that the reaction does indeed represent long-range electron transfer and not some rate-limiting conformational process due to the assembly. This point can be clarified by examination of the effect of the driving force on the rate of electron transfer. If the rate is determined by some conformational change, it should be independent of the driving force for the reaction. There is now an extensive body of information available on driving force effects in the cyt c system obtained by taking advantage of the fact that replacement of Zn for Fe in the heme center allows reactions of the excited state zinc porphyrin to be investigated (eq. (4.40)), and under suitable circumstances, the subsequent thermal reaction can also be investigated (eq. (4.41)) [82]. The triplet excited state of zinc substituted cytochrome c, cyt c(*Zn^{II}), lies 1.7 eV above the ground state and has a substantial lifetime [83] which makes it ideal for photoinduced electron transfer studies [84]. It acts both as an oxidant in $[(NH_3)_5Ru^{II}his\text{-}33\text{-}cyt\ c(*Zn^{II})]$ and as a reductant in $[(NH_3)_5Ru^{III}his\text{-}33\text{-}cyt\ c(*Zn^{II})]$. Studies have also been carried out by modification of the ruthenium derivative. The results of the different reactions and their driving forces are presented in Table 4.9.

$$[(NH_3)_5Ru^{III}his\text{-}33\text{-}cyt\ c(*Zn^{II})] \underset{k_2}{\overset{k_1}{\rightleftharpoons}} [(NH_3)_5Ru^{II}his\text{-}33\text{-}cyt\ c(Zn^{II})^+]$$

$$(4.40)$$

$$[(NH_3)_5Ru^{II}his\text{-}33\text{-}cyt\ c(Zn^{II})^+] \underset{k_2}{\overset{k_1}{\rightleftharpoons}} [(NH_3)_5Ru^{III}his\text{-}33\text{-}cyt\ c(Zn^{II})]$$

(4.41)

Table 4.9. Rate constants for intramolecular electron transfer within cyt c derivatives

Protein	$E°$ (V)	k (s^{-1})	$\Delta H^{\ddagger}$ (kJ mol^{-1})	$\Delta S^{\ddagger}$ (J K^{-1} mol^{-1})	Ref.
$[(NH_3)_5Ru^{II}his\text{-}33\text{-}cyt\ c(Fe^{III})]$	0.18	3.0×10^1	8	-180	78
$[(NH_3)_5Ru^{II}his\text{-}33\text{-}cyt\ c(Fe^{III})]$	0.18	5.3×10^1	15	-163	80
$[(NH_3)_5Ru^{II}his\text{-}33\text{-}cyt\ c(*Zn^{II})]$	0.36	2.4×10^2	9	-171	84
$[(NH_3)_5Ru^{III}his\text{-}33\text{-}cyt\ c(*Zn^{II})]$	0.70	7.7×10^5	7	-113	84
$[(NH_3)_5Ru^{II}his\text{-}33\text{-}cyt\ c(Zn^{II})^+]$	1.01	1.6×10^6	—	—	84
$[(NH_3)_4(isn)Ru^{II}his\text{-}33\text{-}cyt\ c(*Zn^{II})]$	1.05	2.9×10^6	<2	-126	85
$[(NH_3)_4(py)Ru^{III}his\text{-}33\text{-}cyt\ c(*Zn^{II})]$	0.97	3.3×10^6	9	-92	85
$[(NH_3)_4(isn)Ru^{II}his\text{-}33\text{-}cyt\ c(Zn^{II})^+]$	0.66	2.0×10^5	<2	-146	85
$[(NH_3)_4(py)Ru^{II}his\text{-}33\text{-}cyt\ c(Zn^{II})^+]$	0.74	3.5×10^5	<2	-142	85
$[(bpy)_2(Im)Ru^{III}his\text{-}33\text{-}cyt\ c(Fe^{II})]$	0.74	2.6×10^6	—	—	—

The analysis of these data in terms of eq. (4.22) is shown in Fig. 4.16 where log k_{et} is plotted against $\Delta G°$ for the series of reactants. While λ_{out} and the contributions to λ_{in} from the ruthenium complexes are expected to be substantially constant (Chapter 2), the reorganization at the porphyrin differs for cyt c(Fe^{II}) ($\lambda = 1.2$ eV), cyt c(Zn^{II})$^+$ ($\lambda = 1.19$ eV), and cyt c(*Zn^{II}) ($\lambda = 1.10$ eV) [87]. The electronic coupling H_{AB} is approximately 0.03 cm^{-1} in the iron–porphyrin system and approximately 0.12 cm^{-1} in the zinc system. These values are much smaller than those found in the long-range electron transfer in the oligoproline complexes.

Further light on this long-range electron transfer has been shed by studies with ruthenated derivatives of the heme-containing oxygen transport protein myoglobin [87]. Again analysis involves driving force studies with zinc-substituted derivatives, and a summary of the results is presented in Table 4.10. In attempting to understand the effect on H_{AB} of changing the distance between the oxidant and the reductant, one must first address the question of how to measure the distance. To this point, metal center to metal center distance has provided an adequate measure of the separation of the reaction centers, however, it is clear that r_o, the distance at which electron transfer is adiabatic, is finite. The convention has been adopted that r_o is taken as the metal and its immediate ligand shell, which includes the whole porphyrin ligand where applicable. The distance r is then the closest distance between these ligand shells.

Table 4.10. Rate constants for electron transfer in myoglobin derivatives [88]

	Distance (Å)	k (s^{-1})
[(NH$_3$)$_5$Ru^{III}his-48-Mb(*Zn^{II})]	12.7	7.2×10^4
[(NH$_3$)$_5$Ru^{III}his-81-Mb(*Zn^{II})]	19.3	1.5×10^2
[(NH$_3$)$_5$Ru^{III}his-116-Mb(*Zn^{II})]	20.1	3.0×10^1
[(NH$_3$)$_5$Ru^{III}his-12-Mb(*Zn^{II})]	22.0	1.4×10^2

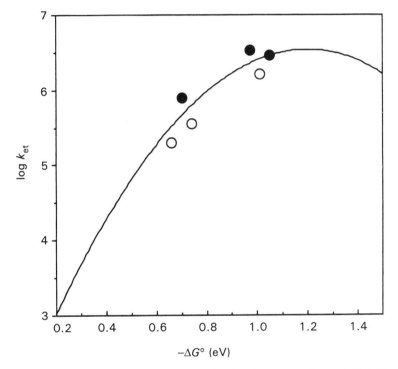

Fig. 4.16. Dependence of log k_{et} for intramolecular electron transfer on $\Delta G°$ in His-33 ruthenium modified Zn cytochrome c. The solid line is the best fit to eq. (4.22) using the parameters $\lambda = 1.2$ eV and $H_{AB} = 0.12$ cm^{-1}. Photoinduced reactions are shown as closed circles, recombination reactions as open circles. From Ref. [87].

It is interesting to note that for myoglobin which has no natural electron transfer function, λ is larger than with the cytochromes despite a similar porphyrin core. A similar conclusion has been reached in studies involving intermolecular electron transfer [89]. There is a distance dependence with $\beta = 0.79$ Å^{-1}. However, H_{AB} is much smaller than values obtained for reactions over comparable distances where the reactant centers are linked by a continuous series of bonds. The strong dependence on the nature of the intervening medium reveals that electronic coupling is promoted by a continuous series of bonding interactions. In these long-range electron transfer

210 Intramolecular electron trnasfer [Ch. 4

reactions of the ruthenated proteins, the peptide backbone provides the only continuous series of covalent bonds between the reactants and the number of bonds is not related to r, the closest distance of approach. Detailed calculations are now able to pinpoint likely pathways for the electron based on through bond, hydrogen bond and through space interactions with limited success [90].

In conclusion, the factors which affect electron transfer in biological systems have received a great deal of attention and this is reflected in the detail with which the experimental results can be interpreted. It is significant that Chapters 2, 3, and 4 have dealt primarily with the simplest of charge transfer process, electron transfer. In Chapter 5 more complex processes will be encountered involving transfer of charge in multiple units.

QUESTIONS

4.1 Discuss two examples of behavior which require an explanation in terms of non-adiabatic electron transfer.

4.2 Plant ferredoxins have molecular weight $\approx$ 10 500 and contain a {2Fe-2S} active site which cycles between the Fe_2^{III} and $Fe^{III}Fe^{II}$ states with a reduction potential of approximately -0.40 V. Pseudo-first-order rate constants for the oxidation of a reduced ferredoxin from spinach by $[Co(NH_3)_6]^{3+}$ and $[Cr(phen)_3]^{3+}$ at 25.0°C and 0.10 M ionic strength are presented in the table.

Pseudo-first-order rate constants at pH 8.0.

$[Co(NH_3)_6]^{3+}$ (M)	$[Cr(NH_3)_6]^{3+}$ (M)	$[Cr(phen)_3]^{3+}$ (M)	k_{obsd} (s^{-1})
2.4×10^{-4}			4.87
3.0×10^{-4}	2.5×10^{-4}		2.90
3.0×10^{-4}	5.0×10^{-4}		2.84
3.0×10^{-4}	1.0×10^{-3}		2.09
3.0×10^{-4}	2.0×10^{-3}		1.47
3.0×10^{-4}	4.0×10^{-3}		0.95
4.9×10^{-4}			7.33
9.9×10^{-4}			10.4
1.98×10^{-3}			14.6
3.93×10^{-3}			16.5
4.87×10^{-3}			17.4
		2.5×10^{-4}	2.45
		5.0×10^{-4}	4.42
		1.0×10^{-3}	7.1
		2.0×10^{-3}	9.8
		4.0×10^{-3}	12.2
		5.0×10^{-3}	13.3

Determine rate laws for the reactions which account for the concentration dependencies, suggest mechanisms and speculate whether these reactions are likely to obey the Marcus relationship.
(Lloyd, E.; Tomkinson, N. P.; Sykes, A. G. *J. Chem. Soc., Dalton Trans.* 1992, 753–756.)

4.3 The self-exchange rate for $[Fe(TPP)(MeIm)_2]^{+/0}$ (TPP = tetraphenyl porphyrin, MeIm = 1-methylimidazole) is 8×10^7 at $-21\,°C$ in CD_2Cl_2 solution. Discuss the relevance of this observation for reactions of the cytochrome c.
(Shirazi, A.; Barbush, M.; Ghosh, S.; Dixon, D. W. *Inorg. Chem.* 1985, **24**, 2495–2502.)

REFERENCES

[1] Robin, M. B.; Day, P. *Adv. Inorg. Chem. Radiochem.* 1967, **10**, 247–422.
[2] Day, P. *Comments Inorg. Chem.* 1981, **1**, 155–167.
[3] Marcus, R. A.; Sutin, N. *Comments Inorg. Chem.* 1986, **5**, 119–133.
[4] Creutz, C. *Prog. Inorg. Chem.* 1983, **30**, 1–73.
[5] Creutz, C.; Taube, H. *J. Am. Chem. Soc.* 1969, **91**, 3988–3989.
[6] Taube, H. *Ann. N.Y. Acad. Sci.* 1978, **313**, 481–495.
[7] Sutton, J. E.; Taube H. *Inorg. Chem.* 1981, **20**, 3125–3134.
[8] Callahann, R. W.; Keene, F. R.; Meyer, T. J.; Salmon, D. J. *J. Am. Chem. Soc.* 1977, **99**, 1064–1073.
[9] Powers, M. J.; Meyer, T. J. *J. Am. Chem. Soc.*, 1980, **102**, 1289–1297.
[10] Felix, F.; Ludi, A. *Inorg. Chem.* 1978, **17**, 1782.
[11] Stein, C. A.; Taube, H. *J. Am. Chem. Soc.* 1978, **100**, 1635–1637.
[12] Stein, C. A.; Lewis, N.; Seitz, G. *J. Am. Chem. Soc.* 1982, **104**, 2596–2599.
[13] Creutz, C. *Inorg. Chem.* 1978, **17**, 3723–3725.
[14] Meyer, T. J. *Chem. Phys. Letters* 1979, **64**, 417–420.
[15] Beratran, D. A.; Hopfield, J. J. *J. Am. Chem. Soc.* 1984, **106**, 1584–1594.
[16] Lewis, N. A.; Obeng, Y. S. *J. Am. Chem. Soc.* 1989, **111**, 7624–7625.
[17] Lewis, N. A.; Obeng, Y. S. *J. Am. Chem. Soc.* 1988, **110**, 2306–2307.
[18] Callahan, R. W.; Brown, G. M.; Meyer, T. J. *J. Am. Chem. Soc.* 1974, **96**, 7829–7830.
[19] Callahan, R. W.; Brown, G. M.; Meyer, T. J. *Inorg. Chem.* 1975, **14**, 1443–1453.
[20] Yeh, A.; Haim, A. *J. Am. Chem. Soc.* 1985, **107**, 369–376.
[21] Yeh, A.; Haim, A.; Tanner, M.; Ludi, A. *Inorg. Chim Acta* 1979, **33**, 51–56.
[22] Vogler, A.; Kisslinger, J. *J. Am. Chem. Soc.* 1982, **104**, 2311.
[23] Burewicz, A.; Haim, A. *Inorg. Chem.* 1988, **27**, 1611–1614.
[24] Walker, G. C.; Barbara, P. F.; Doorn, S. K.; Dong, Y.; Hupp, J. T. *J. Phys. Chem.* 1991, **95**, 5712–5715.
[25] Doorn, S. K.; Stoutland, P. O.; Dyer, R. B.; Woodruff, W. H. *J. Am. Chem. Soc.* 1992, **114**, 3133–3134.
[26] Curtis, J. C. Meyer, T. J. *Inorg. Chem.* 1982, **21**, 1562–1571.

[27] Khoshtariya, D. E.; Kjaer, A. M.; Marsagishvili, T. A.; Ulstrup, J. *J. Phys. Chem.* 1991, **95**, 8797–8804.
[28] Isied, S. S.; Taube, H. *J. Am. Chem. Soc.* 1973, **95**, 8198–8200.
[29] Jwo, J.-J.; Haim, A. *J. Am. Chem. Soc.* 1976, **98**, 1172–1176.
[30] Zawacky, S. K. S.; Taube, H. *J. Am. Chem. Soc.* 1981, **103**, 3379–3387.
[31] Reider, K.; Taube, H. *J. Am. Chem. Soc.* 1977, **99**, 7891–7894.
[32] Fischer, H.; Tom, G. M.; Taube, H. *J. Am. Chem. Soc.* 1976, **98**, 5512–5517.
[33] Geselowitz, D. A. *Inorg. Chem.* 1987, **26**, 4135–4137.
[34] Isied, S. S.; Vassilian, A. *J. Am. Chem. Soc.* 1984, **106**, 1726–1732.
[35] Isied, S. S.; Vassilian, A. *J. Am. Chem. Soc.* 1984, **106**, 1732–1736.
[36] Isied, S. S.; Ogawa, M. Y.; Wishart, J. F. *Chem. Rev.* 1992, **92**, 381–394.
[37] Isied, S. S.; Vassilian, A.; Magnuson, R. H.; Schwarz, H. A. *J. Am. Chem. Soc.* 1985, **107**, 7432–7438.
[38] Vassilian, A.; Wishart, J. F.; van Hemelryck, B; Schwarz, H. A.; Isied, S. S. *J. Am. Chem. Soc.* 1990, **112**, 7278–7286.
[39] Hopfield, J. J. *Proc. Natl. Acad. Sci., USA* 1974, **71**, 3640–3644.
[40] Isied, S. S.; Vassilian, A.; Wishart, J. F.; Creutz, C.; Schwarz, H.; Sutin, N. *J. Am. Chem. Soc.* 1988, **110**, 635–637.
[41] Closs, G. L.; Piotrowiak, P.; MacInnis, J. M.; Fleming, G. R. *J. Am. Chem. Soc.* 1988, **110**, 2652–2653.
[42] Siddarth, P.; Marcus, R. A. *J. Phys. Chem.* 1990, **94**, 2985–2989.
[43] Siddarth, P.; Marcus, R. A. *J. Phys. Chem.* 1992, **96**, 3213–3217.
[44] Moore, G. R.; Eley, C. G. S.; Williams, G. *Adv. Inorg. Bioinorg. Mech.* 1984, **3**, 1–96.
[45] Williams, G.; Moore, G. R.; Williams, R. J. P. *Comments Inorg. Chem.* 1985, **4**, 55–98.
[46] Dickinson, R. E.; Takano, T.; Eisenberg, D.; Kallai, O. B.; Samson, L; Cooper, A.; Margoliash, E. *J. Biol. Chem.* 1971, **246**, 1511–1535.
[47] Gupta, R. K.; Koenig, S. H.; Redfield, A. G. *J. Magn. Reson.* 1972, **7**, 66–73.
[48] Barlow, G. H.; Margoliash, E. *J. Biol. Chem.* 1966, **241**, 1473–1477.
[49] Wherland, S.; Gray, H. B. *Proc. Natl. Acad. Sci., USA* 1976, **73**, 2950–2954.
[50] Rush, J. D.; Lan, J.; Koppenol, W. H. *J. Am. Chem. Soc.* 1987, **109**, 2679–2682.
[51] Cheddar, G. Meyer, T. E.; Cusanovich, M. A.; Stout, C. D.; Tollin, G. *Biochemistry,* 1989, **28**, 6318–6322.
[52] Butler, J.; Davies, D. M.; Sykes, A. G.; Koppenol, W. H.; Osheroff, N.; Margoliash, E. *J. Am. Chem. Soc.* 1981, **103**, 469–471.
[53] Holwerda, R. A.; Knaff, D. B.; Gray, H. B.; Clemmer, J. D.; Crowley, R.; Smith, J. M.; Mauk, A. G. *J. Am. Chem. Soc.* 1980, **102**, 1142–1146.
[54] McArdle, J. V.; Gray, H. B.; Creutz, C.; Sutin, N. *J. Am. Chem. Soc.* 1974, **96**, 5737–5741.
[55] Cummins, D.; Gray, H. B. *J. Am. Chem. Soc.* 1977, **99**, 5158–5167.
[56] Carney, M. J.; Lesniak, J. S.; Likar, M. D., Pladziewicz, J. R. *J. Am. Chem. Soc.* 1984, **106**, 2565–2569.
[57] Sutin, N. *Adv. Chem.* 1977, **162**, 156–172.

[58] Hodges, H. L.; Holwerda, R. A.; Gray, H. B. *J. Am. Chem. Soc.* 1974, **96**, 3132–3137.
[59] Eley, C. G. S.; Moore, G. R.; Williams, G.; Williams, R. J. P. *Eur. J. Biochem.* 1982, **124**, 295–303.
[60] Armstrong, G. D.; Chambers, J. A.; Sykes, A. G. *J. Chem. Soc., Dalton Trans.* 1986, 755–758.
[61] Drake, P. L.; Hartshorn, R. T.; McGinnis, J.; Sykes, A. G. *Inorg. Chem.* 1989, **28**, 1361–1366.
[62] Wherland, S.; Gray, H. B. *Biological Aspects of Inorganic Chemistry*, Wiley, New York, 1977, 289–360.
[63] Sykes, A. G. *Structure and Bonding* 1991, **75**, 175–224.
[64] Sykes, A. G. *Chem. Soc. Rev.* 1985, **14**, 283–315.
[65] Guss, J. M.; Freeman, H. C. *J. Mol. Biol.* 1983, **169**, 521–563.
[66] Guss, J. M.; Harrowell, P. R.; Murata, M.; Norris, V. A.; Freeman, H. C. *J. Mol. Biol.* 1986, **192**, 361–387.
[67] Sinclair-Day, J. D.; Sisley, M. J.; Sykes, A. G.; King, G. C.; Wright, P. E. *J. Chem. Soc., Chem. Commun.* 1985, 505–507.
[68] Segal, M. G.; Sykes, A. G. *J. Am. Chem. Soc.* 1978, **100**, 4585–4592.
[69] McGinnis, J.; Sinclair-Day, J. D.; Sykes, A. G. *J. Chem. Soc., Dalton Trans.* 1986, 2007–2009.
[70] Pladziewicz, J. R.; Brenner, M. S. *Inorg. Chem.* 1987, **26**, 3629–3634.
[71] Cookson, D. J.; Hayes, M. T.; Wright, P. E. *Biochim. Biophys. Acta* 1980, **591**, 162–176.
[72] Sinclair-Day, J. D.; Sykes, A. G. *J. Chem. Soc., Dalton Trans.* 1986, 2069–2073.
[73] McGinnis, J.; Sinclair-Day, J. D.; Sykes, A. G. *J. Chem. Soc., Dalton Trans.* 1986, 2011–2012.
[74] Mauk, A. G.; Bordignon, E.; Gray, H. B. *J. Am. Chem. Soc.* 1982, **104**, 7654–7657.
[75] Brunschwig, B. S.; DeLaive, P. J.; English, A. M.; Goldberg, M.; Gray, H. B.; Mayo, S. L.; Sutin, N. *Inorg. Chem.* 1985, **24**, 3743–3749.
[76] Armstrong, F. A. *Adv. Inorg. Bioinorg. Mech.* 1982, **1**, 65–120.
[77] Winkler, J. R.; Nocera, D. G.; Yocom, K. M.; Bordignon, E.; Gray, H. B. *J. Am. Chem. Soc.* 1982, **104**, 5798–5800.
[78] Nocera, D. G.; Winkler, J. R.; Yocom, K. M.; Bordignon, E.; Gray, H. B. *J. Am. Chem. Soc.* 1984, **106**, 5145–5150.
[79] Isied, S. S.; Worosila, G.; Atherton, S. J. *J. Am. Chem. Soc.* 1982, **104**, 7659–7661.
[80] Isied, S. S.; Kuehn, C.; Worosila, G. *J. Am. Chem. Soc.* 1984, **106**, 1722–1726.
[81] Wishart, J. F.; van Eldick, R.; Sun, J.; Su, C.; Isied, S. S. *Inorg. Chem.* 1992, **31**, 3986–3989.
[82] Winkler, J. R.; Gray, H. B. *Chem. Rev.* 1992, **92**, 369–379.
[83] Dixit, B. P. S. N.; Moy, V. T.; Vanderkooi, J. M. *Biochemistry* 1984, **23**, 2103–2107.
[84] Elias, H.; Chou, M. H.; Winkler, J. R. *J. Am. Chem. Soc.* 1988, **110**, 429–434.
[85] Meade, T. J.; Gray, H. B.; Winkler, J. R. *J. Am. Chem. Soc.* 1989, **111**, 4353–4356.
[86] Chang, I.-J.; Gray, H. B.; Winkler, J. R. *J. Am. Chem. Soc.* 1991, **113**, 7056–7057.

[87] Winkler, J. R.; Gray, H. B. *Chem. Rev.* 1992, **92,** 369–379.
[88] Cowan, J. A.; Upmacis, R. K.; Beratan, D. N.; Onuchic, J. N.; Gray, H. B. *Ann. New York Acad. Sci.* 1988, **550,** 68–84.
[89] Mauk, A. G.; Gray, H. B. *Biochem. Biophys. Res. Commun.* 1979, **86,** 206–210.
[90] Beratan, D. N.; Onuchie, J. N.; Betts, J. N.; Bowler, B. E.; Gray, H. B. *J. Am. Chem. Soc.* 1990, **112,** 7915–7921.

5

Multiple electron transfers

5.1 INTRODUCTION

In this final chapter, the focus of the discussion changes from simple electron transfer to more complex reactions where multiple electron transfer is involved. Many of the concepts are the same. The distinction between processes which involve substitution as an essential component of the charge transfer and processes which do not remains valid. However, anion and cation transfer provide a direct link with the more conventional concepts of substitution mechanisms. The chapter begins with an examination of multiple electron transfer between metal ion complexes where different mechanistic types are introduced. These mechanistic types are expanded in a discussion of electron transfer reactions involving metal ion complexes and non-metallic substrates and finally some of the complexities inherent in reactions between non-metallic species are explored.

The transfer of more than one electron or its equivalent in a reaction involves mechanisms which are intrinsically more complex than those involved in the transfer of a single electron. Changes in oxidation state by two or more units result in large structural changes and frequently in changes in coordination number for the species involved. For example, oxidation of $[V(H_2O)_6]^{2+}$ by one electron leads to $[V(H_2O)_6]^{3+}$, which has a similar six-coordinate structure, but oxidation by two electrons leads to $[VO(H_2O)_4]^{2+}$, which is predominantly square pyramidal with one oxo ligand. Changes in coordination number during the course of an electron transfer reaction carry with them a requirement that substitution processes are also involved. There is frequently such close coupling between transfer of an electron and the substitution process that it is difficult to distinguish whether substitution precedes or succeeds electron transfer or whether they are synchronous and if so, whether they are better described as atom, anion or cation transfer processes. Questions such as these will arise frequently in this chapter.

A distinction must be raised at this point between reactions which are *complementary* such as the reaction between $[Tl_{aq}]^{3+}$ and $[U_{aq}]^{4+}$, (eq. (5.1)), where more than one electron is transferred but the stoichiometry is 1 : 1 [1] and those which are *non-complementary* such as the reaction in eq. (5.2) where the stoichiometry differs from 1 : 1 [2]. In both reactions, $[Tl_{aq}]^{3+}$ acts as a net acceptor of two electrons.

$$[Tl_{aq}]^{3+} + [U_{aq}]^{4+} \rightarrow [Tl_{aq}]^{+} + [(UO_2)_{aq}]^{2+} \tag{5.1}$$

$$[Tl_{aq}]^{3+} + 2\,[Fe(H_2O)_6]^{2+} \rightarrow [Tl_{aq}]^{+} + 2\,[Fe(H_2O)_6]^{3+} \tag{5.2}$$

The complementary reaction, eq. (5.1), can proceed in a single bimolecular elementary step involving transfer of two electrons, or in a more complex series of reactions. However, the non-complementary reaction, eq. (5.2), must take place by a more complex mechanism involving a minimum of two elementary steps with the formation of reaction intermediates. Whilst unravelling the intricacies of complex mechanisms can be challenging, the increased complexity of the experimental rate law greatly enhances the amount of mechanistic information available. In addition to the rate law, product analysis and the detection of reaction intermediates either by direct spectrophotometric observation or by chemical reaction can provide significant information. Consequently, a knowledge of the chemistry of intermediate species is very important.

In order to explore the complexities of reactions involving multiple electron transfer, a few examples will be considered to illustrate different facets of behavior rather than any attempt made to present comprehensive coverage [3, 4]. The first two examples involve metal ion complexes where a two-electron change is involved and the intermediate oxidation state is thermodynamically unstable with respect to disproportionation. The reactivity patterns differ because of differences in the nature of the lability of the reagents. While multiple electron transfer can be exhibited by reagents where the intermediate oxidation state is thermodynamically stable, this behavior is much less common and electron transfer in single units prevails. Reactions where multiple electron transfer is accompanied by transfer of a bridging group are also considered and the implications of these as atom or group transfer processes are discussed.

5.2 REACTIONS OF THALLIUM(III)/(I)

The main group metal ion reagent $[Tl_{aq}]^{3+}$ is a powerful two-electron oxidant. In 1.0 M $HClO_4$, the potential of eq. (5.3) is 1.25 V [5]. Whereas $[Tl_{aq}]^{3+}$ is susceptible to hydrolysis with pK_h values of 1.14 and 2.61 (eqs (5.4) and (5.5)), which means the $[TlOH_{aq}]^{2+}$ is the dominant species in moderately acidic solution, the lower oxidation state, $[Tl_{aq}]^{+}$, shows no evidence for hydrolysis in the acidic pH range [6]. As with most main group metals, substitution rates are generally high for both oxidation states so that the assignment of inner-sphere and outer-sphere mechanisms is problematic.

$$[Tl_{aq}]^{3+} + 2e^{-} \rightleftharpoons [Tl_{aq}]^{+} \tag{5.3}$$

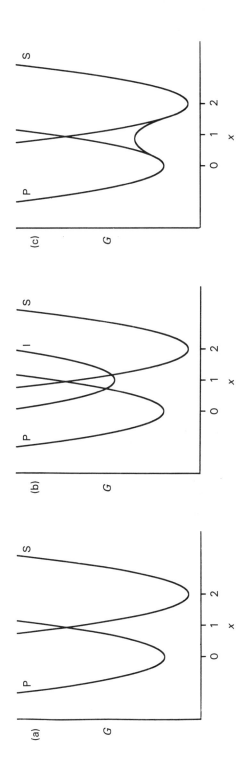

Fig. 5.1. Energetic requirements for two-electron transfer. In (a) an outer-sphere two-electron transfer is depicted. The reorganizational energies, λ, are much larger than for single electron transfer. In (b) a thermodynamically unfavorable intermediate oxidation state provides a lower energy reaction pathway, while in (c) an inner-sphere mechanism operates.

$$[Tl_{aq}]^{3+} \overset{K_{h1}}{\rightleftharpoons} [TlOH_{aq}]^{2+} + H^+ \tag{5.4}$$

$$[TlOH_{aq}]^{2+} \overset{K_{h2}}{\rightleftharpoons} [(Tl(OH)_2)_{aq}]^+ + H^+ \tag{5.5}$$

The complementary reduction of $[Tl_{aq}]^{3+}$ by $[U_{aq}]^{4+}$ obeys a simple two-term rate law which is first-order in both reactants and independent of the reaction products (eq. (5.6)) in strongly acidic media, where the subscript T refers to total concentration of the reagent [7]. Scheme 5.1 represents the simpler of two mechanisms which are possible according to the rate-law. Electron transfer takes place by parallel pathways which involve single two-electron steps (eq. (5.7)).

$$\frac{-d[[U_{aq}]^{4+}]}{dt} = \frac{k_1[[U_{aq}]^{4+}][[Tl_{aq}]^{3+}]_T}{[H^+]} + \frac{k_2[[U_{aq}]^{4+}][[Tl_{aq}]^{3+}]_T}{[H^+]^2} \tag{5.6}$$

$$[Tl_{aq}]^{3+} + [U_{aq}]^{4+} \rightarrow [Tl_{aq}]^+ + [(UO_2)_{aq}]^{2+} \tag{5.7}$$

Scheme 5.1.

The energetic requirements of this situation are illustrated by Fig. 5.1. If the reaction is considered to be outer-sphere in nature, a large Frank–Condon barrier can be anticipated since the changes in geometry and coordination involved in reaction (5.7) are large. Simultaneous transfer of more than one electron in a single step by an outer-sphere mechanism is not necessarily a forbidden process; however, competing mechanisms can intrude. Two of the more common competitive mechanisms are also illustrated in Fig. 5.1. The reaction can proceed by an inner-sphere mechanism where there is strong electronic coupling between the donor and acceptor states as the result of formation of a bridge between the two centers. In fact this is the likely mechanism in this instance. The domination of terms in the rate expression with inverse acid dependencies is a consequence of the changes in protonation of the reagents during the electron transfer process and may also be indicative of the formation of inner-sphere hydroxo bridges. Alternatively, electron transfer may be by sequential one electron transfers (Scheme 5.2).

$$[Tl_{aq}]^{3+} + [U_{aq}]^{4+} \underset{k_{-1}}{\overset{k_1}{\rightleftharpoons}} [Tl_{aq}]^{2+} + [(UO_2)_{aq}]^+ \tag{5.8}$$

$$[Tl_{aq}]^{2+} + [(UO_2)_{aq}]^+ \overset{k_2}{\rightarrow} [Tl_{aq}]^+ + [(UO_2)_{aq}]^{2+} \tag{5.9}$$

$$[Tl_{aq}]^{3+} + [(UO_2)_{aq}]^+ \rightarrow [Tl_{aq}]^{2+} + [(UO_2)_{aq}]^{2+} \tag{5.10}$$

$$[Tl_{aq}]^{2+} + [U_{aq}]^{4+} \rightarrow [Tl_{aq}]^+ + [(UO_2)_{aq}]^+ \tag{5.11}$$

$$2\,[\text{Tl}_{\text{aq}}]^{2+} \rightarrow [\text{Tl}_{\text{aq}}]^{+} + [\text{Tl}_{\text{aq}}]^{3+} \tag{5.12}$$

$$2\,[(\text{UO}_2)_{\text{aq}}]^{+} \rightarrow [(\text{UO}_2)_{\text{aq}}]^{2+} + [\text{U}_{\text{aq}}]^{4+} \tag{5.13}$$

Scheme 5.2.

The mechanism in Scheme 5.2 is also consistent with the rate law in eq. (5.6). However, this rate law represents only one limiting form, where k_1 in eq. (5.8) is rate-determining. Single-electron transfer results in the production of the thermodynamically unfavorable oxidation states $[\text{Tl}_{\text{aq}}]^{2+}$ and $[(\text{UO}_2)_{\text{aq}}]^{+}$, and while structural rearrangement is presumably less important than for the two-electron change, the energetics for a rate-determining single-electron outer-sphere process are also very unfavorable (Fig. 5.1). Complications can arise from the presence of the intermediate oxidation states and a chain mechanism where the reaction (5.8) is the initiation step, reactions (5.10) and (5.11) are the propagation steps, and reactions (5.9), (5.12), and (5.13) are termination steps, is possible. The chain mechanism is expected to result in more complex kinetic behavior and to be susceptible to interruption by scavengers for the chain carriers. A knowledge of the chemistry of the intermediate oxidation states is therefore desirable for detailed mechanistic studies.

In spite of these arguments which favor Scheme 5.1 as the mechanism, there remains ambiguity in the nature of the rate-limiting step. In a limiting case in Scheme 5.2, where the initial electron transfer, k_1, is rate-limiting and k_2 exceeds the rate at which the products $\{[\text{Tl}_{\text{aq}}]^{2+},[(\text{UO}_2)_{\text{aq}}]^{+}\}$ can diffuse apart, the second electron transfer will take place within the same solvent cage as the initial transfer. No free single-electron oxidized or reduced products are produced and the reaction is indistinguishable from the two-electron mechanism in Scheme 5.1. Note, however, that the energetic requirements for the two mechanisms differ. Those for Scheme 5.2 should reflect the energetics of single-electron transfer, and as such should be predictable using the Marcus relationship.

Whereas a complementary reaction can occur in a single step since it involves the assembly of only two molecular species, clearly the non-complementary process cannot since at least three species must be assembled. The reduction of $[\text{Tl}_{\text{aq}}]^{3+}$ by $[\text{Fe}(\text{H}_2\text{O})_6]^{2+}$ shows complex behavior. The kinetics of this reaction have been investigated in HClO_4 and show a marked inhibition by $[\text{Fe}(\text{H}_2\text{O})_6]^{3+}$ following the rate law (eq. (5.15)), [2]. The rate law is consistent with a rate-limiting step involving a combination of one $[\text{Tl}_{\text{aq}}]^{3+}$ and two $[\text{Fe}(\text{H}_2\text{O})_6]^{2+}$ in a mechanism in which an intermediate, scavengable by both $[\text{Fe}(\text{H}_2\text{O})_6]^{2+}$ and $[\text{Fe}(\text{H}_2\text{O})_6]^{3+}$, is formed from one $[\text{Tl}_{\text{aq}}]^{3+}$ and one $[\text{Fe}(\text{H}_2\text{O})_6]^{2+}$. It is proposed that the intermediate is $[\text{Tl}_{\text{aq}}]^{2+}$, Scheme 5.3, and the rate constants are $k_1 = 1.39 \times 10^{-2}$ M^{-1} s^{-1} and $k_{-1}/k_2 = 5.1 \times 10^{-2}$ in 1.0 M HClO_4 at 25.0 °C [8].

$$[\text{Tl}_{\text{aq}}]^{3+} + 2[\text{Fe}(\text{H}_2\text{O})_6]^{2+} \rightarrow [\text{Tl}_{\text{aq}}]^{+} + 2[\text{Fe}(\text{H}_2\text{O})_6]^{3+} \tag{5.14}$$

$$\text{Rate} = \frac{k_1 k_2 [[\text{Tl}_{aq}]^{3+}][[\text{Fe}(H_2O)_6]^{2+}]^2}{k_{-1}[[\text{Fe}(H_2O)_6]^{3+}] + k_2[[\text{Fe}(H_2O)_6]^{2+}]} \quad (5.15)$$

$$[\text{Tl}_{aq}]^{3+} + [\text{Fe}(H_2O)_6]^{2+} \underset{k_{-1}}{\overset{k_1}{\rightleftharpoons}} [\text{Tl}_{aq}]^{2+} + [\text{Fe}(H_2O)_6]^{3+} \quad (5.16)$$

$$[\text{Tl}_{aq}]^{2+} + [\text{Fe}(H_2O)_6]^{2+} \overset{k_2}{\rightarrow} [\text{Tl}_{aq}]^{+} + [\text{Fe}(H_2O)_6]^{3+} \quad (5.17)$$

Scheme 5.3.

The acidity dependence of k_1 indicates that it is a composite term, $k_1 = \{K_{h1}/(K_{h1} + [H^+])\}\{k_1' + K_{h2}k_1''/[H^+]\}$, where k_1' and k_1'' are 1.4×10^{-2} M^{-1} s^{-1} and approximately 4 M^{-1} s^{-1} for reactions of $[\text{TlOH}_{aq}]^{2+}$ and $[\text{Tl(OH)}_{2aq}]^{+}$ respectively. The term k_{-1}/k_2 also has an acidity dependence which suggests the hydrolyzed complexes are the dominant reactants in both k_{-1} and k_2 [2]. The absence of pathways involving unhydrolyzed complexes suggests that the reactions may be inner-sphere with hydroxy-bridged intermediates.

There is an alternative mechanism for the non-complementary reduction of $[\text{Tl}_{aq}]^{3+}$ which avoids the formation of Tl^{2+}. In Scheme 5.4, the initial step is a two-electron process yielding $[\text{Fe}_{aq}]^{4+}$ as a transient, and the rate law (eq. (5.20), derived by the application of the steady-state approximation to this species, should show a term inhibitory in $[\text{Tl}_{aq}]^{+}$.

$$[\text{Tl}_{aq}]^{3+} + [\text{Fe}(H_2O)_6]^{2+} \rightleftharpoons [\text{Tl}_{aq}]^{+} + [\text{Fe}_{aq}]^{4+} \quad (5.18)$$

$$[\text{Fe}_{aq}]^{4+} + [\text{Fe}(H_2O)_6]^{2+} \rightarrow 2[\text{Fe}(H_2O)_6]^{3+} \quad (5.19)$$

Scheme 5.4.

$$\frac{-d[[\text{Tl}_{aq}]^{3+}]}{dt} = \frac{k_1 k_2 [[\text{Tl}_{aq}]^{3+}][[\text{Fe}(H_2O)_6]^{2+}]^2}{k_{-1}[[\text{Tl}_{aq}]^{+}] + k_2[[\text{Fe}(H_2O)_6]^{2+}]} \quad (5.20)$$

The two mechanisms shown in Schemes 5.3 and 5.4 can be distinguished on the basis of rate law and clearly Scheme 5.3 is correctly identified in the reduction of $[\text{Fe}(H_2O)_6]^{2+}$ by the term inhibitory in $[\text{Fe}(H_2O)_6]^{3+}$ rather than $[\text{Tl}_{aq}]^{+}$. However, under conditions where $k_2[\text{Fe}^{II}] \gg k_{-1}[\text{Fe}^{III}]$ for Scheme 5.3 or $k_2[\text{Fe}^{II}] \gg k_{-1}[[\text{Tl}_{aq}]^{+}]$ for Scheme 5.4, the rate laws are indistinguishable. In such cases, distinction between the mechanisms can be made on the basis of scavenging the reactive intermediates and the effect which this has on the reaction products. For example, $[\text{Tl}_{aq}]^{3+}$ oxidizes $[\text{V}(H_2O)_6]^{3+}$ to give predominantly $[\text{VO}_{aq}]^{2+}$ (eq.

(5.21)), and the rate law shows no inhibitory terms [9,10]. The second-order rate constant, 350 M^{-1} s^{-1} at 0.5°C and in 1.0 M HClO$_4$ is much more rapid than the reaction with [Fe(H$_2$O)$_6$]$^{2+}$. However, rapid induction of one equivalent of [Fe(H$_2$O)$_6$]$^{3+}$ results from the addition of each equivalent of [V(H$_2$O)$_6$]$^{3+}$ to [Tl$_{aq}$]$^{3+}$, indicating that the reactive intermediate, [Tl$_{aq}$]$^{2+}$, is formed in the reaction (eqs (5.22)–(5.23)).

$$[Tl_{aq}]^{3+} + 2\,[V(H_2O)_6]^{3+} \overset{fast}{\rightleftharpoons} [Tl_{aq}]^{+} + 2\,[VO_{aq}]^{2+} \qquad (5.21)$$

$$[Tl_{aq}]^{3+} + [V(H_2O)_6]^{3+} \rightarrow [Tl_{aq}]^{2+} + [VO_{aq}]^{2+} \qquad (5.22)$$

$$[Tl_{aq}]^{2+} + [Fe(H_2O)_6]^{2+} \rightarrow [Tl_{aq}]^{+} + [Fe(H_2O)_6]^{3+} \qquad (5.23)$$

It is possible to rule out an alternative mechanism in which a complementary two-electron reaction is followed by comproportionation to give the [VO]$^{2+}$ product since the comproportionation reaction is an order of magnitude too slow to participate in the overall reaction. [VO$_{aq}$]$^{2+}$ is oxidized only slowly by [Tl$_{aq}$]$^{3+}$ (eq. (5.24)), but note that reaction with [Tl$_{aq}$]$^{2+}$ is more rapid, and induction of the minor product [(VO$_2$)$_{aq}$]$^{+}$ is observed in the presence of added [VO$_{aq}$]$^{2+}$ as noted in the case of added [Fe(H$_2$O)$_6$]$^{2+}$ [11].

$$[Tl_{aq}]^{3+} + 2\,[VO_{aq}]^{2+} \overset{slow}{\rightarrow} [Tl_{aq}]^{+} + 2\,[(VO_2)_{aq}]^{+} \qquad (5.24)$$

The mechanism in Scheme 5.4 has been proposed for the reduction of [Tl$_{aq}$]$^{3+}$ by [Cr(H$_2$O)$_6$]$^{2+}$ [12]. The reaction is too fast to allow kinetic studies but the chromium(III) product is identified as a bis μ-hydroxy-dimer (eq. (5.25)), rather than monomeric [Cr(H$_2$O)$_6$]$^{3+}$, which is found in the reaction with one-electron oxidants such as [Fe(H$_2$O)$_6$]$^{3+}$.

$$[[Tl_{aq}]^{3+} + [Cr(H_2O)_6]^{2+} \rightarrow [Tl_{aq}]^{+} + [(H_2O)_4Cr(\mu\text{-}OH)_2Cr(OH_2)_4]^{4+} \qquad (5.25)$$

Since chromium(III) is inert to substitution on the timescale of the electron transfer process, the dimeric and monomeric forms do not interconvert rapidly and [(H$_2$O)$_4$Cr(μ-OH)$_2$Cr(OH$_2$)$_4$]$^{4+}$ must result from the comproportionation reaction (eq. (5.27)), involving substitution at labile [Cr(H$_2$O)$_6$]$^{2+}$.

$$[Tl_{aq}]^{3+} + [Cr(H_2O)_6]^{2+} \rightleftharpoons [Tl_{aq}]^{+} + [CrO_{aq}]^{2+} \qquad (5.26)$$

$$[CrO_{aq}]^{2+} + [Cr(H_2O)_6]^{2+} \rightarrow [(H_2O)_4Cr(\mu\text{-}OH)_2Cr(OH_2)_4]^{4+} \qquad (5.27)$$

The oxidation of $[V(H_2O)_6]^{2+}$ is also thought to be a complementary two-electron process with formation of $[VO_{aq}]^{2+}$ [10].

Electron transfer behavior of thermodynamically unstable intermediates such as $[Tl_{aq}]^{2+}$ can be examined independently. Rapid generation of $[Tl_{aq}]^{2+}$ is possible by either flash photolysis or pulse radiolysis with ·OH radicals (eq. (5.28)), and studies involving this reagent have given considerable insight in understanding the chemistry of the $[Tl_{aq}]^{3+/+}$ change [8, 13]. The rate of the reaction between $[Tl_{aq}]^{2+}$ and $[Fe(H_2O)_6]^{2+}$, k_2 in eq. (5.17), is determined to be 6.7×10^6 M^{-1} s^{-1} in 1.0 M HClO$_4$. From this value, it can be calculated that $k_{-1} = 3.4 \times 10^5$ M^{-1} s^{-1} and hence that $K_1\ (=k_1/k_{-1}) = 4.1 \times 10^{-8}$ (eq. (5.16)), corresponding to a difference in the reduction potential between $[Fe(H_2O)_6]^{3+/2+}$ and $[Tl_{aq}]^{3+/2+}$ of 0.44 V. In 1.0 M HClO$_4$, the reduction potential of $[Tl_{aq}]^{3+/+}$ is 1.26 V from which the potential for $[Tl_{aq}]^{2+/+}$ can be calculated to be 2.22 V.

$$[Tl_{aq}]^+ + \cdot OH \rightarrow [Tl_{aq}]^{2+} \tag{5.28}$$

The evaluation of the reduction potential of $[Tl_{aq}]^{2+}$ has provided a key for the interpretation of the mechanism of the self-exchange process between $[Tl_{aq}]^{3+}$ and $[Tl_{aq}]^+$ (eq. (5.29)). This is a formal two-electron process and the rate law is first-order in each reagent with a second-order rate constant of 4×10^{-3} M^{-1} s^{-1} at 25.0°C in 1.0 M HClO$_4$ [14–16]. The reaction is very sensitive to the reaction medium and there is some question regarding the [H$^+$] dependence (eq. (5.30)), which is interpreted in terms of two pathways, one involving $[Tl_{aq}]^{3+}$ and the other $[TlOH_{aq}]^{2+}$. However, the major question whether the mechanism involves $[Tl_{aq}]^{2+}$ or not can be answered. The equilibrium constant for the comproportionation reaction (eq. (5.31)), is 4×10^{-33}, and since the rate constant for the comproportionation reaction is 2×10^9 M^{-1} s^{-1} [17], the rate constant for the exchange reaction proceeding by $[Tl_{aq}]^{2+}$ is 8×10^{-24} M^{-1} s^{-1}, too slow to participate in the observed reaction. The mechanism is a two-electron process and since the ions are labile on the timescale of the redox process, an inner-sphere reaction is most likely.

$$[Tl_{aq}]^{3+} + [*Tl_{aq}]^+ \rightleftharpoons [*Tl_{aq}]^{3+} + [Tl_{aq}]^+ \tag{5.29}$$

$$\text{Rate} = k_1\,[[Tl_{aq}]^{3+}][[*Tl_{aq}]^+] + \frac{k_2\,[[Tl_{aq}]^{3+}][[*Tl_{aq}]^+]}{[H^+]} \tag{5.30}$$

$$[Tl_{aq}]^{3+} + [Tl_{aq}]^+ \rightleftharpoons 2\,[Tl_{aq}]^{2+} \tag{5.31}$$

In conclusion, the reactions of the $[Tl_{aq}]^{3+/+}$ system reveal evidence for both one- and two-electron processes. The distinction between these two processes is blurred since two-electron transfer may occur by two consecutive one-electron events without the detection of the intermediate state. The high substitution lability of the ions in both oxidation states prevents comments about group transfer. In the next section this latter question is explored for reactions where the oxidation states are inert to substitution.

5.3 REACTIONS OF PLATINUM(IV)/(II)

Platinum(II) complexes such as $[Pt(NH_3)_4]^{2+}$ are substitution-inert low-spin d^8 square-planar complexes whereas platinum(IV) complexes such as $[Pt(NH_3)_4Cl_2]^{2+}$ prefer a substitution-inert low-spin d^6 six-coordinate geometry. The intermediate oxidation state, platinum(III) is thermodynamically unstable, and the substantial geometry change between the two thermodynamically stable oxidation states focuses interest on electron transfer mechanisms.

The kinetics of the formal two-electron self-exchange process between [*trans*-$Pt(en)_2Cl_2]^{2+}$ and $[Pt(en)_2]^{2+}$, monitored [18] by ^{36}Cl isotope methods show that the rate is catalyzed by chloride ion (eq. (5.32)), with $k = 15\ M^{-2}\ s^{-1}$ at 25.0°C and 0.01 M ionic strength. Rates determined by ^{195}Pt, ^{14}C exchange give similar values.

$$\text{Rate} = k\,[[\textit{trans}\text{-Pt(en)}_2\text{Cl}_2]^{2+}]\,[[\text{Pt(en)}_2]^{2+}]\,[\text{Cl}^-] \tag{5.32}$$

The rate law suggests a mechanism in which a $[[Pt(en)_2]^{2+},Cl^-]$ adduct reacts by an inner-sphere pathway involving a bridging Cl^- from the platinum(IV) complex (eqs (5.33a)–(5.33b)).

$$[\text{Pt(en)}_2]^{2+} + \text{Cl}^- \rightleftharpoons [[\text{Pt(en)}_2]^{2+},\text{Cl}^-] \tag{5.33a}$$

$$[\textit{trans}\text{-*Pt(en)}_2\text{Cl}_2]^{2+} + [[\text{Pt(en)}_2]^{2+},\text{Cl}^-] \longrightarrow [\textit{trans}\text{-Pt(en)}_2\text{Cl}_2]^{2+} + [\text{*Pt(en)}_2]^{2+} + \text{Cl}^- \tag{5.33b}$$

The bridge is transferred in the reaction, and the transition state in eq. (5.34) is proposed. The axial interactions of the platinum(II) are weak and it is unclear when bond formation takes place. One consequence of the mechanism is that the reactions are subject to steric effects; the corresponding reaction of the Me_4en complexes is negligibly slow.

(5.34)

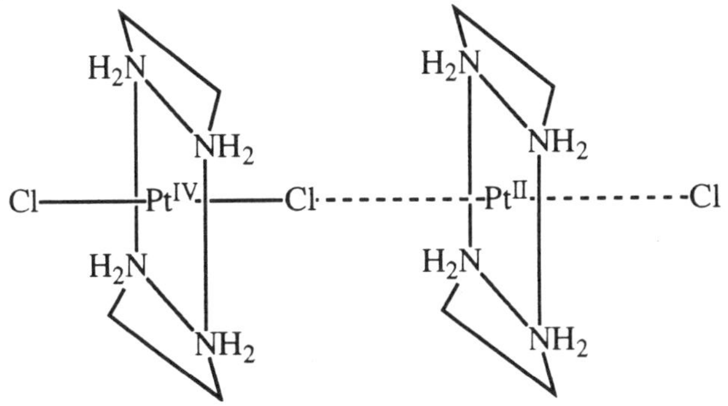

Reactions of cis-[Pt(en)$_2$Cl$_2$]$^{2+}$ with [Pt(en)$_2$]$^{2+}$ and of [Pt(NH$_3$)$_5$Cl]$^{3+}$ with [Pt(NH$_3$)$_4$]$^{2+}$ are also slow. These latter reactions are not self-exchange processes but result in trans-[Pt(en)$_2$Cl$_2$]$^{2+}$ and trans-[Pt(NH$_3$)$_4$Cl$_2$]$^{2+}$ respectively and follow a similar halide-ion-catalyzed rate law. Reduction involves cleavage of a trans Pt—N bond which is energetically less favorable than cleavage of the Pt—halide bond [19]. Rate data are presented in Table 5.1.

Table 5.1. Rate and activation parameters for the reduction of platinum(IV) by Platinum(II) at 25.0°C

Reaction		μ (M)	k (M^{-1} s^{-1})	$\Delta H^{\ddagger}$ (kJ mol^{-1})	$\Delta S^{\ddagger}$ (J K^{-1} mol^{-1})	Ref.
trans-[Pt(NH$_3$)$_4$Cl$_2$]$^{2+}$	[Pt(NH$_3$)$_4$]$^{2+}$ Cl$^-$	1.0	4×10^2			18
cis-[Pt(NH$_3$)$_4$Cl$_2$]$^{2+}$	[Pt(NH$_3$)$_4$]$^{2+}$ Cl$^-$	1.0	1.5×10^{-1}			18
[Pt(NH$_3$)$_5$Cl]$^{3+}$	[Pt(NH$_3$)$_4$]$^{2+}$ Cl$^-$	1.0	4×10^{-2}			18
[Pt(NH$_3$)$_5$I]$^{3+}$	[Pt(NH$_3$)$_4$]$^{2+}$ I$^-$	0.016	3.9×10^2	25	−121	19
[Pt(NH$_3$)$_5$I]$^{3+}$	[Pt(NH$_3$)$_4$]$^{2+}$ Br$^-$	0.016	1.2×10^4	33	−63	19
[Pt(NH$_3$)$_5$I]$^{3+}$	[Pt(NH$_3$)$_4$]$^{2+}$ Cl$^-$	0.016	5.6×10^2	46	−42	19
[Pt(NH$_3$)$_5$Br]$^{3+}$	[Pt(NH$_3$)$_4$]$^{2+}$ Br$^-$	0.32	12	42	−79	19
[Pt(NH$_3$)$_5$Cl]$^{3+}$	[Pt(NH$_3$)$_4$]$^{2+}$ Cl$^-$	0.32	1.2×10^{-3}	75	−54	19

It seems likely that these complementary reactions do not involve platinum(III), but it is difficult to decide whether they are best described as inner-sphere processes with the transfer of two electrons through a halide bridge or as outer-sphere processes with halogen cation transfer. Energetically the latter would appear to be unlikely and whereas equatorial substitution on the square-planar platinum(II) complexes is sluggish, the weak association of ligands in the axial positions can be rapid so that an inner-sphere two-electron transfer is preferred. Note however, that platinum(III) has been invoked to explain inhibition by one-electron transfer reagents in the self-exchange between [PtCl$_6$]$^{2-}$ and [PtCl$_4$]$^{2-}$ which is also catalyzed by Cl$^-$ [20].

Non-complementary reductions of [trans-Pt(en)$_2$Cl$_2$]$^{2+}$ by single equivalent reductants show a variety of mechanisms. The reaction of [Pt(NH$_3$)$_5$Cl]$^{3+}$ with [Cr(H$_2$O)$_6$]$^{2+}$ is rapid and results in the formation of equal amounts of [Cr(H$_2$O)$_5$Cl]$^{2+}$ and [Cr(H$_2$O)$_6$]$^{3+}$ in acidic solutions, whereas at higher pH, the latter product is diminished and appreciable amounts of the dimer [(Cr(OH)$_2$Cr)$_{aq}$]$^{4+}$ are formed [21]. The mechanism involves an inner-sphere transfer of two electrons through a Cl$^-$ bridge to give a transient chromium(IV) complex, [CrCl$_{aq}$]$^{3+}$, which subsequently reacts with [Cr(H$_2$O)$_6$]$^{2+}$ to give the tell-tale dimeric species (eqs (5.35)–(5.37)).

$$[Pt(NH_3)_5Cl]^{3+} + [Cr(H_2O)_6]^{2+} \rightarrow [Pt(NH_3)_4]^{2+} + [CrCl_{aq}]^{3+} + NH_3 \quad (5.35)$$

fast
$$[CrCl_{aq}]^{3+} + [Cr(H_2O)_6]^{2+} \rightarrow [(H_2O)_5CrClCr(H_2O)_5]^{5+} \quad (5.36)$$

$$[(H_2O)_5CrClCr(H_2O)_5]^{5+} + [Cr(H_2O)_6]^{2+} \rightarrow \text{products} \quad (5.37)$$

Interestingly the mechanism of reduction by $[V(H_2O)_6]^{2+}$ is quite different [22]. The reaction is acid independent and there is no formation of $[VO_{aq}]^{2+}$ which would otherwise be detectable under the conditions of the experiments. Single electron transfer is proposed with the formation of transient platinum(III) intermediates which subsequently react with $[V(H_2O)_6]^{2+}$ (eqs (5.38)–(5.39)). The reaction rates show a correlation (eq. (5.40)) with those for the corresponding reactions of $[Ru(NH_3)_6]^{2+}$ which proceed by a similar mechanism involving outer-sphere electron transfer (Fig. 5.2). The slope of the correlation is close to that predicted by the Marcus relationship for outer-sphere reactions and this is the suggested mechanism. In partial confirmation, for $[PtCl_6]^{2-}$ as oxidant, the second-order rate constant exceeds the normal rate of substitution at $[V(H_2O)_6]^{2+}$. Rate constants for one-electron reduction of the platinum(IV) complexes are presented in Table 5.2.

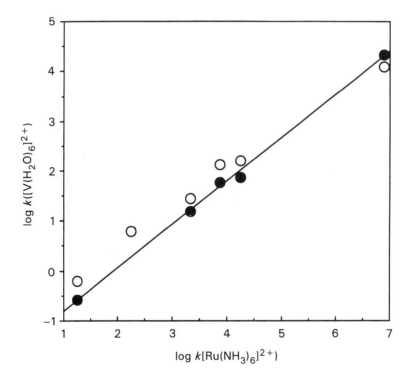

Fig. 5.2. Correlation of the rate constants $k([VH_2O)_6]^{2+})$ and $k([Ru(NH_3)_6]^{2+})$ for reductions of platinum(IV) complexes at $\mu = 1.00$ M (LiClO$_4$) (open circles) and 0.10 M (LiClO$_4$) (closed circles), from Ref. [22].

$$[Pt(NH_3)_5Cl]^{3+} + [V(H_2O)_6]^{2+} \rightarrow [Pt(NH_3)_5Cl]^{2+} + [V(H_2O)_6]^{3+} \quad (5.38)$$

$$\text{Pt(NH}_3)_5\text{Cl}]^{2+} + [\text{V(H}_2\text{O})_6]^{2+} \rightarrow$$
$$[\text{Pt(NH}_3)_4]^{2+} + [\text{V(H}_2\text{O})_6]^{3+} + \text{NH}_3 + \text{Cl}^- \quad (5.39)$$

$$\log k_{[\text{V(H}_2\text{O})_6]^{2+}} = 0.89 \log k_{[\text{Ru(NH}_3)_6]^{2+}} - 1.68 \quad (5.40)$$

Table 5.2. Rate constants and activation parameters for the one-electron reduction of platinum(IV) at 25.0°C.

Reaction		μ (M)	k ($M^{-1} s^{-1}$)	$\Delta H^{\ddagger}$ (kJ mol^{-1})	$\Delta S^{\ddagger}$ (J K^{-1} mol^{-1})	Ref.
$[\text{Pt(NH}_3)_5\text{Cl}]^{3+}$	$[\text{Cr(H}_2\text{O})_6]^{2+}$	1.0	5×10^4			23
$[\text{Pt(NH}_3)_5\text{Cl}]^{3+}$	$[\text{V(H}_2\text{O})_6]^{2+}$	1.0	0.61	41	-105	24
$[\text{Pt(NH}_3)_5\text{Br}]^{3+}$	$[\text{V(H}_2\text{O})_6]^{2+}$	1.0	6.0	31	-120	22
$trans$-$[\text{Pt(NH}_3)_4\text{Cl}_2]^{2+}$	$[\text{V(H}_2\text{O})_6]^{2+}$	1.0	28.3	45	-61	22
$[\text{PtCl}_6]^{2-}$	$[\text{V(H}_2\text{O})_6]^{2+}$	1.0	1.23×10^4	19	-100	24
$[\text{Pt(NH}_3)_5\text{Cl}]^{3+}$	$[\text{Ru(NH}_3)_6]^{2+}$	0.1^a	17.7			24
$[\text{Pt(NH}_3)_5\text{Br}]^{3+}$	$[\text{Ru(NH}_3)_6]^{2+}$	0.1^a	1.8×10^2			22
$trans$-$[\text{Pt(NH}_3)_4\text{Cl}_2]^{2+}$	$[\text{Ru(NH}_3)_6]^{2+}$	0.1^a	2.14×10^3			22
$[\text{PtCl}_6]^{2-}$	$[\text{Ru(NH}_3)_6]^{2+}$	0.1^a	7.9×10^6			22

^{a} 15°C.

The rate law for the oxidation of $[\text{Pt(en)}_2]^{2+}$ by $[\text{IrCl}_6]^{2-}$ is complex (eq. (5.41)), and shows strong inhibition by $[\text{IrCl}_6]^{3-}$ confirming that it proceeds by consecutive one-electron steps (eqs (5.42)–(5.43)) [25]. The halide dependence of the major pathway suggests that the initial reactant is the ion pair $[[\text{Pt(en)}_2]^{2+},\text{Cl}^-]$, and that there is no transfer of Cl$^-$ from the oxidant.

$$\frac{-d[[\text{IrCl}_6]^{2-}]}{dt} = \frac{2k_1 k_2 [[\text{Pt(en)}_2]^{2+}][[\text{IrCl}_6]^{2-}]^2[\text{Cl}^-]^2}{k_{-1}[[\text{IrCl}_6]^{3-}] + k_2[[\text{IrCl}_6]^{2-}][\text{Cl}^-]} \quad (5.41)$$

$$[\text{Pt(en)}_2]^{2+} + [\text{IrCl}_6]^{2-} + \text{Cl}^- \underset{k_1}{\overset{k_{-1}}{\rightleftharpoons}} [\text{Pt(en)}_2\text{Cl}]^{2+} + [\text{IrCl}_6]^{3-} \quad (5.42)$$

$$[\text{Pt(en)}_2\text{Cl}]^{2+} + [\text{IrCl}_6]^{2-} + \text{Cl}^- \overset{k_2}{\rightarrow} [\text{Pt(en)}_2\text{Cl}]^{2+} + [\text{IrCl}_6]^{3-} \quad (5.43)$$

Interestingly, a slightly different mechanism is revealed by the rate law (eq. (5.44)), for the anionic reductant $[\text{PtCl}_4]^{2-}$ [26]. Again, the initial step, formation of platinum(III) is rate-limiting, but in this case the first-order dependence on [Cl$^-$] indicates that significant amounts of a reactive ion pair between $[\text{PtCl}_4]^{2-}$ and Cl$^-$ are not present (eq. (5.45)). The second electron transfer involves incorporation of free halide ion with the intermediate platinum(III) complex and is followed by rapid formation of $[PtCl_6]^{2-}$ product (eqs (5.46)–(5.47)).

$$\frac{-d[[IrCl_6]^{2-}]}{dt} = \frac{2k_1k_2[[PtCl_4]^{2-}][[IrCl_6]^{2-}]^2[Cl^-]}{k_{-1}[[IrCl_6]^{3-}] + k_2[[IrCl_6]^{2-}][Cl^-]} \qquad (5.44)$$

$$[PtCl_4]^{2-} + [IrCl_6]^{2-} \underset{k_{-1}}{\overset{k_1}{\rightleftharpoons}} [PtCl_4]^- + [IrCl_6]^{3-} \qquad (5.45)$$

$$[PtCl_4]^- + [IrCl_6]^{2-} + Cl^- \overset{k_2}{\longrightarrow} [PtCl_5]^- + [IrCl_6]^{3-} \qquad (5.46)$$

$$[PtCl_5]^- + Cl^- \overset{fast}{\longrightarrow} [PtCl_6]^{2-} \qquad (5.47)$$

Interesting insight into the energetics of these reactions is provided by a study of the oxidation of $[Pt(NH_3)_4]^{2+}$ by $[Fe(CN)_6]^{3-}$ which is first-order in both reactants and results in the mixed-valence trinuclear complex $[(CN)_5Fe^{II}CNPt^{IV}(NH_3)_4NCFe^{II}(CN)_5]^{4-}$ by an inner-sphere mechanism [27]. The product exhibits an intervalence transfer band at 470 nm in aqueous solution and the reaction can be reversed photochemically by excitation into this band. Initial excitation gives $[(CN)_5Fe^{III}CNPt^{III}(NH_3)_4NCFe^{II}(CN)_5]^{4-}$ which can further react. Analysis of the band shape together with estimates of the reduction potentials for the metal centers gives a value for λ, the reorganizational energy, of 175 kJ mol^{-1} which is large but consistent with substantial stuctural reorganization at the platinum center [28].

In the reactions of both thallium and platinum, single-step two-electron transfer pathways are noted. In the case of platinum, these are shown to involve transfer of a bridging ligand, and may be inner-sphere in nature since the platinum(II) is coordinatively unsaturated. It is likely also that the thallium reactions are inner-sphere since pathways involving inverse [H$^+$] dependencies, consistent with hydroxy-bridged intermediates, predominate, although there is no proof for transfer of the bridging ligand. Both series of reactions are adequately described within the framework developed in Chapters 2 and 3 as electron transfer processes, albeit with the added complication that in some instances two electrons are involved. The energetics of these processes are difficult to probe, in part because the reactions are complex but also because driving forces differ from reaction to reaction. The exceptions are the self-exchange reactions since for those $\Delta E° = 0;$, however data for two-electron self-exchange reactions are limited.

5.4 OXO-ION REAGENTS

An important group of multiple-electron transfer reagents in aqueous solution is formed by higher oxidation state species which exist as oxy anions and cations as a result of extensive hydrolysis. Reduction potentials for some of the more common reagents are presented in Table 5.3, and discussion will center on the chemistry of chromium(VI), which is a three-electron reagent. This example is chosen because the intermediate oxidation states chromium(V) and chromium(IV) are thermodynamically unstable in the absence of stabilizing ligands. The chemistry of some of the other species where intermediate oxidation states are thermodynamically stable has been discussed in Chapters 2 and 3.

Table 5.3. Reduction potentials for oxo-ion reagents in acid solution at 25.0°C [5]

Half-reaction	E° (V)
$[TiO_{aq}]^{2+} + 2\ H^+ + e^- \rightleftharpoons [Ti_{aq}]^{3+} + H_2O$	0.10
$[(VO_2)_{aq}]^+ + 2\ H^+ + e^- \rightleftharpoons [VO_{aq}]^{2+} + H_2O$	1.00
$[VO_{aq}]^{2+} + 2\ H^+ + e^- \rightleftharpoons [V_{aq}]^{3+} + H_2O$	0.34
$[HCrO_4]^- + 7\ H^+ + 3e^- \rightleftharpoons [Cr_{aq}]^{3+} + 4\ H_2O$	1.38
$[HCrO_4]^- + 2\ H^+ + e^- \rightleftharpoons [H_3CrO_4]$	≈0.55
$[MnO_4]^- + 8\ H^+ + 5e^- \rightleftharpoons [Mn_{aq}]^{2+} + 4\ H_2O$	1.51
$[MnO_4]^- + e^- \rightleftharpoons [MnO_4]^{2-}$	0.56
$[MnO_4]^{2-} + e^- \rightleftharpoons [MnO_4]^{3-}$	0.27
$[RuO_4]^- + e^- \rightleftharpoons [RuO_4]^{2-}$	0.59
$[(UO_2)_{aq}]^{2+} + e^- \rightleftharpoons [(UO_2)_{aq}]^+$	0.16
$[(UO_2)_{aq}]^{2+} + 4\ H^+ + 2e^- \rightleftharpoons [U_{aq}]^{4+} + 2\ H_2O$	0.27

The reduction of chromium(VI) is a three-electron process which involves a substantial change in coordination around the metal center (eq. (5.48)). Intermediate oxidation states are known but they are thermodynamically unstable with respect to disproportionation. Estimated reduction potentials are 1.34 V and 2.10 V in acid solution for the $[Cr_{aq}]^{5+/4+}$ and $[Cr_{aq}]^{4+/3+}$ changes so that chromium(IV) state is the least stable oxidation state [29].

$$[HCrO_4]^- + 7\ H^+ + 3e^- \rightleftharpoons [Cr(H_2O)_6]^{3+} \tag{5.48}$$

The chemistry of $[HCrO_4]^-$ is dominated by the protic equilibria (eqs (5.49) and (5.50)), and dimerization to form $[Cr_2O_7]^{2-}$ (eq. (5.51)) so that in dilute aqueous acid, $[HCrO_4]^-$ is the dominant species. The ion is moderately labile.

$$[H_2CrO_4] \rightleftharpoons [HCrO_4]^- + H^+ \quad K_{a1} = 0.21 \tag{5.49}$$

$$[HCrO_4]^- \rightleftharpoons [CrO_4]^{2-} + H^+ \quad K_{a2} = 1.04 \times 10^{-6} \tag{5.50}$$

$$[HCrO_4]^- \rightleftharpoons [Cr_2O_7]^{2-} + H_2O \quad K_d = 98\ M^{-1} \tag{5.51}$$

Non-complementary reduction of [HCrO$_4$]$^-$ by reagents which are incapable of inner-sphere electron transfer such as [Fe(bpy)$_3$]$^{2+}$ show rate laws which are first order in each reagent and are subject to acid catalysis (eq. (5.52)) with k_2 dominant [30, 31]. The initial step, formation of chromium(V), is rate-limiting and the acid catalysis reflects the greater reduction potential for the protonated oxidant since there is a net addition of seven protons in the overall reaction. The proposed mechanism is given in eqs (5.53)–(5.56).

$$-\frac{d[[Fe(bpy)_3]^{2+}]}{dt} = \{k_0 + k_1[H^+] + k_2[H^+]^2\} [[HCrO_4]^-] [[Fe(bpy)_3]^{2+}] \quad (5.52)$$

$$[HCrO_4]^- + [Fe(bpy)_3]^{2+} \xrightarrow{k_0} [HCrO_4]^{2-} + [Fe(bpy)_3]^{3+} \quad (5.53)$$

$$[H_2CrO_4] + [Fe(bpy)_3]^{2+} \xrightarrow{k_1} [H_2CrO_4]^- + [Fe(bpy)_3]^{3+} \quad (5.54)$$

$$[H_2CrO_4] + [Fe(bpy)_3]^{2+} + H^+ \xrightarrow{k_2} [H_3CrO_4] + [Fe(bpy)_3]^{3+} \quad (5.55)$$

$$[H_3CrO_4] + 2\,[Fe(bpy)_3]^{2+} + 5\,H^+ \xrightarrow{\text{fast}} [Cr(H_2O)_6]^{3+} + 2\,[Fe(bpy)_3]^{3+} \quad (5.56)$$

With reductants which contain ligands such as CN$^-$ and Cl$^-$, capable of bridge formation, the rate laws indicate that the initial production of chromium(V) is rate-determining and the [H$^+$] dependencies are similar [30]. In the reaction of [Fe(CN)$_6$]$^{4-}$, however, product analysis reveals the formation of a binuclear product [(CN)$_5$FeCNCr(H$_2$O)$_5$] which must result from inner-sphere reduction of chromium(IV), since substitution at [Cr(H$_2$O)$_6$]$^{3+}$ is sluggish. With [IrCl$_6$]$^{3-}$, the product is [Cr(H$_2$O)$_6$]$^{3+}$ but the participation of inner-sphere steps cannot be eliminated because the overall reaction is slow [32].

More labile one-electron reductants, such as [Fe(H$_2$O)$_6$]$^{2+}$, show a second-order dependence on the reductant (eq. (5.57)), revealing that the rate-limiting step is reduction of chromium(V) to chromium(IV) under conditions of low reductant concentrations (eqs (5.58)–(5.59)) [31]. Again the acidity dependence signifies acid catalysis and it can be deduced that the chromium(V) species has the formula [H$_3$CrO$_4$]. It has been suggested that the major coordination change in going from four-coordinate chromium(VI) to six-coordinate chromium(III) occurs between chromium(V) and chromium(IV) and that, while this latter step is faster than the initial reduction of chromum(VI) for outer-sphere reactions, for inner-sphere reactions, where the chromium(V) is bound to a labile metal ion, this coordination change is not facilitated. This has been borne out in studies of chromium(V) and chromium(IV) stabilized in 2-ethyl-2-hydroxybutanoate complexes (eq. (5.60)), where the self-exchange rate for the chromium(V/IV) change is at least six orders of magnitude smaller than that for chromium(IV/III) [33–36].

$$\frac{-d[[HCrO_4]^-]}{dt} = \frac{k_1 k_2 [[HCrO_4]^-][[Fe(H_2O)_6]^{2+}]^2 [H^+]^3}{k_{-1}[[Fe(H_2O)_6]^{3+}] + k_2[[Fe(H_2O)_6]^{2+}][H^+]} \qquad (5.57)$$

$$[HCrO_4]^- + [Fe(H_2O)_6]^{2+} + 2H^+ \underset{k_{-1}}{\overset{k_1}{\rightleftharpoons}} [H_3CrO_4] + [Fe(H_2O)_6]^{3+} \qquad (5.58)$$

$$[H_3CrO_4] + [Fe(H_2O)_6]^{2+} + H^+ \overset{k_2}{\longrightarrow} [(Cr^{IV}O_2)_{aq}] + [Fe(H_2O)_6]^{3+} + 2H_2O \qquad (5.59)$$

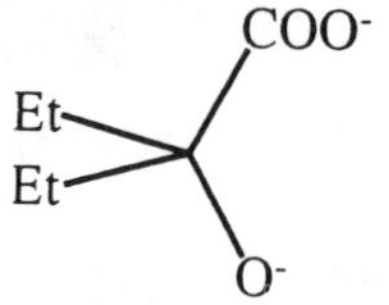

$$\text{Et} \underset{\text{Et}}{\overset{\text{COO}^-}{\underset{O^-}{\diagdown\diagup}}} \qquad (5.60)$$

An important class of reactions involves reduction with reagents where the oxidized product retains an oxo ligand. There are both one-electron reagents such as $[VO]^{2+}$ and two-electron reagents such as $[U_{aq}]^{4+}$. In these reactions, a common distinguishing feature is a reduced dependence on $[H^+]$. The reduction by $[VO]^{2+}$ is independent of $[H^+]$ with the rate law in eq. (5.61) indicating that reduction of chromium(V) is rate-limiting [37–39]. In the proposed mechanism, there is a match in the number of H^+ lost by the reductant and the number of H^+ gained by the oxidant in the initial step (eq. (5.62)). In an elegant experiment where $[VO_2]^+$ is scavenged by the addition of $[V(H_2O)_6]^{3+}$, the initial step k_1 becomes rate-limiting and the rate law is simplified to eq. (5.65), confirming that the sequence of steps is similar to that observed in the reaction with $[Fe(H_2O)_6]^{2+}$.

$$\frac{-d[[HCrO_4]^-]}{dt} = \frac{k_1 k_2 [[HCrO_4]^-][[VO]^{2+}]^2}{k_{-1}[[VO_2]^+]} \qquad (5.61)$$

$$[HCrO_4]^- + [VO]^{2+} + H_2O \underset{k_{-1}}{\overset{k_1}{\rightleftharpoons}} [H_3CrO_4] + [VO_2]^+ \qquad (5.62)$$

$$[H_3CrO_4] + [VO]^{2+} + H^+ \overset{k_2}{\rightarrow} [(Cr^{IV}O_2)_{aq}] + [VO_2]^+ + H_2O \qquad (5.63)$$

$$[(Cr^{IV}O_2)_{aq}] + [VO]^{2+} \overset{fast}{\rightarrow} products \qquad (5.64)$$

$$\frac{-d[[HCrO_4]^-]}{dt} = k_1[[HCrO_4]^-][[VO]^{2+}] \qquad (5.65)$$

The balance of [H$^+$] in the rate law, indicates that the proton demand for protonation or deprotonation of the oxo-ligands is satisfied. The facilitation of reactions of this sort by [H$^+$] catalysis has been noted in section 2.15. However, the reactions may take place by direct transfer of the oxo-ligand from the oxidant to the reductant, most probably in an inner-sphere process. This is not oxygen atom transfer since a one-electron process requires transfer of O$^-$. Isotopic tracer studies have revealed a formal O atom transfer [40] with the two-electron reductant [U$_{aq}$]$^{4+}$, and again the rate law has a reduced [H$^+$] dependence with a dominant term in [H$^+$]$^{-1}$ (eq. (5.66) [41, 42]. Uranium(V) is not formed in the reaction and at least one of the oxygens of [UO$_2$]$^{2+}$ is derived from the oxidant so that a transition state as shown in eq. (5.67) can be proposed.

$$\frac{-d[[HCrO_4]^-]}{dt} = \left\{k_1 + \frac{k_2}{[H^+]}\right\}[[U_{aq}]^{4+}][[HCrO_4]^-] \qquad (5.66)$$

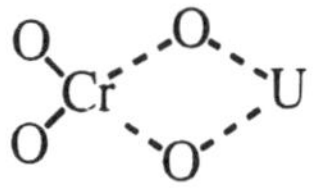

$$\begin{array}{c} O \diagdown \quad\diagup O \diagdown \\ Cr U \\ O \diagup \quad\diagdown O \diagup \end{array} \qquad (5.67)$$

This pattern of reactivity is found in many reactions involving oxo ions. It is the two-electron analogy of the one-electron inner-sphere reaction with group transfer and can be viewed in those terms as transfer of O^{2-} from the oxidant to the reductant concomitant with the transfer of 2e$^-$ from the reductant to the oxidant. Alternatively, it can be thought of as O atom transfer. Consideration of the energetics of the reactions clearly indicates that formation of isolated O atoms is not possible and it is better to think of the O atom as stabilized by coordination to the metal centers. The detailed energetics of these reactions have been difficult to probe, predominantly because they are very complex processes and trends from structural modifications

232 Multiple electron transfer [Ch. 5

are difficult to maintain without gross changes in mechanism. This situation is changing as simpler 'atom-transfer' reactions are examined in detail.

5.5 'ATOM TRANSFER' REACTIONS

The range of two-electron self-exchange processes has been augmented by the processes presented in eqs (5.68) and (5.69). These self-exchange reactions involve a formal change of two electrons explicitly accompanied by transfer of an atom or group, and the intermediate oxidation state is thermodynamically unstable.

$$[CpM(CO)_3]^- + [CpM(CO)_3X] \rightarrow [CpM(CO)_3X] + [CpM(CO)_3]^- \quad (5.68)$$

$$[(Cp)_2M] + [(Cp)_2MX]^+ \rightarrow [(Cp)_2MX]^+ + [(Cp)_2M] \quad (5.69)$$

The rates for the processes have been investigated mainly by NMR methods in acetonitrile solution and are first-order in both reagents. Rate constants (Table 5.4) are considerably in excess of the normal rates of substitution at both metal centers so that the classical inner-sphere mechanism involving formation of a bridge common to the inner-coordination spheres of both reactant metal can be ruled out. Although the one-electron self-exchange rates involving the radical species are rapid inner-sphere processes [43], a stepwise mechanism with two single-electron transfer steps can be excluded on thermodynamic grounds.

Table 5.4. Rate and activation parameters for the self-exchange rates of atom transfer reactions

Reaction	k ($M^{-1} s^{-1}$)	$\Delta H^{\ddagger}$ (kJ mol^{-1})	$\Delta S^{\ddagger}$ (J K^{-1} mol^{-1})	$\Delta V^{\ddagger}$ (cm^3 mol^{-1})	Ref.
$[CpCr(CO)_3]^- + [CpCr(CO)_3H]$	1.8×10^4	20.5	−95		44
$[CpMo(CO)_3]^- + [CpMo(CO)_3I]$	1.5×10^4	26.8	−75		45
$[CpMo(CO)_3]^- + [CpMo(CO)_3Br]$	16	50.6	−50		45
$[CpMo(CO)_3]^- + [CpMo(CO)_3Cl]$	9.0×10^{-2}	79.1	0		45
$[CpMo(CO)_3]^- + [CpMo(CO)_3H]$	2.5×10^3	22.2	−106		44
$[CpW(CO)_3]^- + [CpW(CO)_3I]$	4.5×10^3	31.4	−70		45
$[CpW(CO)_3]^- + [CpW(CO)_3Br]$	1.5	63.2	−29		45
$[CpW(CO)_3]^- + [CpW(CO)_3Cl]$	2.1×10^{-3}	74.0	−46		45
$[CpW(CO)_3]^- + [CpW(CO)_3H]$	6.5×10^2	21.7	−117		44
$[CpMo(CO)_3]^- + [CpMo(CO)_3CH_3]$	$\approx 1 \times 10^{-5}$				45
$[(Cp)_2Ru] + [(Cp)_2RuI]^+$	7.4×10^3	28.0	−36.4	−7.7	46
$[(Cp)_2Ru] + [(Cp)_2RuBr]^+$	1.6×10^3	34.3	−65.7	−2.9	46
$[(Cp)_2Os] + [(Cp)_2OsI]^+$	1.8×10^4	31.4	−56.1	−7.6	46

These reactions can be considered as X^+ transfers (eq. (5.70)), combined $1e^-$ and X atom transfers (eq. (5.71)), or $2e^-$ and X^- transfers (eq. (5.72)), as shown for the reaction of $[CpMo(CO)_3]^-$ with $[CpMo(CO)_3Cl]$ where X = Cl. Details of the mechanisms are not well understood; however, an attempt has been made to probe the energetics with use of Scheme 5.5 [45].

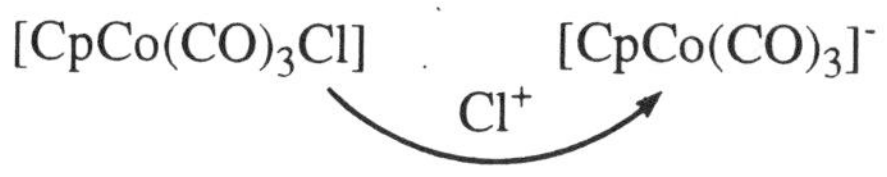

$$[CpCo(CO)_3Cl] \underset{Cl^+}{\curvearrowright} [CpCo(CO)_3]^- \quad (5.70)$$

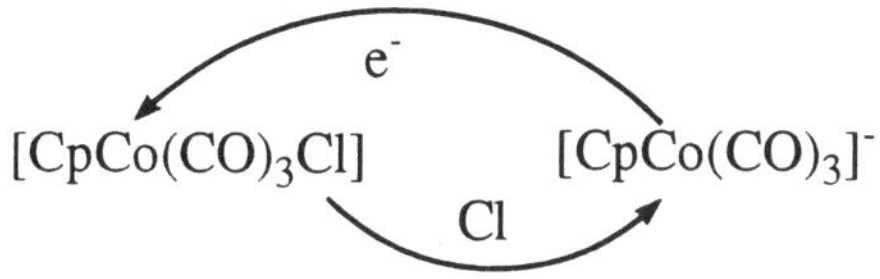

$$[CpCo(CO)_3Cl] \underset{Cl}{\overset{e^-}{\curvearrowright}} [CpCo(CO)_3]^- \quad (5.71)$$

$$[CpCo(CO)_3Cl] \underset{Cl^-}{\overset{2e^-}{\curvearrowright}} [CpCo(CO)_3]^- \quad (5.72)$$

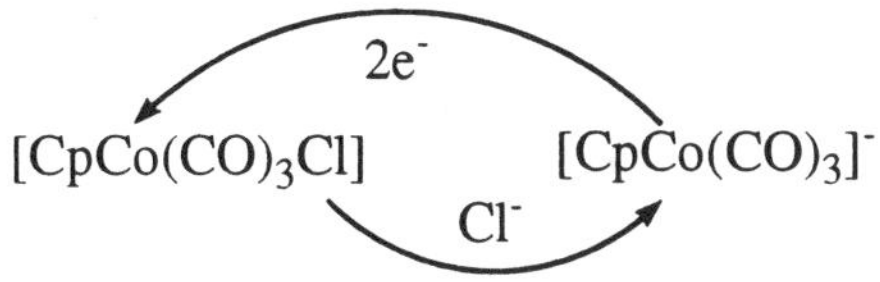

The reactants form a weakly interacting outer-sphere complex (eq. (5.73)), and the energetics of the electron transfer process are determined by stretching of the Mo—Cl bond until the Cl atom is equidistant between the two metal centers, at which point electron transfer takes place. The transition state is a tautomer comprising both structures in eq. (5.75). The isovalent state $\{[CpMo^I(CO)_3\cdots Cl \cdots CpMo^I(CO)_3]\}^-$ is of higher energy but can be admixed to the transition state, contributing to a lower activation energy (Fig. 5.3). Activation volume data obtained in a number of solvents suggest that there is also a significant solvent reorganization involved with the charge transfer process [46].

$$[CpMo(CO)_3Cl] + [CpMo(CO)_3]^- \rightleftharpoons$$
$$\{[CpMo(CO)_3Cl],[CpMo(CO)_3]\}^- \quad (5.73)$$

$$\{[CpMo(CO)_3Cl],[CpMo(CO)_3]\}^- \rightarrow$$
$$\{[CpMo(CO)_3\cdots Cl],[CpMo(CO)_3]\}^- \quad (5.74)$$

$$\{[CpMo(CO)_3\cdots Cl],[CpMo(CO)_3]\}^- \rightarrow$$
$$\{[CpMo(CO)_3],[Cl\cdots CpMo(CO)_3]\}^- \quad (5.75)$$

$$\{[CpMo(CO)_3],[Cl\cdots CpMo(CO)_3]\}^- \rightarrow$$
$$\{[CpMo(CO)_3],[CpMo(CO)_3Cl]\}^- \quad (5.76)$$

$$\{[CpMo(CO)_3],[CpMo(CO)_3Cl]\}^- \rightleftharpoons$$
$$[CpMo(CO)_3]^- + [CpMo(CO)_3Cl] \quad (5.77)$$

Scheme 5.5.

The reaction involving $[CpMo(CO)_3]^-/[CpMo(CO)_3H]$ shows an isotope effect $k_H/k_D = 3.7$ when $[CpMo(CO)_3]^-/[CpMo(CO)_3D]$ is used and this is consistent with rate-limiting H· or H$^+$ transfer. A transition state with a symmetric Mo···H···Mo arrangement yields a calculated k_H/k_D of 3.4.

The description of these reactions as cation transfer processes serves to underscore the gradation between electron and atom transfer. Probing the sequential arrangement of these processes is of considerable importance in presenting a fuller description of the reactions. Mechanistic differentiation has proved elusive. Isotope effects have

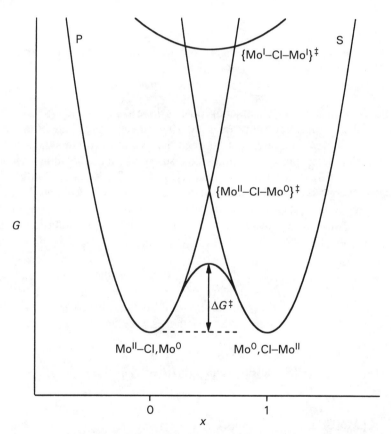

Fig. 5.3. Reaction coordinate diagram for the atom transfer reaction between $[CpMo(CO)_3Cl]$ and $[CpMo(CO)_3]^-$. The reaction coordinate is the stretching of the Mo–Cl bond and the intersection of the two parabolas is described by the tautomer $\{Mo^{II}$–Cl–$Mo^0 \longleftrightarrow Mo^0$–Cl–$Mo^{II}\}^{\ddagger}$. Admixing of the isovalent state $\{Mo^I$–Cl–$Mo^I\}^{\ddagger}$ with this state contributes to a lower activation energy.

been useful, as have correlations with bond energies, but there are many gaps in the data and more studies are required.

5.6 THE ROLE OF ADDUCT FORMATION IN NON-COMPLEMENTARY REACTIONS

In the examples of non-complementary reactions presented in sections 5.2–5.5, some reference has been made to the role of inner-sphere processes in facilitating multiple-electron transfer. Further examples of this behavior are to be found in the remaining sections of the chapter. Besides this stoichiometric effect in controlling reactivity, adduct formation also plays an energetic role. Difficulties in assessing this role have been encountered in the examination of inner-sphere mechanisms in Chapter 3. However, there is one particular case involving non-complementary reactions which has led to a limited understanding of a significant class of these reactions.

The class of reactions to which this discussion can be applied are those where a two-electron reagent, generally the reductant, has a very unstable intermediate oxidation state and where adduct formation with a single-electron oxidant produces an intermediate with discrete spectroscopic properties of a charge transfer nature. Complexes of this sort are particularly prevalent in sluggish redox interactions involving organic and organometallic substrates. The mechanistic implications of electron transfer have assumed an important role in organic chemistry and there are many parallels with the work described in this book but the subject cannot be treated with any detail here. The reader is referred to reviews elsewhere for information on these reactions [47, 48].

The initial step in the irreversible oxidation of $[SnMe_2Et_2]$ by $[Fe(phen)_3]^{3+}$ in acetonitrile solution is the formation of a cation radical (eq. (5.78)), which subsequently decays with loss of either ·Et or ·Me (eq. (5.79)). This reaction is thought to be outer-sphere in nature since the rates for a variety of compounds, $[SnR_4]$, show a Marcus-type correlation with the ionization potentials, I, for the reductants, and these can be directly related to the reduction potentials (Fig. 5.4) [49]. Reduction potentials for the cation radicals in solution are not known but are expected to be quite large so that the first step in the reaction is very endoenergetic [50, 51].

$$[Fe(phen)_3]^{3+} + [SnMe_2Et_2] \xrightarrow{k} [Fe(phen)_3]^{2+} + [SnMe_2Et_2]^+ \qquad (5.78)$$

$$[SnMe_2Et_2]^+ \begin{cases} \xrightarrow{k_{Et}} [SnMe_2Et]^+ + \cdot Et \rightarrow \rightarrow \\ \xrightarrow{k_{Me}} [SnMeEt_2]^+ + \cdot Me \rightarrow \rightarrow \end{cases} \qquad (5.79)$$

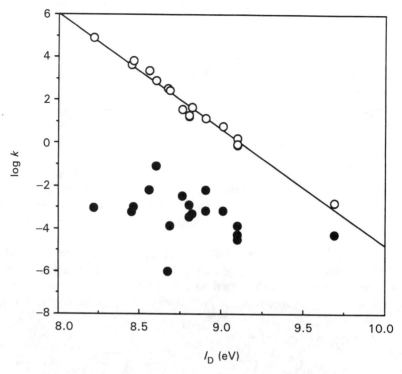

Fig. 5.4. Contrasting behavior of the dependence of the rates for $[IrCl_6]^{2-}$ (closed circles), and $[Fe(phen)_3]^{3+}$ (open circles) oxidations of alkylmetals on the ionization potentials of alkylmetals, a measure of the thermodynamic driving force for the reactions. The slope of the $[Fe(phen)_3]^{3+}$ correlation is −0.5. From Ref. [49].

The behavior with $[IrCl_6]^{2-}$ is quite different. Firstly the product analysis indicates that the reaction selectivity, $S\ (= k_{Et}/k_{Me})$, is 11 rather than 27 found for $[Fe(phen)_3]^{3+}$, so that the reactivity of the cation radical is modified. Secondly there is no correlation of the rates with ionization potentials, I (Fig. 5.4). Instead the rates show evidence for steric inhibition. It is concluded that while the $[Fe(phen)_3]^{3+}$ reactions are outer-sphere in nature, the $[IrCl_6]^{2-}$ reactions involve adduct formation and are best considered a limiting form of inner-sphere reaction.

The energetics of this inner-sphere process can be described in eqs (5.80)–(5.82), by two work terms, precursor and successor respectively, and the activation energy for the intramolecular electron transfer. In a very endoenergetic process, such as eq. (5.81), the transition state is very product-like so that the reverse reaction is close to being activationless and the overall reaction is driven by the subsequent decomposition of the cation radical. Addition of eqs (5.80)–(5.82) gives the relationship between $\Delta G^{\ddagger}$ and $\Delta G°$ in eq. (5.83) where $\omega_{AB} \approx 0$ for the interaction of $[IrCl_6]^{2-}$ and a neutral molecule.

$$[\text{IrCl}_6]^{2-} + [\text{SnMe}_2\text{Et}_2] \underset{}{\overset{\omega_{AB}}{\rightleftharpoons}} \{[\text{IrCl}_6]^{2-},[\text{SnMe}_2\text{Et}_2]\} \quad (5.80)$$

$$\{[\text{IrCl}_6]^{2-},[\text{SnMe}_2\text{Et}_2]\} \overset{k}{\rightarrow} \{[\text{IrCl}_6]^{3-},[\text{SnMe}_2\text{Et}_2]^+\} \quad (5.81)$$

$$\{[\text{IrCl}_6]^{3-},[\text{SnMe}_2\text{Et}_2]^+\} \underset{\omega_{BA}}{\rightleftharpoons} [\text{IrCl}_6]^{3-} + [\text{SnMe}_2\text{Et}_2]^+ \quad (5.82)$$

$$\Delta G^{\ddagger} = \Delta G^{\circ} + \omega_{BA} \quad (5.83)$$

The quantity ω_{BA} depends on the strength of the interaction between the $[\text{IrCl}_6]^{3-}$ and $[\text{SnMe}_2\text{Et}_2]^+$ and will reflect steric factors as required of the experimental observations. Where the charge transfer spectra of the intermediate $\{[\text{IrCl}_6]^{2-},[\text{SnMe}_2\text{Et}_2]\}$ can be detected (eqs (5.84)–(5.85)), $\omega_{AB} + \omega_{BA}*$ may be estimated by using Mulliken theory, which relates the energy of the charge transfer to the ionization potential of the electron donor and the electron affinity of the electron acceptor by a constant which includes $\omega_{BA}*$.

$$\{[\text{IrCl}_6]^{2-},[\text{SnMe}_2\text{Et}_2]\} \overset{h\nu_{CT}}{\rightleftharpoons} \{[\text{IrCl}_6]^{3-},[\text{SnMe}_2\text{Et}_2]^+\}* \quad (5.84)$$

$$[\text{IrCl}_6]^{3-},[\text{SnMe}_2\text{Et}_2]^+\}* \underset{\omega_{AB}*}{\rightleftharpoons} [\text{IrCl}_6]^{3-} + [\text{SnMe}_2\text{Et}_2]^+ \quad (5.85)$$

By choosing the same donor and by referencing the charge transfer spectra to a standard acceptor for which $\omega_{BA}* = \omega^{\circ}_{BA}*$, then $\Delta\omega_{BA} = (\omega_{BA}* - \omega^{\circ}_{BA}*) = \Delta h\nu_{CT} - \Delta I$ may be used to correlate the rate data (eq. (5.86)). This very simple relationship relates the activation energy for the electron transfer reaction to the driving force and the strength of the inner-sphere interaction and is illustrated in (Fig. 5.5) [52]. It is this latter factor which includes the steric and other factors which have proved difficult to incorporate into less empirical relationships.

$$\Delta G^{\ddagger} - \Delta \omega_{BA} = \Delta G^{\circ} \tag{5.86}$$

The insight into the energetics of electron transfer within adducts which these observations provide is tempered by the applicability, which is restricted to reactions which are highly endoenergetic such as non-complementary processes in organic and organometallic chemistry. Besides a role in modifying the energetics of electron transfer, adducts can also provide a role in modifying the stoichiometry of the reaction transition state, and it is this aspect which will be amplified in the remainder of the chapter.

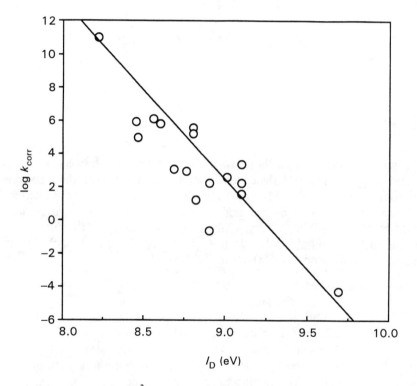

Fig. 5.5. Rates for the $[IrCl_6]^{2-}$ oxidations of alkylmetals from Fig. 5.4, corrected for steric effects according to eq. (5.86). The line is drawn with unit slope.

5.7 NON-METALLIC REAGENTS

Most non-metallic reagents require a two-electron change on altering from one thermodynamically stable state to another, and therefore form a special set of reagents for multiple-electron transfer. Besides these inorganic species, there is extensive reduction and oxidation chemistry involving organic substrates which is beyond the scope of the present treatment, although many of the trends and reactivity patterns are mirrored with organic substrates.

Some common non-metallic redox couples are illustated in Table 5.5. There are two main types: those in which oxidized and reduced forms are monomeric, differing in charge or the degree of protonation, and the second type where the oxidized form is a dimer of the reduced form. Again, charge or the degree of protonation differs between the two oxidation states. Where the degree of protonation differs, the reduction potential is pH-dependent.

Table 5.5. Reduction potentials of selected non-metallic redox couples in acidic solution at 25°C

Reaction	$E°$ (V)
$SO_4^{2-} + 4\,H^+ + 2e^- \rightleftharpoons H_2SO_3 + H_2O$	0.16
$Cl_2 + 2e^- \rightleftharpoons 2\,Cl^-$	1.40
$Br_2 + 2e^- \rightleftharpoons 2\,Br^-$	1.09
$I_2 + 2e^- \rightleftharpoons 2\,I^-$	0.62
$S_2O_8^{2-} + 2e^- \rightleftharpoons 2\,SO_4^{2-}$	1.96
$S_2O_6^{2-} + 4\,H^+ + 2e^- \rightleftharpoons 2\,H_2SO_3$	0.57
$S_4O_6^{2-} + 2e^- \rightleftharpoons 2\,S_2O_3^{2-}$	0.08
$(SCN)_2 + 2e^- \rightleftharpoons 2\,SCN^-$	0.77

There are many parallels between reactions involving non-metallic substrates and those involving metal ions. For example, there are two distinct mechanisms: inner-sphere and outer-sphere. The preferred pathway will depend on the characteristics of the oxidants and reductants. Substitution occurs most frequently at the metal center and in such cases the inner-sphere mechanism requires that the metal be substitution-labile and that the substrate have donor atoms which have lone pairs available for coordination. If these requirements are not met then the mechanisms are generally outer-sphere. A reaction which takes place with a metal ion which has no readily accessible oxidation levels differing by two equivalents is non-complementary and may lead to complex rate dependencies, with the formation of radicals as unstable intermediates. The thermodynamic fate of the radical species can have an important role in determining the course of the reaction. If, on the other hand, the metal ion has such oxidation levels, then both complementary and non-complementary processes are possible.

In the following section, oxidation and reduction reactions of a number of different non-metallic reagents with metal ion complexes will be considered. The coverage is not exhaustive but an attempt has been made to include examples which illustrate the major features and reactivity patterns. The coverage begins with reactions of molecular oxygen, a ubiquitous oxidant in aqueous solution which proves to be problematic in the handling of solutions of reductants. The chemistry of the important intermediate oxidation states is examined, and the reactivity patterns are developed with reference to other oxidizing and reducing substrates.

5.8 MOLECULAR OXYGEN AND HYDROGEN PEROXIDE

Molecular oxygen has a solubility of 1.1×10^{-3} M at one atmosphere in aqueous solution at 25°C. The standard reduction potential for the complete four-electron reduction to water (eq. (5.87)) is 1.229 V and falls with increasing pH to 0.401 V for the reaction in basic solution (eq. (5.88)); however, for solution studies, formal potentials based on 1.0 M concentrations of the gas are more useful and are 1.273 V and 0.445 V respectively.

$$O_2 + 4 H^+ + 4e^- \rightleftharpoons 2 H_2O \tag{5.87}$$

$$O_2 + 2 H_2O + 4e^- \rightleftharpoons 4 OH^- \tag{5.88}$$

The oxidation states intermediate between O_2 and H_2O are of considerable kinetic importance and some dicussion of the thermodynamics and kinetics of their reactions is essential for understanding the chemistry of O_2. One-electron reduction yields the superoxide ion O_2^-, with a formal reduction potential of -0.16 V (eq. (5.89)). It can be protonated with $pK_a = 4.8$.

$$O_2 + e^- \rightleftharpoons O_2^- \tag{5.89}$$

This initial reduction is thermodynamically unfavorable, but addition of a second electron to give hydrogen peroxide is very favorable and is coupled to the addition of two H^+ (eq. (5.90)). Hydrogen peroxide can be deprotonated with $pK_a = 11.7$.

$$O_2^- + 2 H^+ + e^- \rightleftharpoons H_2O_2 \tag{5.90}$$

Although H_2O_2 with a two-electron reduction potential of 1.76 V is a more potent oxidant than O_2, it tends to be slow to react since addition of an electron carries with it a requirement for cleavage of the O—O bond. Consequently, in reactions of O_2 with many one-electron reductants, especially those which react by outer-sphere mechanisms such as $[Cr(bpy)_3]^{2+}$ and $[Co(sep)]^{2+}$, the overall reaction is a two-electron process (eq. (5.91)), and the rate-limiting step is the initial formation of the superoxide ion (eq. (5.92)) which then reacts with a second equivalent of reductant or disproportionates (eqs (5.93)–(5.95)).

$$2\ [Cr(bpy)_3]^{2+} + 2\ H^+ + O_2 \rightarrow H_2O_2 + 2\ [Cr(bpy)_3]^{3+} \tag{5.91}$$

$$[Cr(bpy)_3]^{2+} + O_2 \rightarrow O_2^- + [Cr(bpy)_3]^{3+} \tag{5.92}$$

$$[Cr(bpy)_3]^{2+} + O_2^- + 2\ H^+ \rightarrow H_2O_2 + [Cr(bpy)_3]^{3+} \tag{5.93}$$

$$H^+ + O_2^- \rightleftharpoons HO_2 \tag{5.94}$$

$$2HO_2 \rightarrow H_2O_2 + O_2 \tag{5.95}$$

A general rate law for this reaction is given by eq. (5.96), where k is the second-order rate constant for the initial electron transfer. In suitable cases, superoxide ion may be detected directly by its characteristic EPR (g_I = 2.008, g_{II} 2.083 at 77 K) and absorption spectra (ε_{245} = 2350 M^{-1} cm^{-1}), or by the effects of chemical traps such as tetranitromethane, but this is not common and, in general, formation of the intermediate can be inferred only from the rate law. Rates of a number of reductions of O_2 by outer-sphere reagents are presented in Table 5.6.

$$\frac{-d[[Cr(bpy)_3]^{2+}]}{dt} = 2k[O_2][[Cr(bpy)_3]^{2+}] \tag{5.96}$$

Table 5.6. Rate constants for outer-sphere oxidation by O_2 and derived self-exchange rates for the $O_2^{0/-}$ reaction at 25°C

Reductant	k_{AB} (M^{-1} s^{-1})	k_{AA} (M^{-1} s^{-1})	Ref.
[Cr(bpy)$_3$]$^{2+}$	6.0 × 10^5	0.8	53
[Cr(phen)$_3$]$^{2+}$	1.5 × 10^5	2.0	53
[Ru(NH$_3$)$_6$]$^{2+}$	6.3 × 10	8.2	54
[Ru(NH$_3$)$_4$(phen)]$^{2+}$	7.7 × 10^{-3}	1.4	55
[Ru(NH$_3$)$_5$(isn)]$^{2+}$	1.1 × 10^{-1}	5.1	55
[Ru(en)$_3$]$^{2+}$	3.6 × 10	7.7	54
[Co(sep)]$^{2+}$	4.3 × 10	1.2	56
[Fe(4,7-(HO)$_2$phen)$_3$]$^{2+}$	5.1 × 10^2		57
[Cu(phen)$_2$]$^+$	5.0 × 10^4		58
[Cu(bpy)$_2$]$^+$	6.0 × 10^5		59

The Marcus linear free-energy relationship has been applied to the rate data to evaluate a self-exchange rate for the O_2/O_2^- couple. A reasonably consistent value around 2 M^{-1} s^{-1} is obtained from the data for O_2 reduction [53]. However, the rates of the reverse of this reaction, reduction of metal ion complexes by O_2^-, give significant discrepancies in the self-exchange rate for reasons which are not adequately understood [60]. Other mechanisms have been noted in reactions of O_2^-. For example, the reaction of O_2^- with [Co(sep)]$^{2+}$ shows an isotope effect of 2.1 when the N—H protons are deuterated and a mechanism involving proton abstraction followed by protonation of the cobalt(III) product may be in operation [61].

$$[Co(sep)]^{2+} + O_2^- \rightarrow [Co(H_{-1}sep)]^+ + HO_2^-$$

$$\downarrow 2H^+$$

$$[Co(sep)]^{3+} + H_2O_2 \tag{5.97}$$

The reduction of H_2O_2 is thermodynamically more favorable than that of O_2, with a formal potential of 1.763 V in acidic media (eq. (5.98)) and 0.867 V in basic solution (eq. (5.99)).

$$H_2O_2 + 2 H^+ + 2e^- \rightleftharpoons 2 H_2O \tag{5.98}$$

$$O_2^- + H_2O + 2e^- \rightleftharpoons 3 OH^- \tag{5.99}$$

Like the reduction of O_2, the reduction of H_2O_2 is also dominated by single-electron transfer processes, but in this case reactions with substitution-inert complexes which may take place by outer-sphere mechanisms are very slow. A primary reason for this is that addition of a single electron requires cleavage of the O—O bond and is very thermodynamically unfavorable (eq. (5.100)). The product of this reaction is OH·, which has an important and extensive chemistry as a powerful one-electron oxidant (eq. (5.101)), reacting mainly by H-abstraction mechanisms.

$$HO_2^- + H_2O + e^- \rightleftharpoons OH\cdot + 2OH^- \tag{5.100}$$

$$OH\cdot + e^- \rightleftharpoons OH^- \qquad E° = 1.91 V \tag{5.101}$$

Consequently outer-sphere reduction of H_2O_2 generally represents a minor pathway [57]. Reaction of the powerful reductant $[Cr(CN)_6]^{4-}$ follows the rate law (eq. (5.102)) with a rate constant $k_0 = 3.3 \times 10^{-2}$ M^{-1} s^{-1}, where the bulk of the reaction occurs by a competing inner-sphere pathway involving $[Cr(CN)_5(OH_2)]^{3-}$ with a rate constant $k_1 = 3.6 \times 10^3$ M^{-1} s^{-1} [62]. Similarly, the reductants $[Ru(NH_3)_6]^{2+}$ and $[Ru(NH_3)_5(OH_2)]^{2+}$ react very slowly and limiting kinetic behavior suggests that an inner-sphere mechanism with a seven-coordinate intermediate may be involved [63].

$$-\frac{d[[Cr(CN)_5(OH_2)]^{3-}]_T}{dt} = 2\left\{\frac{k_0 K_1 [CN^-] + k_1}{K_1 [CN^-] + 1}\right\} [H_2O_2][[Cr(CN)_5(OH_2)]^{3-}]_T \tag{5.102}$$

Reactions of O_2 and H_2O_2 with labile reductants allow more complex mechanisms and some important studies are noted below. The reductions of $[Cr(H_2O)_6]^{2+}$ and $[V(H_2O)_6]^{2+}$ result in the complete reduction of O_2 to H_2O (eq. (5.103)). The mechanism which has been proposed for the $[Cr(H_2O)_6]^{2+}$ reaction involves the initial formation of a metal peroxo complex (eq. (5.104)), absorbing at 245 and 290 nm, with a formation rate constant 1.6×10^8 M^{-1} s^{-1}, typical of substitution at chromium(II).

$$O_2 + 4[Cr(H_2O)_6]^{2+} + 4 H^+ \rightarrow 4[Cr(H_2O)_6]^{3+} + 2 H_2O \tag{5.103}$$

$$[Cr(H_2O)_6]^{2+} + O_2 \rightarrow [Cr(O_2)_{aq}]^{2+} \tag{5.104}$$

Subsequent reactions depend on a variety of factors. Under conditions of excess O_2, this intermediate is long-lived [64] but in the presence of excess $[Cr(H_2O)_6]^{2+}$, it reacts to form a μ-peroxo dimer which in turn consumes two further equivalents of

[Cr(H$_2$O)$_6$]$^{2+}$. The product is a dimeric complex of chromium(III) which contains both atoms from the O$_2$ [65]. As a result of the dimer formation, the formation of O$_2^-$ is avoided. In contrast, the reduction of H$_2$O$_2$ by [Cr(H$_2$O)$_6$]$^{2+}$ gives predominantly monomeric product [66] and a radical mechanism is proposed. The initial step is formation of OH· radical with a rate constant of 7.06×10^4 M^{-1} s^{-1} at 1.0 M ionic strength followed by the rapid reaction of [Cr(H$_2$O)$_6$]$^{2+}$ with OH·, $k = 1.2 \times 10^{10}$ M^{-1} s^{-1} [67, 68].

The initial step in the reaction of O$_2$ with [V(H$_2$O)$_6$]$^{2+}$ is also adduct formation (Scheme 5.6), with a formation rate constant of 2.0×10^3 M^{-1} s^{-1} in 0.12 M HClO$_4$ [69]. However a more complex mechanism results since the reductant, [V(H$_2$O)$_6$]$^{2+}$, is capable of transfer of more than one electron. In the presence of excess reductant, dimer formation ensues yielding two equivalents of [VO(H$_2$O)$_4$]$^{2+}$ directly, while the intermediate [V(O$_2$)$_{aq}$]$^{2+}$ also reacts by a first-order two-electron decomposition to give [VO(H$_2$O)$_4$]$^{2+}$ and H$_2$O$_2$.

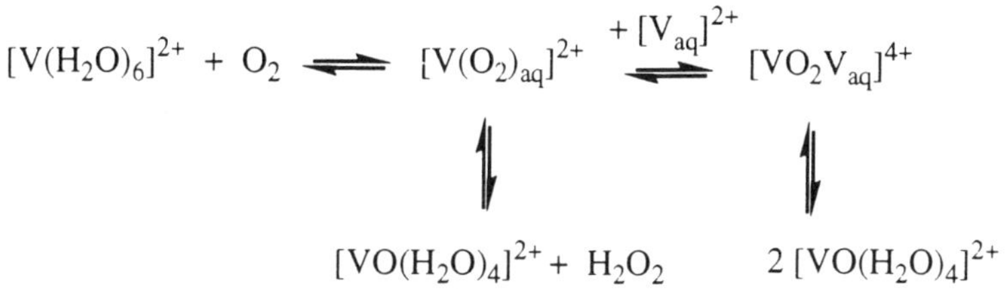

$$[V(H_2O)_6]^{2+} + O_2 \rightleftharpoons [V(O_2)_{aq}]^{2+} \xrightarrow{+ [V_{aq}]^{2+}} [VO_2V_{aq}]^{4+}$$

$$\downarrow \qquad\qquad\qquad \downarrow$$

$$[VO(H_2O)_4]^{2+} + H_2O_2 \qquad 2\,[VO(H_2O)_4]^{2+}$$

(5.105)

Scheme 5.6.

These inner-sphere mechanisms offer pathways which avoid the formation of high-energy species such as O$_2^-$ in the reduction of O$_2$. Cleavage of the O—O bond is readily achieved in systems where oxo complexes are formed. In the reduction of H$_2$O$_2$, single-electron transfer and the formation of OH· is a preferred pathway and again O—O bond cleavage is promoted by the formation of oxo complexes [68].

The reverse of these reactions, oxidations of H$_2$O and H$_2$O$_2$, have also been investigated. Oxidation of H$_2$O to give O$_2$ is most efficiently carried out by powerful aquo-ion oxidants such as [Co(H$_2$O)$_6$]$^{3+}$ and [Mn(H$_2$O)$_6$]$^{3+}$ which react by complex mechanisms which are thought to be inner-sphere in nature [70, 71]. The reactions are kinetically complex but are inhibited by acid. Studies with the strong oxidants [Fe(bpy)$_3$]$^{3+}$, [Ru(bpy)$_3$]$^{3+}$, [Os(bpy)$_3$]$^{3+}$ and derivatives have also attracted much attention. In such reactions production of O$_2$ is not stoichiometric and ligand oxidation competes [73–75]. However, the simple rate law observed in each case (eq. (5.106)), points to a common rate-limiting step for which mechanisms ranging from direct electron transfer from OH$^-$ to give OH·, to deprotonation or covalent hydration of the imine ligand, and nucleophilic attack by OH$^-$ at the metal center have been proposed [76].

$$\frac{-d[[\mathrm{Ru(bpy)_3}]^{3+}]}{dt} = k\,[[\mathrm{Ru(bpy)_3}]^{3+}][\mathrm{OH}^-] \tag{5.106}$$

The rates exceed the rate of dissociation of the bpy ligand from the oxidant (Table 5.7), and the latter mechanism involving formation of a seven-coordinate intermediate seems most plausible. Subsequent steps involve substitution of one of the bpy ligands or the formation of oxo complexes on further oxidation (eqs (5.107)–(5.108)) [77]. The minor pathway involving water oxidation is thought to involve formation of an oxo-bridged dimer, $[(\mathrm{bpy})_2(\mathrm{H_2O})\mathrm{Ru}(\mu\text{-}\mathrm{O})\mathrm{Ru}(\mathrm{OH_2})(\mathrm{bpy})]^{4+}$, identified as an effective catalyst in the electrochemical oxidation of water [78]. The active species is the formal ruthenium(V) complex $[(\mathrm{bpy})_2\mathrm{ORu}(\mu\text{-}\mathrm{O})\mathrm{RuO}(\mathrm{bpy})]^{4+}$ and studies with $^{18}\mathrm{O}$ oxo derivatives suggest that the $\mathrm{O_2}$ produced in the reaction derives predominantly from the coordinated oxygen [79].

$$[\mathrm{Ru(bpy)_3}]^{3+} + \mathrm{OH}^- \xrightarrow{k} [\mathrm{Ru(bpy)_3OH}]^{2+} \tag{5.107}$$

$$[\mathrm{Ru(bpy)_3}]^{3+} + [\mathrm{Ru(bpy)_3OH}]^{2+} \rightarrow [\mathrm{Ru(bpy)_3O}]^{2+} + [\mathrm{Ru(bpy)_3}]^{2+} + \mathrm{H}^+ \tag{5.108}$$

Table 5.7. Rate constants for the reaction of OH$^-$ with $[\mathrm{Ru(bpy)_3}]^{3+}$ and derivatives [76]

Oxidant	μ (M)	k (M^{-1} s^{-1})	$\Delta H^{\ddagger}$ (kJ mol^{-1})	$\Delta S^{\ddagger}$ (J K^{-1} mol^{-1})
$[\mathrm{Fe(bpy)_3}]^{3+}$	1.0	16	62	−17
$[\mathrm{Ru(bpy)_3}]^{3+}$	1.0	148	64	12
$[\mathrm{Os(bpy)_3}]^{3+}$	1.0	4.7	65	−13
$[\mathrm{Fe(phen)_3}]^{3+}$	1.0	420	47	−42
$[\mathrm{Os(phen)_3}]^{3+}$	1.0	156	50	−35

The stoichiometric oxidation of the intermediate state, $\mathrm{H_2O_2}$, is more straightforward and can be accomplished with more modest oxidants, but the mechanisms are thought to be predominantly inner-sphere and the rate laws show a strong dependence on $[\mathrm{H}^+]^{-1}$ [80]. That $\mathrm{H_2O_2}$ can act as both oxidant and reductant with a number of complexes allows them to function as catalysts for disproportionation to give $\mathrm{H_2O}$ and $\mathrm{O_2}$ (eq. (5.109)), and studies of this peroxidase activity have attracted some interest [81].

$$2\,\mathrm{H_2O_2} \rightarrow 2\,\mathrm{H_2O} + \mathrm{O_2} \tag{5.109}$$

Outer-sphere reactions have also been examined and are much less common, occurring with powerful outer-sphere reagents such as $[\mathrm{Ni(bpy)_3}]^{3+}$ [82]. There is no inhibition by $[\mathrm{H}^+]$ in the rate law (eq. (5.110)), where $k = 0.34$ M^{-1} s^{-1} at 25.0°C and 2.0 M ionic strength with $\Delta H^\ddagger = 40$ kJ mol^{-1} and $\Delta S^\ddagger = -126$ J K^{-1} mol^{-1} suggesting the mechanism in eqs (5.111) and (5.112).

$$-\frac{d[[Ni(bpy)_3]^{3+}]}{dt} = k\,[H_2O_2][[Ni(bpy)_3]^{3+}] \qquad (5.110)$$

$$[Ni(bpy)_3]^{3+} + H_2O_2 \rightarrow [Ni(bpy)_3]^{2+} + O_2^- + H^+ \qquad (5.111)$$

$$[Ni(bpy)_3]^{3+} + O_2^- \xrightarrow{\text{fast}} [Ni(bpy)_3]^{2+} + O_2 \qquad (5.112)$$

Reactions with the aqua-ion oxidants such as $[Co(H_2O)_6]^{3+}$ [83], $[Fe(H_2O)_6]^{3+}$ [84], and cis-$[Co(NH_3)_2(H_2O)_4]^{3+}$ [85], exhibit the rate law shown in eq. (5.113). The mechanisms are inner-sphere involving complexes of the type $[Co(NH_3)_2(H_2O)_2(OH)(O_2H_2)]^{2+}$. However, mechanistic interpretation of these reactions is challenging because of the proton ambiguity which leaves the pathways (5.114) and (5.115) kinetically indistinguishable when the kinetics are examined over a limited range of [H$^+$].

$$\frac{-d[[Co(NH_3)_2(H_2O)_4]^{3+}]}{dt} = \frac{k\,[[Co(NH_3)_2(H_2O)_4]^{3+}]\,[H_2O_2]}{[H^+]} \qquad (5.113)$$

$$[Co(NH_3)_2(H_2O)_3(OH)]^{2+} + H_2O_2 \rightarrow [Co(NH_3)_2(H_2O)_2(OH)(O_2H_2)]^{2+} \qquad (5.114)$$

$$[Co(NH_3)_2(H_2O)_4]^{3+} + HO_2^- \rightarrow [Co(NH_3)_2(H_2O)_3(O_2H)]^{2+} \qquad (5.115)$$

Substitution-controlled reactions of HO$_2^-$ are thought to predominate but a great deal relies on the proton balance of the reagents where electron transfer is accompanied by the transfer of H$^+$. In such conditions atom transfer processes may be competitive. A detailed study of the oxidation of H$_2$O$_2$ by [(bpy)$_2$pyRuO]$^{2+}$ involves a process where the disposition of protons is crucial to the thermodynamic driving force (eq. (5.116) [86, 87]. Reduction of the oxidant involves addition of two protons whereas oxidation of H$_2$O$_2$ involves loss of two protons. The rate law shows two terms (eq. (5.117)), one of which is independent of acid and the mechanism (eqs (5.118)–(5.122)) requires two consecutive single-electron transfer reactions since [(bpy)$_2$pyRuOH]$^{2+}$ is detected as an intermediate and it cannot be formed by comproportionation of [(bpy)$_2$pyRuO]$^{2+}$ and [(bpy)$_2$pyRuOH$_2$]$^{2+}$ on the timescale of the reactions.

$$[(bpy)_2pyRuO]^{2+} + H_2O_2 \rightarrow [(bpy)_2pyRuOH_2]^{2+} + O_2 \qquad (5.116)$$

$$-\frac{d[[(bpy)_2pyRuO]^{2+}]}{dt} = \{k_1 + k_2K_a/[H^+]\} [H_2O_2][[(bpy)_2pyRuO]^{2+}] \quad (5.117)$$

$$H_2O_2 \rightleftharpoons HO_2^- + H^+ \quad K_a = 2.28 \times 10^{-12} \quad (5.118)$$

$$[(bpy)_2pyRuO]^{2+} + H_2O_2 \xrightarrow{k_1} [(bpy)_2pyRuOH]^{2+} + HO_2 \quad (5.119)$$

$$[(bpy)_2pyRuO]^{2+} + HO_2 \xrightarrow{fast} [(bpy)_2pyRuOH]^{2+} + O_2 \quad (5.120)$$

$$[(bpy)_2pyRuOH]^{2+} + H_2O_2 \xrightarrow{k_3} [(bpy)_2pyRuOH_2]^{2+} + HO_2 \quad (5.121)$$

$$[(bpy)_2pyRuO]^{2+} + HO_2^- \xrightarrow{k_2} [(bpy)_2pyRuOH]^+ + O_2 \quad (5.122)$$

The key point of interest is a large solvent isotope effect, k_{H_2O}/k_{D_2O}, of 22 for the acid independent pathway, k_1, and is interpreted as evidence for atom transfer. The acid dependent pathway, k_2, also shows an isotope effect but the magnitude is comparable to the thermodynamic effect (eq. (5.118)) where $K_{aH_2O}/K_{aD_2O} = 7.28$. Similar observations have been made in the reduction of $[(bpy)_2pyRuOH]^{2+}$. Labelling studies with $[(bpy)_2pyRu^{18}O]^{2+}$ give no incorporation of the label in the O_2 produced in the reaction, suggesting that a peroxy adduct is not formed.

Reactions of cobalt(II) amine complexes with O_2 form the final topic in this section and are important because they have provided a rich chemistry of redox active adducts [88, 89]. In the oxidation of these species it is found that O_2 reacts much faster with the more labile $[Co(NH_3)_5OH_2]^{2+}$ than $[Co(NH_3)_6]^{2+}$ and that the initial reaction involves formation of a peroxo adduct (eq. (5.123)). Subsequent reaction with a second metal complex leads to the formation of a μ-peroxo dimer (eq. (5.124)), characterized by a O—O bond distance of 1.47 Å. The general mechanism is followed by a number of cobalt(II) species and rate data are given in Table 5.8.

$$[Co(NH_3)_5OH_2]^{2+} + O_2 \underset{k_{-1}}{\overset{k_1}{\rightleftharpoons}} [Co(NH_3)_5O_2]^{2+} \quad (5.123)$$

$$[Co(NH_3)_5OH_2]^{2+} + [Co(NH_3)_5O_2]^{2+} \underset{k_{-2}}{\overset{k_2}{\rightleftharpoons}} [(NH_3)_5Co(\mu\text{-}O_2)Co(NH_3)_5]^{2+}$$

(5.124)

Table 5.8. Rate constants for the reactions of cobalt(II) complexes with O_2 at 25°C

Complex	μ (M)	k ($M^{-1} s^{-1}$)	Ref.
$[Co(NH_3)_5OH_2]^{2+}$	2.0	2.5×10^4	90
$[Co(en)_2(OH_2)_2]^{2+}$	0.2	4.7×10^5	91
$[Co(trien)(OH_2)_2]^{2+}$	0.2	2.5×10^4	92
$[Co(trien)(OH_2)(OH)]^+$	0.2	2.8×10^5	92
$[Co([14]aneN_4)(H_2O)_2]^{2+}$	0.1	5.0×10^5	93
$[Co(Me_6[14]aneN_4)(H_2O)_2]^{2+}$	0.1	5.0×10^6	94
$[Co(Me_6[14]aneN_4)(H_2O)(OH)]^+$	0.1	8.9×10^5	95

In the case of the macrocycle complexes such as $[Co([14]aneN_4)(H_2O)_2]^{2+}$, the two formation steps can be distinguished [93], and the 1:1 adduct exists in equilibrium with the μ-peroxo complex [96]. The values for k_2 and k_{-2} are 4.9×10^5 M^{-1} s^{-1} and 0.6 s^{-1} respectively at 25°C. Use of the sterically demanding macrocycle $[Co(Me_6[14]aneN_4)(H_2O)_2]^{2+}$ results in inhibition of μ-peroxo formation [94]. Where the μ-peroxo complex has labile ligands in positions *cis* to the peroxo linkage, formation of additional μ-hydroxy- or μ-amido-bridges is possible and this can be detected as a slower ring closure step. These double bridged complexes have significantly higher thermodynamic stability than the singly bridged species.

The chemistry of the μ-peroxo complexes is extensive. In the presence of excess O_2 or $S_2O_8^{2-}$, the complexes can be oxidized to form μ-superoxo species. Reduction potentials for the complexes have been measured and are generally in the vicinity of 1.0 V, indicating significant stabilization of the superoxo-ligand [97]. Reduction reactions with metal ion complexes are predominantly outer-sphere in nature and yield the μ-peroxo complex (section 2.6). By contrast, inner-sphere mechanisms are preferred for reductions of the μ-peroxo complexes.

When $[([14]aneN_4)Co(\mu\text{-}O_2)Co([14]aneN_4)]^{4+}$ is treated with acid, reactions of the transient $[Co([14]aneN_4)(H_2O)(O_2)]^{2+}$ can be examined. The $[Co([14]aneN_4)(H_2O)(O_2)]^{2+}$ is scavenged by both oxidants and reductants, and rate constants for a variety of outer-sphere reactions are presented in Table 5.9. Both oxidation and reduction are limited by μ-peroxo decomposition with a limiting rate constant of 0.57 s^{-1} at 25.0°C and 0.10 M ionic strength. An estimate of 0.3 V for the $[Co([14]aneN_4)(H_2O)(O_2)]^{2+/+}$ reduction potential yields a self-exchange rate for the couple of approximately 10^3 M^{-1} s^{-1} [98].

Table 5.9. Rate constants for reactions of $[Co([14]aneN_4)(H_2O)(O_2)]^{2+}$ at 25°C

Reductants	μ (M)	k (M^{-1} s^{-1})
$[Ru(NH_3)_6]^{2+}$	0.1	2.3×10^5
$[Co(sep)]^{2+}$	0.3	$\approx 1 \times 10^6$
$[V(OH_2)_6]^{2+}$	0.1	1.8×10^5
Oxidant		
$[Ru(NH_3)_4(phen)]^{3+}$	0.1	$\approx 1 \times 10^7$

The behavior observed in the reactions of metal complexes with O_2, H_2O_2 and H_2O represents a number of features. Firstly, outer-sphere one-electron reactions are observed and can be understood in terms of Marcus Theory. Secondly, outer-sphere reactions for these non-complementary reactions can be very slow when the single electron radical is unstable or production of the radical involves large changes in geometry resulting in slow self-exchange rates. High-energy intermediate species frequently react by alternative mechanisms such as atom abstraction. Thirdly, inner-sphere and atom transfer mechanisms provide alternative facile pathways for multiple-electron transfer and bond cleavage. These observations are quite general and can be illustrated with examples from other oxidants and reductants.

5.9 REACTIONS OF HALOGENS, PSEUDOHALOGENS AND RELATED SPECIES

The uniting theme of the reagents in this section is that formation of the oxidized forms requires dimerization of the reduced forms. In this respect they are expected to show parallels with the behavior of $H_2O_2/2H_2O$ system. The reduction of $[S_2O_8]^{2-}$, like the reductions of H_2O_2 requires concomitant cleavage of the peroxo O—O bond. Studies of non-complementary reactions with outer-sphere reagents suggest a mechanism (eqs (5.125)–(5.127)) involving the radical SO_4^- as an intermediate [99, 100].

$$[Fe(bpy)_3]^{2+} + S_2O_8^{2-} \overset{K_0}{\rightleftharpoons} \{[Fe(bpy)_3]^{2+}, S_2O_8^{2-}\} \quad (5.125)$$

$$\{[Fe(bpy)_3]^{2+}, S_2O_8^{2-}\} \overset{k_1}{\rightarrow} [Fe(bpy)_3]^{3+} + SO_4^{2-} + SO_4^- \quad (5.126)$$

$$[Fe(bpy)_3]^{2+} + SO_4^- \xrightarrow{fast} [Fe(bpy)_3]^{3+} + SO_4^{2-} \quad (5.127)$$

Rate constants, K_0k_1, for these reactions are accumulated in Table 5.10. In suitable cases with highly charged oxidants such as the dimeric $[(NH_3)_5Ru(pz)Ru(NH_3)_5]^{5+}$, there is kinetic evidence for ion pair formation with $K_0 = 2.7 \times 10^2$ M^{-1} and $k_1 = 47$ s^{-1} [101].

Table 5.10. Rate constants for the reductive cleavage of $S_2O_8^{2-}$ by metal complexes at 25 °C.

Reductant	μ (M)	K_0k_1 (M^{-1} s^{-1})	Ref.
$[Fe(bpy)_3]^{2+}$	0.03	0.41	99
$[Fe(phen)_3]^{2+}$	0.03	0.11	99
$[Fe(terpy)_2]^{2+}$	0.03	0.51	99
$[Fe(bpy)_3]^{2+}$	0.03^a	0.6	100
$[Ru(NH_3)_5(pz)]^{2+}$	0.1	3.6×10^3	101
$[Ru(NH_3)_4(bpy)]^{2+}$	0.1	3.6×10^3	101
$[Ru_2(NH_3)_{10}(pz)]^{5+}$	0.1	1.3×10^4	101
$[Ru_2(NH_3)_{10}(pz)]^{4+}$	0.1	4.3×10^5	101

^{a}35 °C

Application of a free-energy relationship based on Marcus Theory suggests that these reagents react by outer-sphere mechanisms. An estimate for the reduction potential for $S_2O_8^{2-}/SO_4^{2-},SO_4^-$ of 1.45 V yields a self-exchange rate for the couple of approximately 10^{-18} M^{-1} s^{-1}, not surprising in view of the large structural reorganization involved. An interesting issue is raised by reaction of the dimeric reductant, $[(NH_3)_5Ru(pz)Ru(NH_3)_5]^{4+}$, which is capable of a complementary two-electron transfer (eq. (5.128)). The absorbance change is biphasic, indicating that $[(NH_3)_5Ru(pz)Ru(NH_3)_5]^{5+}$ is formed but the amount of this species indicates that the majority of it (80%) is formed by the comproportionation reaction (eq. (5.129)), rather than (eq. (5.130)). Evidence from the effect of $[Fe(H_2O)_6]^{2+}$ as a radical scavenger also pinpoints that 80% of the reaction takes place through a complementary two-electron process. However, the preferred explanation involves two consecutive electron transfers within a single encounter complex.

$$[(NH_3)_5Ru(pz)Ru(NH_3)_5]^{4+} + S_2O_8^{2-} \rightarrow$$
$$[(NH_3)_5Ru(pz)Ru(NH_3)_5]^{6+} + 2\,SO_4^{2-} \quad (5.128)$$

$$[(NH_3)_5Ru(pz)Ru(NH_3)_5]^{4+} + [(NH_3)_5Ru(pz)Ru(NH_3)_5]^{6+} \rightarrow$$
$$2\,[(NH_3)_5Ru(pz)Ru(NH_3)_5]^{5+} \quad (5.129)$$

$$[(NH_3)_5Ru(pz)Ru(NH_3)_5]^{4+} + S_2O_8^{2-} \rightarrow$$
$$[(NH_3)_5Ru(pz)Ru(NH_3)_5]^{5+} + SO_4^{2-} + SO_4^- \quad (5.130)$$

Reduction of $[S_2O_8]^{2-}$ by $[Cr(H_2O)_6]^{2+}$ is more rapid than would be expected for an outer-sphere process, 2.5×10^4 M^{-1} s^{-1} at 25°C and 1.0 M ionic strength, and leads to the formation of $[Cr(H_2O)_6]^{3+}$ and $[Cr(H_2O)_5(SO_4)]^+$ in equal amounts (eq. (5.131)). The initial electron transfer (eq. (5.132)), is inner-sphere and the subsequent step (eq. (5.133)) may involve outer-sphere electron transfer or H-atom abstraction by the reactive SO_4^- ($E° = 2.43$ V) [102]. Reduction by $[Fe(H_2O)_6]^{2+}$ with $k = 26$ M^{-1} s^{-1} at 25°C and 0.10 M ionic strength is also thought to be inner-sphere [101].

$$[S_2O_8]^{2-} + [Cr(H_2O)_6]^{2+} \rightarrow$$
$$[Cr(H_2O)_6]^{3+} + [Cr(H_2O)_4(SO_4)]^+ + [SO_4]^{2-} \quad (5.131)$$

$$[S_2O_8]^{2-} + [Cr(H_2O)_6]^{2+} \rightarrow [SO_4]^- + [Cr(H_2O)_4(SO_4)]^+ \quad (5.132)$$

$$[SO_4]^- + [Cr(H_2O)_6]^{2+} \rightarrow [SO_4]^{2-} + [Cr(H_2O)_6]^{3+} \quad (5.133)$$

A further series of reactions which involves cleavage of a bond in the oxidant is the oxidation of metal complexes by halogens. These reactions are complicated by hydrolysis and adduct formation, illustrated for Br_2 in eqs (5.134) and (5.135). Reaction kinetics show a dependence on both pH and $[Br^-]$ from which the rate constants for reactions of all three oxidizing species, Br_2, HOBr, and Br_3^-, can be evaluated.

$$Br_2 + OH^- \rightleftharpoons HOBr + Br^- \quad (5.135)$$

$$Br_2 + Br^- \rightleftharpoons Br_3^- \quad (5.136)$$

Rate data for a number of reactions of metal complexes with halogens are presented in Table 5.11.

The list includes both complementary and non-complementary processes. In some instances, for example $[Fe(phen)_3]^{2+}$ and $[Co(sep)]^{2+}$, the reactions must take place by outer-sphere mechanisms since the rate of the electron transfer process exceeds the rate of substitution at the metal center. These reactions are found to follow linear free energy relationships as predicted by the Marcus relationship. In other reactions, inner-sphere mechanisms can be demonstrated. In the reactions of $[Co(edta)]^{2-}$ with Br_2, the initial product is $[Co(edta)Br]^{2-}$ revealing that the mechanism is inner-sphere since substitution must take place prior to oxidation of the metal center to the substitution-inert low-spin d^6 configuration. The relative reactivity of the various oxidizing species is also an important clue to mechanism. Empirically it is observed that HOBr is more reactive than Br_2 for inner-sphere reactions. Reactions of the hypohalous acids are slow where an outer-sphere mechanism is assigned.

Table 5.11. Rate constants and activation parameters for reactions of metal complexes with halogens at 25.0°C

Reaction	μ (M)	k_{X_2} (M^{-1} s^{-1})	$\Delta H^{\ddagger}$ (kJ mol^{-1})	$\Delta S^{\ddagger}$ (J K^{-1} mol^{-1})	$k_{X_3^-}$ (M^{-1} s^{-1})	Ref.
[Fe(phen)$_3$]$^{2+}$ + Cl$_2$	1.0	2.4				103–105
[Rh$_2$(OAc)$_4$] + Cl$_2$	1.0	0.12				106
[Fe(phen)$_3$]$^{2+}$ + Br$_2$	1.0	1.6	64	−25		105
[Co(edta)]$^{2-}$ + Br$_2$	1.0	7.3 × 10^{-2}			5.4 × 10^{-2}	107
[Co(cdta)]$^{2-}$ + Br$_2$	1.0	2.0 × 10^{-3}			1.1 × 10^{-3}	107
[Fe(cdta)]$^{2-}$ + Br$_2$	1.0	9.3 × 10^6			1.2 × 10^6	108
[Mn(edta)]$^{2-}$ + Br$_2$	1.0	7.8 × 10^{-3}			2.5 × 10^{-4}	108
[Ni(Me$_2$L)] + Br$_2$	1.0	2.1 × 10^7			≈1 × 10^4	109
[Ni(Me$_2$LH)]$^+$ + Br$_2$	1.0	7.1 × 10^5			2 × 10^2	109
[Ni(Me$_2$L)]$^+$ + Br$_2$	1.0	1.2 × 10^5				109
[Co(edta)]$^{2-}$ + I$_2$	1.0	5.2 × 10^{-4}			3.1 × 10^{-5}	110
[Co(cdta)]$^{2-}$ + I$_2$	1.0	6.4 × 10^{-5}			2.2 × 10^{-5}	110
[Fe(edta)]$^{2-}$ + I$_2$	1.0	1.6 × 10^6			5.0 × 10^3	108
[Fe(cdta)]$^{2-}$ + I$_2$	1.0	1.2 × 10^6			4.4 × 10^3	108
[Co(sep)]$^{2+}$ + I$_2$	0.1	5.9 × 10^4			3.93 × 10^4	111

Three examples of complementary reactions with the halogens are considered and the detailed mechanisms show remarkable variation. With the labile reductant $[U_{aq}]^{4+}$, the basic rate law (eq. (5.136)) shows a first-order dependence on both reagents and is inhibited by a term in $[H^+]^{-2}$ [112–114]. This strong acid dependence is a reflection of the proton balance in the reaction (eq. (5.137)), and is a characteristic of reactions of oxo anions. The reactive reductant is $[U(OH)_{2aq}]^{2+}$.

$$\frac{-d[Br_2]}{dt} = \frac{k[[U_{aq}]^{4+}][Br_2]}{[H^+]^2} \tag{5.136}$$

$$[U_{aq}]^{4+} + Br_2 \rightarrow [(UO_2)_{aq}]^{2+} + 2\,Br^- + 4\,H^+ \tag{5.137}$$

Although species such as Br_3^- and HOBr participate in the reactions, there is no evidence for formation of intermediates and an inner-sphere two-electron transfer mechanism is consistent with the small variation in reaction rate when the halide is varied from I_2 to Cl_2.

Reaction of the halogens with complexes of platinum(II) is quite different. The halogen is incorporated into the platinum(IV) product, a formal oxidative addition (eq. (5.138)). The mechanism is complex and at least two steps have been noted [115–118]. The first step is the rapid formation of a platinum(IV) aqua-halo intermediate (eq. (5.139)), and is followed by a slower anation to give trans-$[Pt(CN)_4Cl_2]^{2-}$ (eq. (5.140)) [119]. A parallel pathway for HOCl as oxidant is a factor of 10^5 slower and leads to trans-$[Pt(CN)_4(OH)_2]^{2-}$ as the initial product. The mechanism of electron transfer involves simultaneous transfer of two electrons by an inner-sphere mechanism. In the case of HOCl and OCl$^-$, it is the oxygen of the oxidant which coordinates and is transferred in the electron transfer process, rather than the halide, accounting for the differences in products.

$$[Pt(CN)_4]^{2-} + Cl_2 \rightarrow \textit{trans-}[Pt(CN)_4Cl_2]^{2-} + H_2O \tag{5.138}$$

$$[Pt(CN)_4]^{2-} + Cl_2 \rightarrow \textit{trans-}[Pt(CN)_4Cl(H_2O)]^- + Cl^- \tag{5.139}$$

$$\textit{trans-}[Pt(CN)_4Cl(H_2O)]^- + Cl^- \rightarrow \textit{trans-}[Pt(CN)_4Cl_2]^{2-} + H_2O \tag{5.140}$$

An interesting example of a complementary reaction is reported observation in the reaction of the nickel(II) complex, $[Ni(Me_2L)]$, with Br_2 [109]. Under conditions of excess reductant where $[Ni(Me_2L)]^+$ is the thermodynamic product, the nickel(IV) complex $[Ni(Me_2L)]^{2+}$ is detected as a transient which subsequently comproportionates (eqs (5.141)–(5.142)). Reactions with Br_3^-, HOBr and OBr$^-$ are unimportant suggesting an outer-sphere mechanism.

$$[Ni(Me_2L)] + Br_2 \rightarrow [Ni(Me_2L)]^{2+} + 2\,Br^- \tag{5.141}$$

$$[Ni(Me_2L)]^{2+} + [Ni(Me_2L)] \rightarrow 2\,[Ni(Me_2L)]^+ \tag{5.142}$$

It is estimated that at least 60% of the reaction proceeds by this mechanism. Like the outer-sphere two-elecron mechanism proposed for the oxidation of $[(NH_3)_5Ru(pz)Ru(NH_3)_5]^{4+}$ by $S_2O_8^{2-}$ [101], detailed analysis of the reaction energetics suggest that the best description involves two sequential one-electron steps within the same encounter complex rather than a synchronous two-electron step.

Non-complementary reactions of the halogens have been widely studied. The reactions of Br_2 with $[Co(edta)]^{2-}$ leads to the formation of $[Co(edta)Br]^{2-}$ as the initial product, thereby proving an inner-sphere mechanism (eqs (5.143)–(5.145)) [110]. In the corresponding reactions with HOBr, the product is $[Co(edta)(OH)]^{2-}$ or $[Co(edta)(OH_2)]^-$ depending on the pH, again consistent with an inner-sphere process [107]. In this case the Br· radicals formed do not oxidize $[Co(edta)]^{2-}$ directly since $[Co(edta)Br]^{2-}$ is not a product under these conditions but instead recombine and hydrolyze to regenerate the oxidant.

$$[Co(edta)]^{2-} + Br_2 \rightleftharpoons [Co(edta)Br_2]^{2-} \tag{5.143}$$

$$[Co(edta)Br_2]^{2-} \rightarrow [Co(edta)Br]^{2-} + Br\cdot \tag{5.144}$$

$$[Co(edta)Br_2]^{2-} + Br\cdot \rightarrow [Co(edta)Br]^{2-} + Br_2 \tag{5.145}$$

The reaction of Cl_2 with $[Fe(phen)_3]^{2+}$ is complicated by a change in the reaction product from the blue $[Fe(phen)_3]^{3+}$ formed by an outer-sphere mechanism in acidic conditions to the yellow dimer $[(phen)_2Fe(OH)_2Fe(phen)_2]^{4+}$ at higher pH where HOCl is the oxidant [104]. Under these conditions, the rate law is given in eq. (5.146), where K_a is the acidity constant for HOCl. The terms k_1 and k_2 are respectively the rates of outer-sphere reaction of $[Fe(phen)_3]^{2+}$ with HOCl and OCl$^-$ while the k_0 term is the rate of dissociation of $[Fe(phen)_3]^{2+}$, providing an inner-sphere pathway which, it is proposed, could lead to a two-electron transfer (eqs (5.147)–(5.149)).

$$\frac{-d[[Fe(phen)_3]^{2+}]}{dt} = k_0 + \frac{k_1[H^+] + K_a k_2}{[H^+] + K_a} [HOCl]_T[[Fe(phen)_3]^{2+}] \tag{5.146}$$

$$[Fe(phen)_3]^{2+} \underset{}{\overset{k_0}{\rightleftharpoons}} [Fe(phen)_2]^{2+} + phen \tag{5.147}$$

$$[Fe(phen)_2]^{2+} + HOCl \overset{fast}{\rightarrow} [Fe^{IV}(phen)_2OH]^{3+} + Cl^- \tag{5.148}$$

Table 5.12. Reactivity patterns with halogens and halogenous acids

Reaction	μ (M)	k_{X_2} ($M^{-1} s^{-1}$)	$k_{X_3^-}$ ($M^{-1} s^{-1}$)	k_{HOX} ($M^{-1} s^{-1}$)	k_{OX^-} ($M^{-1} s^{-1}$)	Ref.
$[Fe(phen)_3]^{2+} + Cl_2$	1.0	2.4		2.2×10^{-2}	2.0×10^{-3}	103, 104
$[Fe(phen)_3]^{2+} + Cl_2$	1.0	2.2				105
$[Fe(phen)_3]^{2+} + Br_2$	1.0	1.6				105
$[Rh_2(OAc)_4] + Cl_2$	1.0	0.12		≈ 0	≈ 0	106
$[Co(edta)]^{2-} + Br_2$	1.0	7.3×10^{-2}	5.4×10^{-2}	65	2×10^3	107
$[Co(cdta)]^{2-} + Br_2$	1.0	2.0×10^{-3}	1.1×10^{-3}	0.16	49	107
$[Ni(Me_2L)] + Br_2$	1.0	2.1×10^7	$\approx 1 \times 10^4$	$\leq 1 \times 10^3$	≈ 0	109
$[Pt(CN)_4]^{2-} + Cl_2$	1.0	1.1×10^7	—	98	≈ 10	119

$$[\text{Fe(phen)}_2]^{2+} + [\text{Fe}^{IV}(\text{phen})_2\text{OH}]^{3+} \xrightarrow{\text{fast}} [(\text{phen})_2\text{Fe(OH)}_2\text{Fe(phen)}]^{4+} + \text{H}^+$$

(5.149)

Oxidations of the metal cations, $[\text{Fe(H}_2\text{O})_6]^{2+}$, $[\text{V(H}_2\text{O})_6]^{2+}$ [120], $[\text{V(H}_2\text{O})_6]^{3+}$, and $[\text{Ti(H}_2\text{O})_6]^{3+}$ have been investigated. The bulk of the evidence in the reactions of X_2 and X_3^- suggests that consecutive one-electron mechanisms are operating and distinction of inner-sphere and outer-sphere mechanisms can be made on the basis of whether the rates exceed substitution control at the metal center. In the case of the oxidation of $[\text{Fe(H}_2\text{O})_6]^{2+}$ by HOCl, a dimeric product is identified as a transient which indicates the presence of a two-electron pathway involving iron(IV) [121].

Table 5.13. Rate constants and activation parameters for the oxidation of aqua metal ions by halogens at 1.0 M ionic strength and 25.0°C

Reaction	k (M^{-1} s^{-1})	$\Delta H^\ddagger$ (kJ mol^{-1})	$\Delta S^\ddagger$ (JK^{-1} mol^{-1})	Ref.
$[\text{V(H}_2\text{O})_6]^{2+} + \text{I}_3^-$	9.7×10^2	38	−59	120
$[\text{V(H}_2\text{O})_6]^{2+} + \text{I}_2$	7.5×10^3	25	−88	120
$[\text{V(H}_2\text{O})_6]^{2+} + \text{Br}_2$	3.0×10^2	15	−109	120
$[\text{V(H}_2\text{O})_6]^{2+} + \text{Cl}_2$	>500	—	—	120
$[\text{V(H}_2\text{O})_5(\text{OH})]^{2+} + \text{Br}_2$	1.8×10^3	—	—	122
$[\text{Ti(H}_2\text{O})_5(\text{OH})]^{2+} + \text{Cl}_2$	1.3×10^4	—	—	123
$[\text{Fe(H}_2\text{O})_6]^{2+} + \text{Br}_2$	0.76	—	—	124
$[\text{Fe(H}_2\text{O})_6]^{2+} + \text{Cl}_2$	80	—	—	121

To summarize the reactions of the halogens: outer-sphere reactions involving single-electron transfer are found with X_2 and some X_3^- oxidants with formation of the corresponding radical; but for HOX as oxidant the outer-sphere pathway is disfavored as the radical is too high in energy. Inner-sphere reactions of HOX are favored. Inner-sphere reactions involving two-electron transfer are also found. The mechanistic complexity appears to arise from the energetic requirements of different pathways. Where inner-sphere complex formation is possible, lower-energy pathways are the frequent result. Considerable insight into the reactions of the halogens has been derived from the study of the reverse of the halogen reduction, halide oxidation. Again there are differences between outer-sphere and inner-sphere pathways which appear to be governed by the energetics involved in radical formation.

The general rate law for the outer-sphere oxidation of halides and pseudo-halides is illustrated for the oxidation of I$^-$ by $[\text{Os(phen)}_3]^{3+}$ in eq. (5.150). The reaction is first-order in $[[\text{Os(phen)}_3]^{3+}]$, but the order in [I$^-$] indicates two pathways, one with a simple first-order dependence on [I$^-$], the other with a second-order dependence. A proposed mechanism [125] involves formation of an ion pair, $\{[\text{Os(bpy)}_3]^{3+},\text{I}^-\}$, which can undergo intramolecular electron transfer to give I· atoms with $k_a = K_0 k_1$ or serve as a reactant in reaction with a second I$^-$ to yield the I$_2^-$ radical ion with

$k_b = K_0 k_2$ (eqs (5.151)–(5.155)). The rate constants k_a and k_b for this and a number of related reactions are presented in Table 5.14.

$$\frac{-d[[Os(bpy)_3]^{3+}]}{dt} = \{k_a + k_b[I^-]\}\,[I^-]\,[[Os(bpy)_3]^{3+}] \tag{5.150}$$

$$[Os(bpy)_3]^{3+} + I^- \underset{}{\overset{K_0}{\rightleftharpoons}} \{[Os(bpy)_3]^{3+}, I^-\} \tag{5.151}$$

$$\{[Os(bpy)_3]^{3+}, I^-\} \underset{k_{-1}}{\overset{k_1}{\rightleftharpoons}} [Os(bpy)_3]^{2+} + I\cdot \tag{5.152}$$

$$\{[Os(bpy)_3]^{3+}, I^-\} + I^- \underset{k_{-2}}{\overset{k_2}{\rightleftharpoons}} [Os(bpy)_3]^{2+} + I_2^- \tag{5.153}$$

$$I\cdot + I^- \underset{k_{-3}}{\overset{k_3}{\rightleftharpoons}} I_2^- \tag{5.154}$$

$$[Os(bpy)_3]^{3+} + I_2^- \overset{k_4}{\rightarrow} [Os(bpy)_3]^{2+} + I_2 \tag{5.155}$$

The driving force for the rate law term, second-order in [I$^-$], is the greater thermodynamic stability of the transient radicals formed. Pulse radiolysis studies with I_2^- and $[Os(bpy)_3]^{2+}$ allow the evaluation of the reverse rates $k_{-2} = 1.1 \times 10^8$ M^{-1} s^{-1}, from which reduction potentials for $I_2^-/2$ I$^-$ of 1.06 V can be calculated [134]. The reduction potential of I· is estimated to be 1.33 V (eq. (5.156)), and is stabilized by the addition of I$^-$ with a formation constant of the order of 1×10^5 M^{-1} (eq. (5.157)).

$$I\cdot + e^- \rightleftharpoons I^- \quad E^\circ = 1.33 \text{ V} \tag{5.156}$$

$$I\cdot + I^- \rightleftharpoons I_2^- \tag{5.157}$$

$$I_2^- + e^- \rightleftharpoons 2\,I^- \quad E^\circ = 1.06 \text{ V} \tag{5.158}$$

In view of the central role of the ion pair in the proposed mechanism, it is surprising that the pathway dependent on [I$^-$]^2 can be detected with anionic oxidants such as [IrCl$_6$]$^{2-}$ [126]. An alternative attempt [135] to rationalize these observations involves formation of ion pairs of the type {I$^-$,I$^-$} (eq. (5.159)), which undergo outer-sphere reactions. The association constant for this species estimated by means

Table 5.14. Rate and activation parameters for reactions of halides and pseudohalides at 25.0 °C

Reaction	μ (M)	k_a (M^{-1} s^{-1})	$\Delta H^{\ddagger}$ (kJ mol^{-1})	$\Delta S^{\ddagger}$ (J K^{-1} mol^{-1})	k_b (M^{-2} s^{-1})	$\Delta H^{\ddagger}$ (kJ mol^{-1})	$\Delta S^{\ddagger}$ (J K^{-1} mol^{-1})	Ref.
[Os(bpy)$_3$]$^{3+}$ + I$^-$	1.0	33	73	−31	1.83×10^4	38	−42	125
[Os(phen)$_3$]$^{3+}$ + I$^-$	1.0	52	57	−17	3.43×10^4	31	−54	125
[IrBr$_6$]$^{2-}$ + I$^-$	0.1	28	64	−69	9.3×10^{-1}	36	−123	126
[IrCl$_6$]$^{2-}$ + I$^-$	0.1	4.09×10^2						127
[Mo(CN)$_8$]$^{3-}$ + I$^-$	0.1	7.1×10^{-1}						128
[Ni([9]aneN$_3$)$_2$]$^{3+}$ + I$^-$	1.0	1.91×10^2	67	23				129
[Cu^{III}H$_{-2}$A$_3$] + I$^-$	0.1	4.5			1.28×10^3			130
[Cu^{III}H$_{-2}$Aib$_3$] + I$^-$	0.1	0.02			3.86			130
[Cu^{III}HH$_{-3}$G$_4$]$^-$ + I$^-$	0.1	0.10			1.4			130
[Ni(bpy)$_3$]$^{3+}$ + Br$^-$	5.0	1.01×10^2	60	−3				131
[Os(bpy)$_3$]$^{3+}$ + SCN$^-$	1.0				9.8	44	−75	125
[Os(phen)$_3$]$^{3+}$ + SCN$^-$	1.0				21.6			125
[IrCl$_6$]$^{2-}$ + SCN$^-$	0.1	57.1	33	−100	8.50×10^3	18	−110	126
[Os(phen)$_3$]$^{3+}$ + S$_2$O$_3^{2-}$	0.1	1.90×10^2						132
[IrCl$_6$]$^{2-}$ + S$_2$O$_3^{2-}$	0.1	1.38×10^2						133
[Ni([9]aneN$_3$)$_2$]$^{3+}$ + S$_2$O$_3^{2-}$	0.1	1.40×10^3						133

of the Fuoss expression (eq. (2.13)), is 0.035 M^{-1}, so that the reagent $I_2^-/\{I^-,I^-\}$ can be ascribed unique properties such as a reduction potential which is evaluated as 0.94 V.

$$2\,I^- \rightleftharpoons \{I^-,I^-\} \qquad (5.159)$$

The Marcus free energy expression can be applied to these outer-sphere electron transfer reactions to allow evaluation of self-exchange rates for the process in eqs (5.160) and (5.161). The value for I·/I$^-$ is 2×10^8 M^{-1} s^{-1}, in line with reactions of other radical species whereas that for $I_2^-/\{I^-,I^-\}$ is 3×10^4 M^{-1} s^{-1}. Electron transfer in this latter self-exchange process is coupled to cleavage of the I—I bond and at first sight the energetics involved in stretching the I—I bond from 3.10 Å in I_2^- to the 4.08 Å estimated for $\{I^-,I^-\}$ are prohibitive and are inconsistent with such a high rate. However, solvation of the incipient I$^-$ plays an important role, and provides partial compensation for the energy required. In Table 5.15, reduction potentials and self-exchange rates for a number of radical species of this type are presented.

$$I + {}^*I^- \rightleftharpoons I^- + {}^*I \qquad (5.160)$$

$$I_2^- + \{{}^*I^-,{}^*I^-\} \rightleftharpoons \{I^-,I^-\} + {}^*I_2^- \qquad (5.161)$$

Table 5.15. Reduction potentials and self-exchange rates for inorganic radicals [136, 137]

Reaction	$E°$ (V)	k_{AA} (M^{-1} s^{-1})
I· + e$^-$ $\rightleftharpoons$ I$^-$	1.33	2×10^8
I$_2$ + e$^-$ $\rightleftharpoons$ I$_2^-$	0.21	9×10^4
Br· + e$^-$ $\rightleftharpoons$ Br$^-$	1.92	
Br$_2$ + e$^-$ $\rightleftharpoons$ Br$_2^-$	0.58	85
Cl· + e$^-$ $\rightleftharpoons$ Cl$^-$	2.41	
Cl$_2$ + e$^-$ $\rightleftharpoons$ Cl$_2^-$	0.70	9×10^4
SCN· + e$^-$ $\rightleftharpoons$ SCN$^-$	1.63	
N$_3$· + e$^-$ $\rightleftharpoons$ N$_3^-$	1.33	$\approx 4 \times 10^4$
SO$_2$ + e$^-$ $\rightleftharpoons$ SO$_2^-$	-0.26	$\approx 1 \times 10^4$
NO$_2$ + e$^-$ $\rightleftharpoons$ NO$_2^-$	1.04	1×10^{-2}
ClO$_2$ + e$^-$ $\rightleftharpoons$ ClO$_2^-$	0.94	1×10^2
CO$_2$ + e$^-$ $\rightleftharpoons$ CO$_2^-$	-1.8	$\approx 1 \times 10^{-5}$
SO$_3^-$ + e$^-$ $\rightleftharpoons$ SO$_3^{2-}$	0.63	3×10^4
S$_2$O$_3^-$ + e$^-$ $\rightleftharpoons$ S$_2$O$_3^{2-}$	1.30	2×10^5

Inner-sphere reactions show a number of interesting features. Oxidation of I$^-$ by [Fe(H$_2$O)$_6$]$^{3+}$ follows kinetics which suggest the importance of an inner-sphere complex. As with the outer-sphere reactions, the dominant pathway involves the formation of I_2^- [138]. A similar mechanism is proposed for the reactions of nickel(III) macrocyclic complexes [139, 140]. In the corresponding reactions of the nickel(III) peptide complex [141] the dominant pathway has a second-order dependence on [I$^-$] and a second-order dependence on the oxidant (eq. (5.162)), indicating that an adduct, $\{[Ni^{III}H_{-2}Aib_3],I^-\}$, is formed. Subsequent reaction involves two pathways: $k_c = K_0^2 k_1$, where reaction with a second $\{[Ni^{III}H_{-2}Aib_3],I^-\}$ gives products without recourse to any radical species and $k_b = K_0 k_2 k_3 / k_{-2}$ where I_2^- predominates (eqs (5.163)–(5.166)).

$$\frac{-d[[Ni^{III}H_{-2}Aib_3]]}{dt} = \{ k_c + \frac{k_b}{[[Ni^{II}H_{-2}Aib_3]^-]} \} [I^-]^2 [[Ni^{III}H_{-2}Aib_3]]^2 \quad (5.162)$$

$$[Ni^{III}H_{-2}Aib_3] + I^- \overset{K_0}{\rightleftharpoons} \{[Ni^{III}H_{-2}Aib_3],I^-\} \quad (5.163)$$

$$2\,\{[Ni^{III}H_{-2}Aib_3],I^-\} \overset{k_1}{\rightarrow} 2\,[Ni^{II}H_{-2}Aib_3]^- + I_2 \quad (5.164)$$

$$\{[Ni^{III}H_{-2}Aib_3],I^-\} + I^- \underset{k_{-1}}{\overset{k_2}{\rightleftharpoons}} [Ni^{II}H_{-2}Aib_3]^- + I_2^- \quad (5.165)$$

$$[Ni^{III}H_{-2}Aib_3] + I_2^- \overset{k_3}{\rightarrow} [Ni^{II}H_{-2}Aib_3]^- + I^2 \quad (5.166)$$

Yet another aspect to the non-complementary reaction process is observed in this reaction. In addition to a role in modifying the energetics of the electron transfer reaction, the intermediate, $\{[Ni^{III}H_{-2}Aib_3],I^-\}$, assumes a role in accommodating the stoichiometry of the reaction, facilitating a lower energy pathway which avoids the formation of thermodynamically unstable radicals. Thus while outer-sphere reactions involving non-metallic substrates are readily inserted into the pattern of Marcus Theory, inner-sphere reactions provide energetically and stoichiometrically more favorable pathways to products.

5.10 REACTIONS OF INORGANIC RADICALS

The marked success found for the application of Marcus Theory to outer-sphere reactions between metal ion complexes and non-metallic substrates has prompted investigations in an attempt to characterize the intrinsic activation barriers involved in radical formation. There are a number of relatively long-lived radical reactants which involve a single-electron transfer and a small or moderate change in the geometry. These include $[ClO_2]^{0/-}$, $[SO_2]^{0/-}$ and $[NO_2]^{0/-}$. They have been extensively investigated and show some adherence to the Marcus relationship.

Dithionite, $[S_2O_4]^{2-}$, is an important and widely used reductant where reaction is accompanied by cleavage of the S—S bond. The potential is dependent on pH and is greatly enhanced in basic media.

$$2\,HSO_3^- + 2\,e^- + 2\,H^+ \rightleftharpoons S_2O_4^{2-} + 2\,H_2O \quad E° = 0.099\text{ V} \quad (5.167)$$

$$2\,SO_3^{2-} + 2\,e^- + 2\,H_2O \rightleftharpoons S_2O_4^{2-} + 4\,OH^- \quad E° = -1.13\text{ V} \quad (5.168)$$

Under conditions of excess reductant, the kinetics of reduction of metal ion complexes show the general two-term rate law shown in eq. (5.169) for the reduction of $[Fe(CN)_6]^{3-}$, and interpreted in terms of the mechanism (eqs (5.170)–(5.172)) [142].

$$-\frac{d[[Fe(CN)_6]^{3-}]}{dt} = \{k_2 K_1^{1/2}\,[[S_2O_4]^{2-}]^{1/2} + k_3\,[[S_2O_4]^{2-}]\}[[Fe(CN)_6]^{3-}]$$

$$(5.169)$$

$$[S_2O_4]^{2-} \underset{k_{-1}}{\overset{k_1}{\rightleftharpoons}} 2\,[SO_2]^- \quad (5.170)$$

$$[Fe(CN)_6]^{3-} + [SO_2]^- \xrightarrow{k_2} [Fe(CN)_6]^{3-} + [SO_2] \quad (5.171)$$

$$[Fe(CN)_6]^{3-} + [S_2O_4]^{2-} \xrightarrow{k_3} [Fe(CN)_6]^{3-} + [S_2O_4]^- \quad (5.172)$$

Under conditions of excess $[[Fe(CN)_6]^{3-}]$, k_1 is rate limiting with a value 1.7 s^{-1} at 25.0°C and 0.10 M ionic strength. There are parallel pathways involving reactions of $[S_2O_4]^{2-}$ and the radical $[SO_2]^-$. The fate of the radical products is dependent on the reaction conditions and can lead to variable stoichiometries. The value of K_1 is $1.4 \times 10^{-9}\text{ M}^{-1}$ so that the reduction potential for $[SO_2]^{0/-}$ is -0.7 V and although the reagent, $[SO_2]^-$, is present at low concentrations it contributes significantly to the rate. Rate and activation parameters for a number of reactions are presented in Table 5.16.

Table 5.16. Rate constants for the reduction of metal complexes by dithionite at 25°C

Oxidant	μ (M)	k_2 (M^{-1} s^{-1})	k_3 (M^{-1} s^{-1})	Ref.
[Co(edta)]$^-$	0.4	1.0×10^3	≤ 0.1	143
[Co(ox)$_3$]$^{3-}$	0.4	1.8×10^4	≤ 0.4	143
[Co(terpy)$_2$]$^{3+}$	0.4	$\leq 10^7$	4.3×10^5	143
[Fe(edta)]$^-$	0.4	$\leq 2 \times 10^6$	3.6×10^4	143
[Co(NH$_3$)$_6$]$^{3+}$	0.95	5.3×10^6	$\leq 4 \times 10^{-3}$	144
[Co(bpy)$_3$]$^{3+}$	1.0	2.1×10^7		145
[Co(sep)]$^{3+}$	1.0	71		145

Application of the Marcus linear free energy relationship leads to a value of approximately 10^4 M^{-1} s^{-1} for the self-exchange rate constant for [SO$_2$]$^{-/0}$ which has been confirmed for reactions in the reverse direction where SO$_2$ is reduced by [Cr(bpy)$_3$]$^{2+}$ [146].

The volatile radical ClO$_2$ is readily prepared by oxidation of [ClO$_2$]$^-$ and has a formal reduction potential of 0.94 V. The kinetics and mechanism of oxidation of [Fe(phen)$_3$]$^{2+}$ are reversible [147] and, together with the results of other outer-sphere reactions, lead to a consistent value of 160 M^{-1} s^{-1} for the [ClO$_2$]$^{0/-}$ self-exchange rate [148]. Similar observations lead to a value for of 8×10^{-3} for [NO$_2$]$^{0/-}$, which has a reduction potential of 1.04 V [127, 148]. Detailed analysis of these self-exchange rates in terms of the structural rearrangement involved in electron transfer for the bent triatomic indicates that the large angular deformation in [NO$_2$]$^{0/-}$ presents a barrier and leads to significant nuclear tunneling [148].

Table 5.17. Correlation of structural rearrangement with self-exchange rates for non-metallic radicals

	k_{AA} (M^{-1} s^{-1})	δ angle (°)	δr (Å)
[SO$_2$]$^{0/-}$	10^4	9.5	0.08
[NO$_2$]$^{0/-}$	8×10^{-3}	19	0.04
[ClO$_2$]$^{0/-}$	160	6.5	0.10

Of related interest are reactions of species such as SO$_3^{2-}$ and NO$_3^-$. The aqueous chemistry of SO$_2$ which has a solubility of 1.6 M at 20°C, involves the eqs (5.173)–(5.175)).

$$SO_2 + H_2O \rightleftharpoons HSO_3^- + H^+ \tag{5.173}$$

$$HSO_3^- \rightleftharpoons SO_3^{2-} + H^+ \tag{5.174}$$

$$2HSO_3^- \rightleftharpoons S_2O_5^{2-} + H_2O \tag{5.175}$$

Outer-sphere one-electron oxidation reactions follow the mechanism (5.176)–(5.179) shown for $[Fe(phen)_3]^{3+}$ [127, 149–152]. The reduction potential for the $[SO_3]^{2-/-}$ couple has been determined to be 0.72 V from kinetic studies on the reaction with $[Ru(NH_3)_4(phen)]^{3+/2+}$ which can be examined in both forward and reverse directions. A consistent value for the self-exchange rate of 4 M^{-1} s^{-1} is obtained and this relatively low value can be explained by distortion of the pyramidal geometry [153].

$$[Fe(phen)_3]^{3+} + SO_3^{2-} \rightarrow [Fe(phen)_3]^{2+} + SO_3^- \tag{5.176}$$

$$[Fe(phen)_3]^{3+} + SO_3^- \rightarrow [Fe(phen)_3]^{2+} + SO_3 \tag{5.177}$$

$$SO_3 + H_2O \rightarrow SO_4^{2-} + 2H^+ \tag{5.178}$$

$$2SO_3^- \rightarrow S_2O_6^{2-} \tag{5.179}$$

The reaction stoichiometry is variable from 1:2 when SO_4^{2-} to 1:1 when $S_2O_6^{2-}$ is the product. In outer-sphere reactions the dominant pathways generally favor the formation of SO_4^{2-}. Inner-sphere reactions of SO_3^{2-} are also important and have been studied extensively. In many instances, well-defined complexes are observed and these exert a profound effect on the course of the reaction by providing low-energy pathways which avoid the formation of radicals as intermediates and are unavailable to outer-sphere reactants. Information on such pathways can be obtained from the reaction stoichiometry, the balance between $S_2O_6^{2-}$ and SO_4^{2-} as products. Early work [154] suggested that the reaction stoichiometry was a reflection of whether the oxidant acted as a one-electron acceptor with $S_2O_6^{2-}$ favored, or a two-electron acceptor with SO_4^{2-} favored. However, for metal ion oxidants, there is also a dependence on the substitution lability of the reagent. In outer-sphere reactions, SO_4^{2-} is the dominant product. The formation of $S_2O_6^{2-}$ is found with labile oxidants such as $[Fe(H_2O)_6]^{3+}$ where the reaction (eq. (5.180)) produces the product without formation of SO_3^- radicals [155]. The reaction with $[Fe(H_2O)_6]^{3+}$ is very complex and although it does involve complexes of the type $[Fe(SO_3)_{aq}]^+$ [156–159] SO_3^- radicals are also produced. It may be that such pathways are favored by weaker oxidants [160].

$$2[Fe(SO_3)_{aq}]^+ \rightarrow 2[Fe(H_2O)_6]^{2+} + S_2O_6^{2-} \tag{5.180}$$

The sulfite ion is ambidentate and can coordinate by either O- or S-bonding. Both coordination modes have been observed with substitution inert cobalt(III) complexes where decomposition of the S-bonded complexes is significantly slower than for the O-bonded species [160–162]. The rate law for reaction of $[Co(NH_3)_5(OH_2)]^{3+}$ indicates that the redox reaction takes place by a single pathway to give SO_4^{2-} as product (eq. (5.181)–(5.183)) [164].

$$[Co(NH_3)_5(OH_2)]^{3+} + SO_2 \rightleftharpoons [Co(NH_3)_5(OSO_2)]^+ + 2\,H^+ \qquad (5.181)$$

$$[Co(NH_3)_5(OSO_2)]^+ \rightarrow [Co(H_2O)_6]^{2+} + 5\,NH_3 + SO_3^- \qquad (5.182)$$

$$[Co(NH_3)_5(OSO_2)]^+ + SO_3^- \xrightarrow{\text{fast}} [Co(H_2O)_6]^{2+} + 5\,NH_3 + SO_4^{2-} + SO_2$$

$$(5.183)$$

5.11 REACTIONS OF HALOGENATE IONS

Formal atom transfer reactions have been noted in sections 5.5 and in a number of inner-sphere reactions involving non-metallic substrates. These include reactions of H_2O_2, Cl_2 and $HOCl$ and they are characterized by the necessity of a change in geometry at the non-metallic substrate center. With the exception of $ClO_2^{0/-}$, which is dicussed in section 5.10, the electron transfer chemistry of the oxy-halogen acids and anions is also complicated by changes in coordination number about the central halogen atom, and consequently atom transfer processes are important. In addition, disproportionation and hydrolysis of the reactants and reaction intermediates provide their own complicating features. Such processes contribute to the participation of these reagents in oscillating reactions which have received considerable study, but detailed analysis of oscillatory behavior is beyond the scope of this section and the emphasis is placed on stoichiometric processes. Reduction potentials and acidity constants for a number of the reagents are presented in Table 5.18. Also included in the table are estimates of oxygen exchange rates in acid solution which give some idea of the labilities of the ions [165].

Table 5.18. Reduction potentials and oxygen exchange rates for selected halogenate ions at 25 °C

Half-reaction	$E°$ (V)	k_{ex} (oxidant, s^{-1})
$[ClO_4]^- + 2\,H^+ + 2e^- \rightleftharpoons [ClO_3]^- + H_2O$	1.20	$<10^{-10}$
$[ClO_3]^- + 3\,H^+ + 2e^- \rightleftharpoons [HClO_2] + H_2O$	1.18	$\approx 10^{-3}$
$[HClO_2] + 2\,H^+ + 2e^- \rightleftharpoons [HClO] + H_2O$	1.67	
$[BrO_4]^- + 2\,H^+ + 2e^- \rightleftharpoons [BrO_3]^- + H_2O$	1.85	$<10^{-7}$
$[ClO_3]^- + 12\,H^+ + 10e^- \rightleftharpoons Cl_2 + H_2O$	1.42	
$[BrO_3]^- + 12\,H^+ + 10e^- \rightleftharpoons Br_2 + 6\,H_2O$	1.48	$\approx 10^{-6}$
$[IO_3]^- + 12\,H^+ + 10e^- \rightleftharpoons I_2 + 6\,H_2O$	1.20	

There is a general trend of increasing two-electron reduction potential as the oxidation state of the halogen decreases so that thermodynamically, multi-electron transfer processes predominate. Problems of autocatalysis and interference by lower oxidation state intermediates are common but can be prevented by use of appropriate scavengers. For example, participation of HOCl in reactions of $[ClO_2]^-$ and $[ClO_3]^-$ may be prevented by carrying out the studies in the presence of Cl^-, which converts HOCl to Cl_2, and phenol which reacts rapidly with the Cl_2. Reactions of a number of these oxidants show strong acid catalysis. In addition to causing an increase in the reduction potential of the anion, protonation of the coordinated oxygen provides a suitable leaving group and increases the lability of the central halogen atom. Thus inner-sphere mechanisms involving substitution at the halogen are facilitated.

The perchlorate anion, $[ClO_4]^-$, is a very weak base, frequently used as an inert weakly-coordinating counter-ion in kinetic studies. Although it is a powerful oxidant it derives its inert character from the very slow substitution at the halogen. The rate of O-exchange with ^{18}O-labelled water is $<10^{-10}$ M^{-1} s^{-1} [166]. There are a number of metal ions, however, which do undergo electron transfer reactions with $[ClO_4]^-$, notably titanium(III), ruthenium(II) and molybdenum(III). While reduction by $[Ru(H_2O)_6]^{2+}$ yields $[ClO_3]^-$, all the other reagents involve complete reduction to Cl^-. Where they have been investigated, the rate laws are first-order in both reagents (eq. (5.184)), and the second-order rate constants are presented in Table 5.19. The mechanisms are thought to be inner-sphere even with $[Ru(NH_3)_6]^{2+}$ where a seven-coordinate intermediate is proposed. While reductants such as $[Cr(H_2O)_6]^{2+}$ and $[U_{aq}]^{3+}$ have a greater thermodynamic driving force, they are unreactive. It is notable that for all the reagents which do react, the electrons are removed from a polarizable σ^* orbital.

$$\frac{-d[[ClO_4]^-]}{dt} = k\,[[Ru(NH_3)_6]^{2+}]\,[[ClO_4]^-] \tag{5.184}$$

Table 5.19. Rate constants for the reduction of $[ClO_4]^-$ at 25.0°C.

Reductant	μ (M)	k (M^{-1} s^{-1})	Ref.
$[Ru(H_2O)_6]^{2+}$	0.3	3.2×10^{-3}	167
$[Ru(NH_3)_6]^{2+}$	0.62	3.8×10^{-5}	168
$[Ti(H_2O)_6]^{3+}$	1.0	2.4×10^{-5}	169
$[V(H_2O)_6]^{2+}$	2.5^a	2.8×10^{-6}	170
$[V(H_2O)_6]^{3+}$	2.5^a	5×10^{-7}	170

^{a}50°C.

The $[BrO_4]^-$ ion shows chemistry similar to that of $[ClO_4]^-$ and has been little studied, but $[IO_4]^-$ is a much more useful oxidant on account that it exhibits an expanded coordination shell (eq. (5.185)), and significantly greater substitution lability [171].

$$[IO_4]^- + 2\,H_2O \rightleftharpoons [H_4IO_6]^- \quad 0.025 \tag{5.185}$$

$$[H_5IO_6] \rightleftharpoons [H_4IO_6]^- + H^+ \quad 5 \times 10^{-4}\ M \tag{5.186}$$

$$[H_4IO_6]^- \rightleftharpoons [H_3IO_6]^{2-} + H^+ \quad 2 \times 10^{-7} \text{ M} \tag{5.187}$$

The reactions with metal complexes are predominantly inner-sphere, involving substitution at the metal center or the iodine atom. For example, with $[Co(edta)]^{2-}$, the reaction rate is substantially independent of pH and the rate law provides evidence for a transient $[Co(edta)OIO_3H_4]^{3-}$ with a stability constant of 2.4 M^{-1} [172]. Decomposition of the transient is first-order with $k = 0.1$ s^{-1} and leads to $[Co(edta)OH_2]^-$ as initial product, suggesting that O-atom transfer is important. The oxidation of $[Cr(H_2O)_6]^{3+}$ to $[HCrO_4]^-$ is thought to involve a trinuclear intermediate (eqs (5.188)–(5.190)), formed by substitution at iodine. The rate-limiting step is a two-electron transfer to give chromium(IV), which is further oxidized in a rapid subsequent step [173].

$$2\,[Cr(H_2O)_6]^{3+} + [H_4IO_6]^- \rightleftharpoons [(H_2O)_5Cr\text{-}O\text{-}IO_4H_5\text{-}O\text{-}Cr(H_2O)_5]^{4+} + H^+ \tag{5.188}$$

$$[(H_2O)_5Cr\text{-}O\text{-}IO_4H_5\text{-}O\text{-}Cr(H_2O)_5]^{4+} \rightarrow 2\,[Cr_{aq}]^{4+} + [IO_3]^- + H^+ \tag{5.189}$$

$$[Cr_{aq}]^{4+} + [H_4IO_6]^- \xrightarrow{\text{fast}} [HCrO_4]^- + [IO_3]^- \tag{5.190}$$

Even with reductants such as $[Fe(CN)_6]^{4-}$ which normally react by outer-sphere electron transfer, the [H$^+$] dependence of the rate indicates that the more labile $[H_4IO_6]^-$ is more reactive and an inner-sphere mechanism with coordination of $[Fe(CN)_6]^{4-}$ through a cyanide ligand to iodine is likely [174].

The halates are more reactive, better donors and with greater lability of the halogen—O bond. Reduction of $[ClO_3]^-$ by $[Cr(H_2O)_6]^{2+}$ is acid catalyzed and proceeds by an inner-sphere mechanism [175, 176]. The products are $[Cr(H_2O)_6]^{3+}$, $[Cr(H_2O)_5Cl]^{2+}$ and 30% $[(H_2O)_5CrOCr(H_2O)_5]^{4+}$ dimer, consistent with formation of chromium(IV) as an intermediate. Tracer studies indicate that a significant proportion of ^{18}O from $[Cl^{18}O_3]^-$ is incorporated into the chromium(III) product indicating atom transfer in an inner-sphere process. Reactions with $[Fe(H_2O)_6]^{2+}$ and $[V(H_2O)_6]^{2+}$ are inner-sphere but involve single-electron transfer [177, 178]. Outer-sphere reductants such as $[Fe(phen)_3]^{2+}$ are unreactive [179]. More extensive studies have been reported with $[BrO_3]^-$. The general rate law (eq. (5.191)) indicates two pathways as illustrated for reaction with $[Fe(CN)_6]^{4-}$ (eq. (5.192)): an acid-independent term and a term in $[H^+]^2$, corresponding to reactions of $[BrO_3]^-$ and $[H_2BrO_3]^+$ respectively (eq. (5.193)) [180].

$$\frac{-d[[Fe(CN)_6]^{4-}]}{dt} = 6\,(k_1 + k_2\,[H^+]^2)\,[[BrO_3]^-]\,[[Fe(CN)_6]^{4-}] \tag{5.191}$$

$$5\ [Fe(CN)_6]^{4-} + [BrO_3]^- + 6\ H^+ \rightarrow$$
$$5\ [Fe(CN)_6]^{4-} + 0.5\ Br_2 + 3\ H_2O \quad (5.192)$$

$$[H_2BrO_3]^+ \rightleftharpoons [BrO_3]^- + 2\ H^+ \quad K_a \ll 1 \quad (5.193)$$

It is likely that the latter reaction is inner-sphere and involves substitution at bromine. The small range of rate constants supports this proposal (Table 5.20). The parallel acid-independent term may be inner-sphere or outer-sphere depending on the lability of the reductant. Single-electron transfer processes prevail.

Table 5.20. Rate constants for reduction of $[BrO_3]^-$ by metal ion complexes at 25.0 °C

Reductants	μ (M)	k_1 ($M^{-1}\ s^{-1}$)	k_2 ($M^{-3}\ s^{-1}$)	Ref.
$[Fe(CN)_6]^{4-}$	0.5	0.00125	0.193	181
$[Fe(bpy)(CN)_4]^{2-}$	0.5	0.0062	0.227	181
$[Fe(bpy)_2(CN)_2]$	0.5	0.03	0.755	180
$[Fe(bpy)_3]^{2+}$	0.5	≤0.0042	≤0.15	179
$[Fe(H_2O)_6]^{2+}$	0.5	0.372	—	177
$[VO_{aq}]^{2+}$	2.0	4.9	—	182
$[IrCl_6]^{2-}$	0.5	—	10	183

With $[IO_3]^-$, the lability of the halogen center again dominates the reactivity patterns. The oxidant is quite basic (eq. (5.194)), and again acid catalysis is observed.

$$HIO_3 \quad [IO_3]^- + H^+ \rightleftharpoons K_a = 0.5\ M \quad (5.194)$$

The reaction with $[V(H_2O)_6]^{2+}$ with a rate constant of 360 $M^{-1}\ s^{-1}$ is too fast to involve substitution at the metal center but may proceed by substitution at iodine [184]. The predominant product, $[V(H_2O)_6]^{3+}$, indicates one-electron steps. A transient intermediate is detected in the reaction with $[Fe(H_2O)_6]^{2+}$, where again single-electron transfer dominates [185]. Strong acid catalysis in the reactions of the inert reductants $[Fe(CN)_6]^{4-}$ and $[IrCl_6]^{3-}$ may also indicate inner-sphere reaction with substitution at iodine [186, 187].

$[ClO_2]^-$ and the other halite ions are prone to disproportionation and this places limits on the study of their redox processes. The principal pathways are inner-sphere. For example, the reaction of $[Fe(phen)_3]^{2+}$ is controlled by the rate of substitution at the metal center [179]. Oxygen atom transfer has been demonstrated by tracer experiments in the reaction with $[U_{aq}]^{4+}$ [188].

To summarize, the reactions of the oxyhalogen anions are characterized by facile inner-sphere reactions involving substitution at the metal center or substitution at the halide [189]. Strong dependencies of the rate on $[H^+]$ are found, the result of both thermodynamic and kinetic factors. Although most of the evidence for the formation of adducts is indirect, adduct formation plays a key role in determining the mechanism, satisfying in this case the stoichiometric [H$^+$] demand in the reactions.

5.12 REACTIONS BETWEEN NON-METALLIC REAGENTS

Reactions between non-metallic reagents hold a special position in mechanistic redox chemistry and present some special problems. They are generally two-electron processes and involve significant changes in the coordination environment around the main group elements so that the reactions are complementary and atom transfer is frequently encountered. Distinct complications arise where multiple oxidation states are possible, and where these prevail as reaction intermediates the differences in reactivity lead to the occurrence of autocatalysis and autoinhibition. Radicals are also possible and the result is that complex kinetic behavior is frequently encountered and many reactions involve oscillatory behavior. As mentioned in section 5.11, discussion of the complex kinetics associated with these reactions is beyond the scope of this book and here emphasis is placed on the study of reactions which are better-behaved from a kinetic point of view.

As outlined in section 5.5, in reactions where formal atom transfer occurs, there is ambiguity in describing the nature of the process. A single-electron transfer which occurs with transfer of a halide or a pseudo-halide from the oxidant to the reductant is formally equivalent to halogen atom transfer; two-electron transfer with transfer of a halide is equivalent to halogen cation transfer; and two-electron transfer with transfer of oxide is equivalent to oxygen atom transfer. Within the description of atom transfer there are also diferences in the way reactions are viewed. For example, the oxidation of I$^-$ by BrCl (eq. (5.195)), can be viewed as a redox process in which I(–1) is oxidized to I(+1), or as a nucleophilic attack of I$^-$ on BrCl with Cl$^-$ as leaving group. The difference is clearly semantic, bound with the definition of oxidation state. Oxidants are generally electrophiles with Lewis acid behavior whereas reductants are generally nucleophiles with Lewis base behavior.

$$I^- + BrCl \longrightarrow IBr + Cl^- \qquad (5.195)$$

As a result of the propensity for these reactions between non-metallic reagents to participate in atom transfer, a number of other mechanistic criteria become important in addition to correlations with reduction potentials and self-exchange rates. These include labelling studies, isotope effects, and correlation of the rates with bond-strengths. The next section is devoted to illustrating some of the more important mechanistic patterns in reactions of reagents which are well-characterized. The area has been reviewed in several publications [189–191].

Some of the most extensively studied non-metallic oxidants are the peroxides [192]. In some instances radicals are encountered signifying one-electron transfer mechanisms which should be amenable to Marcus-type correlations, but in others the mechansims are much simpler and atom transfer predominates. Although reaction stoichiometries can be complex, a number of oxidations by H_2O_2 show a simple two-term rate law shown for the reaction with I^- in eq. (5.196), which is interpreted in terms of nucleophilic substitution at peroxide oxygen with OH^- and OH_2 as leaving groups (eqs (5.197)–(5.199)), where I^- is the reductant. Rate constants for reactions with a number of reductants are presented in Table 5.21.

$$\text{Rate} = \{k_1 + k_2[H^+]\}[H_2O_2][I^-] \tag{5.196}$$

$$HOOH_2^+ \underset{}{\overset{10^{4.7}}{\rightleftharpoons}} HOOH + H^+ \tag{5.197}$$

$$I^- + HOOH \rightarrow IOH + OH^- \tag{5.198}$$

$$I^- + HOOH_2^+ \rightarrow IOH + OH_2 \tag{5.199}$$

Table 5.21. Rate constants for the reactions of H_2O_2 at 25°C [192]

Reductant	k_1 (M^{-1} s^{-1})	k_2 (M^{-1} s^{-1})
SO_3^{2-}	0.2	—
$S_2O_3^{2-}$	2.5×10^{-2}	1.7
I^-	0.6	10.5
SCN^-	5.2×10^{-4}	2.5×10^{-2}
NO_2^-	3×10^{-7}	—
Br^-	2.3×10^{-5}	1.4×10^{-2}
Cl^-	1.1×10^{-7}	5.0×10^{-5}

The main evidence for this comes from rate comparisons where the rates of oxidation of a number of substrates show a good correlation with the reactivity of the reagents as nucleophiles at saturated carbon atoms (Fig. 5.6). In suitable cases such as the oxidation of NO_2^-, labelling studies indicate transfer of a single oxygen atom (eq. (5.200)), consistent with this formal substitution mechanism.

$$H_2{}^*O_2 + NO_2^- \rightarrow {}^*ONO_2^- + {}^*OH^- + H^+ \tag{5.200}$$

The reaction with SO_3^{2-} is subject to general acid catalysis (eq. (5.201)), indicating that proton transfer participates in the rate-determining step and that the acid dependence is not solely due to a stoichiometric pre-equilibrium protonation. As a result, formation of an intermediate adduct, $HOOSO_2^-$, must be proposed with an acid-catalyzed decomposition [193].

$$\frac{d[HSO_4^-]_T}{dt} = \left\{ \frac{k_1 K_a (k_2 [H^+] + k_3 [HA])}{k_{-1} + k_2 [H^+] + k_3 [HA]} \right\} [HOOH][HSO_3^-]_T \tag{5.201}$$

Sec. 5.12] Reactions between non-metallic reagents 269

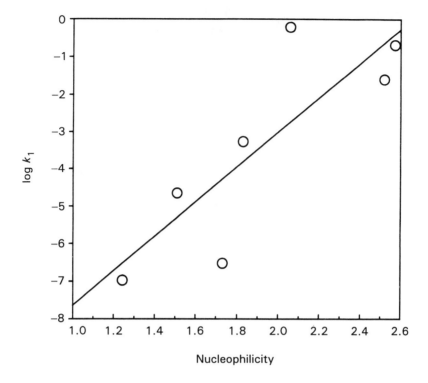

Fig. 5.6. Plot of log k_1 against reductant nucleophilicity for the data in Table 5.21.

$$H_2SO_3 \;\overset{K_a}{\rightleftharpoons}\; HSO_3^- + H^+ \tag{5.202}$$

$$HSO_3^- + HOOH \;\underset{k_{-1}}{\overset{k_1}{\rightleftharpoons}}\; HOOSO_2^- \tag{5.203}$$

$$HOOSO_2^- + H^+ \;\overset{k_2}{\longrightarrow}\; H^+ + HSO_4^- \tag{5.204}$$

$$HOOSO_2^- + HA \;\overset{k_3}{\longrightarrow}\; HA + HSO_4^- \tag{5.205}$$

Corresponding oxidations by the hypohalides such as the oxidation of I^- by OCl^- (eq. (5.206)) would appear to be very similar processes and indeed, until recently, it was thought that this reaction proceeded by a similar mechanism involving displacement of Cl^- by I^- at oxygen, a formal O atom transfer [194]. However, this reaction is also subject to general acid catalysis (eq. (5.207)), signifying that H^+ is incorporated in the transition state. It is argued that this acid catalysis points to the facilitation of OH^- as leaving group in the reaction and that ICl must therefore be an intermediate (eqs (5.208)–(5.210)) where $k_1 k_2/k_{-1} = 4.4 \times 10^{15}$ M^{-2} s^{-1} and $k_2 = 1.4 \times 10^8$ M^{-1} s^{-1} at 25.0°C and 0.5 M ionic strength [195]. This reaction and a number of others (Table 5.22) therefore proceed by Cl^+ transfer rather than by O atom transfer. There is a trend in the rates which depends on the nucleophilicity of the halide towards halogen cation.

$$I^- + OCl^- \rightarrow OI^- + Cl^- \tag{5.206}$$

$$-\frac{d[OCl^-]}{dt} = \{k_H[H^+] + k_{HA}[HA]\}[OCl^-][I^-] \tag{5.207}$$

$$OCl^- + H_2O \underset{k_{-1}}{\overset{k_1}{\rightleftharpoons}} HOCl + OH^- \tag{5.208}$$

$$HOCl + I^- \overset{k_2}{\rightarrow} OH^- + ICl \tag{5.209}$$

$$ICl + 2\, OH^- \overset{k_3}{\rightarrow} OI^- + Cl^- + H_2O \tag{5.210}$$

Table 5.22. Reactions of OCl^- and $HOCl$ involving Cl^+ transfer and associated processes

Reaction	k (M^{-2} s^{-1})	Ref.
$H^+ + OCl^- + Br^-$	3.7×10^{10}	196
$H^+ + OCl^- + I^-$	4.4×10^{15}	195
$H^+ + HOCl + Br^-$	1.3×10^6	196
$H^+ + HOCl + Cl^-$	2.8×10^4	197
$H^+ + HOBr + Br^-$	1.6×10^{10}	198
$H^+ + HOI + I^-$	4.4×10^{12}	198

In the reduction of HOCl by NO_2^-, the rate law is given by eq. (5.211) and again Cl^+ transfer is proposed in the mechanism (eqs (5.212)–(5.217)) with NO_2Cl as an intermediate [199]. Labelling studies when the reaction is carried out in $^{18}OH_2$ indicate that the O atom added in NO_3^- comes from solvent, consistent with the proposal. A similar finding has been shown with oxidation of SO_3^{2-} [200].

$$-\frac{d[OCl^-]}{dt} = \frac{K_h k_1 [NO_2^-]}{k_{-1}[OH^-]^2}(k_4 + k_2[NO_2^-])[OCl^-] \qquad (5.211)$$

$$OCl^- + H_2O \underset{}{\overset{K_h}{\rightleftharpoons}} HOCl + OH^- \qquad (5.212)$$

$$HOCl + NO_2^- \underset{k_{-1}}{\overset{k_1}{\rightleftharpoons}} NO_2Cl + OH^- \qquad (5.213)$$

$$NO_2Cl + NO_2^- \underset{k_{-2}}{\overset{k_2}{\rightleftharpoons}} N_2O_4 + Cl^- \qquad (5.214)$$

$$N_2O_4 + OH^- \overset{k_3}{\rightarrow} NO_3^- + NO_2^- + H^+ \qquad (5.215)$$

$$NO_2Cl \underset{k_{-4}}{\overset{k_4}{\rightleftharpoons}} NO_2^+ + Cl^- \qquad (5.216)$$

$$NO_2^+ + OH^- \overset{k_5}{\rightarrow} NO_3^- + H^+ \qquad (5.217)$$

Halogen reduction reactions represent a further area where halogen cation transfer is anticipated. The hydrolysis reactions as shown for Cl_2 in eq. (5.218) are believed to involve adduct formation as a first step (eq. (5.219)) [201], and the rates, which are much greater for the interhalogens than for the halogens themselves (Table 5.23), reflect the polarity of the complexes rather than the strength of the halogen—halogen bond. On the other hand, formation of the trihalides does not show a trend with the polarity of the halogen but is dependent on size and reflects solvation of the reaction products.

$$Cl_2 + H_2O \rightleftharpoons HOCl + Cl^- + H^+ \qquad (5.218)$$

$$Cl_2 + H_2O \rightarrow H_2OCl_2 \tag{5.219}$$

Table 5.23. Rate constants for reduction of halogens at 25.0°C and 0.10 M ionic strength

Reaction	k (s^{-1})	k (M^{-1} s^{-1})	Ref.
$Cl_2 + H_2O \rightarrow HOCl + Cl^- + H^+$	11.0[a]		202
$Br_2 + H_2O \rightarrow HOBr + Br^- + H^+$	110[a]		202
$I_2 + H_2O \rightarrow HOI + I^- + H^+$	3.0[a]		202
$ICl + H_2O \rightarrow HOI + Cl^- + H^+$	2.4×10^6		203
$IBr + H_2O \rightarrow HOBr + Cl^- + H^+$	8×10^5		204
$I_2 + I^- \rightarrow I_3^-$		6.2×10^9	205
$IBr + I^- \rightarrow I_2Br$		2.0×10^9	204
$ICl + I^- \rightarrow I_2Cl$		5.1×10^8	206

[a] 20°C.

As noted in Section 5.4, oxidation and reduction reactions of oxyanion species are characterized by strong H$^+$ dependencies which reflect the thermodynamic changes in the reactions. As an example, some reactions of BrO_3^- are considered. In the oxidation of halide ions (eq. (5.220)), the rate law shows a dependence on [H$^+$]^2 (eq. (5.221)), and rate constants are presented in Table 5.24. The exact mechanism is not yet clear, but formation of [H$_2$BrO$_3$]$^+$ is unlikely and protonation of an intermediate adduct is considered more likely (eq. (5.222)).

$$BrO_3^- + 5\,Br^- + 6\,H^+ \rightarrow 3\,Br_2 + 3\,H_2O \tag{5.220}$$

$$-\frac{d[BrO_3^-]}{dt} = [BrO_3^-][Br^-][H^+]^2 \tag{5.221}$$

$$BrO_3^- + Br^- + 2\,H^+ \rightarrow Br_2O_2 \tag{5.222}$$

Table 5.24. Rate constants for reduction of BrO_3^- at 25.0°C and 1.0 M ionic strength

Reaction	k (M^{-3} s^{-1})	Ref.
$BrO_3^- + I^- + 2\,H^+$	49	207
$BrO_3^- + Br^- + 2\,H^+$	3.3[a]	208
$BrO_3^- + Cl^- + 2\,H^+$	6.5×10^{-3}[b]	209

[a] 0.11 M ionic strength.
[b] 1.2 M ionic strength.

In the reaction with HSO_3^- (eq. (5.223)), the rate is first-order in both oxidant and reductant and the acidity dependence reveals two pathways involving reactions of H_2SO_3 and HSO_3^- with rate constants 2.2×10^2 and 0.17 M^{-1} s^{-1} respectively at $25.0\,°C$ [210]. Labelling studies with ^{18}O-labelled BrO_3^- in 0.1 M HCl indicate that 2.20 oxygen atoms are transferred for each BrO_3^- and an inner-sphere mechanism with atom transfer is likely in at least two steps in the overall reaction [211].

$$BrO_3^- + 3\,HSO_3^- \longrightarrow 3\,SO_4^{2-} + Br^- + 3\,H^+ \qquad (5.223)$$

While the rate of the corresponding reaction with $[BrO_4]^-$ is independent of $[H^+]$ above pH 4.5 with a rate constant of 5.8×10^{-3} M^{-1} s^{-1} at $25\,°C$, labelling studies reveal transfer of a single oxygen atom (eq. (5.224)) [212].

$$BrO_4^- + SO_3^{2-} \longrightarrow BrO_3^- + SO_4^{2-} \qquad (5.224)$$

There is a consensus that the acidity dependencies in these reactions reveal evidence for the formation of adducts as intermediates in the redox process. This thesis, that replacement is a prerequisite to redox is borne out by the similarity of the rate expressions for redox and O exchange at the oxyanions, and by the strong correlations of the reactivity trends for reduction and lability. Thus the trend in the ease of reduction of the halates is $IO_3^- > BrO_3^- > ClO_3^-$ and for a single halogen is $OCl^- > ClO_2^- > ClO_3^- > ClO_4^-$ [191]. There are a variety of mechanisms for the complementary two-electron processes. Oxygen atom transfer does occur, as does halide cation transfer, and adduct formation facilitates both processes, playing a dominant role in the reactions. However, not all reactions of these oxyanions are as simple. Autocatalysis, oscillatory behavior and radical formation are common and are beyond the scope of this chapter [213, 214]. However, it is equally true that not all the reactions involving radicals show complex behavior. The initial step in the reaction of ClO_2 with NO_2^- is the production of NO_3^- (eq. (5.225)), and is first-order in both reagents with a second-order rate constant of 153 M^{-1} s^{-1} at $25\,°C$. Detection of inhibition by ClO_2^- pinpoints a mechanism involving single-electron transfer (eqs (5.226)–(5.227)), although overall the reaction is complex and involves atom transfer in subsequent steps. The observed rate is three orders of magnitude faster than the rate calculated by the Marcus model and a strong-overlap or inner-sphere mechanism is proposed.

$$2\,ClO_2 + NO_2^- + H_2O \longrightarrow 2\,ClO_2^- + NO_3^- + 2\,H^+ \qquad (5.225)$$

$$ClO_2 + NO_2^- \underset{k_{-1}}{\overset{k_1}{\rightleftharpoons}} ClO_2^- + NO_2 \qquad (5.226)$$

$$2\,ClO_2 + NO_2 + H_2O \overset{k_2}{\longrightarrow} 2\,ClO_2^- + NO_3^- + 2\,H^+ \qquad (5.227)$$

QUESTIONS

5.1 Reduction of $[UO_2]^{2+}$ by $[Ti(H_2O)_6]^{3+}$ follows the stoichiometry:

$$[UO_2]^{2+} + 2[Ti(H_2O)_6]^{3+} \rightarrow [U_{aq}]^{4+} + 2[TiO_{aq}]^{2+}$$

and the rate law

$$-\frac{d[[UO_2]^{2+}]}{dt} = k[[UO_2]^{2+}][[Ti(H_2O)_6]^{3+}]$$

In the presence of $[VO_{aq}]^{2+}$ which, on the timesecale of the reaction, will not oxidize $[U_{aq}]^{4+}$ and $[Ti(H_2O)_6]^{3+}$ or reduce $[UO_2]^{2+}$ and $[TiO_{aq}]^{2+}$, the reaction is catalyzed and the rate law is:

$$-\frac{d[[UO_2]^{2+}]}{dt} = \frac{k_1[[UO_2]^{2+}][[Ti(H_2O)_6]^{3+}]^2}{k_2[[Ti(H_2O)_6]^{3+}] + [[VO_{aq}]^{2+}]}$$

Deduce a mechanism for the reaction.

5.2 The rate law for the oxidation of the tungsten(V) dimer $[W_2(O)_2(\mu\text{-}O)(\mu\text{-}S)(\mu\text{-edta-}N,N')]^{2-}$, abbreviated $[W_2OS]^{2-}$, by $[IrCl_6]^{2-}$ to give monomeric tungsten(VI) at 1.00 M ionic strength and 25 °C is of the form:

$$-\frac{d[[IrCl_6]^{2-}]}{dt} = \frac{2a[[W_2OS]^{2-}][[IrCl_6]^{2-}]}{1 + b[[IrCl_6]^{3-}]}$$

where $a = 8.3 \times 10^2$ M^{-1} s^{-1} and $b = 8.6 \times 10^3$ M^{-1}. Provide a mechanism for this reaction and justify structures for any intermediates which are proposed.
(Ojo, J. F.; Hasegawa, Y.; Sasaki, Y.; Ikari, S. *Inorg. Chem.* **1990**, *29*, 1712–1716.)

5.3 The outer-sphere oxidation of $[IrCl_6]^{3-}$ by ozone, O_3, has a second-order rate constant of 1.7×10^4 M^{-1} s^{-1} at 25.0 °C and 0.2 M ionic strength. The reduction potential of $O_3^{0/-}$ is estimated to be 1.02 V. Calculate a self-exchange rate for the $O_3^{0/-}$ reaction and compare the value with those in Table 5.17.

The rate constant for reaction between O_3 and ClO_2^- to given O_3^- and ClO_2 is 4×10^6 M^{-1} s^{-1}, much faster than the rate calculated for rate-limiting outer-sphere electron transfer 2×10^2 M^{-1} s^{-1}. Propose an explanation which rationalizes the discrepancy.
(Bennett, L. E.; Warlop, P. *Inorg. Chem.* **1990**, *29*, 1975–1981.)

5.4 Consult the following papers dealing with oxidations of the organic substrates, (a) ascorbic acid and (b) 2-mercaptosuccinic acid by $[Fe(H_2O)_6]^{3+}$ and $[Fe(CN)_6]^{3-}$, and explain the role of intermediate complexes in determining the courses of the reactions.

(Xu, J.; Jordan, R. B. *Inorg. Chem.* 1990, **29**, 4180–4184. Bänsch, B.; Martinez, P.; Uribe, D.; Zuluaga, J.; van Eldik, R. *Inorg. Chem.* 1991, **30**, 4555–4559. Bänsch, B.; Martinez, P.; Zuluaga, J.; Uribe, D.; van Eldik, R. *Z. Phys. Chem.* 1991, **170**, 59–71. Bridgart, G. J.; Fuller, M. W.; Wilson, I. R. *J. Chem. Soc., Dalton Trans.* 1973, 1274–1279. Ellis, K. J.; Lappin, A. G.; Alexander McAuley, A. *J. Chem. Soc., Dalton Trans.* 1975, 1930–1934.)

5.5 The reagent methylviologen, MV^{2+}, is a frequently used organic one-electron transfer reagent with a reduction potential of -0.44 V and a self-exchange rate of 8×10^6 M^{-1} s^{-1} at 25 °C and 0.10 M ionic strength. The radical, MV^+, is readily generated by oxidative quenching of $[*Ru(bpy)_3]^{2+}$. Use the Marcus relationship to calculate rates for reaction with $[Fe(CN)_6]^{4-}$ and $[Ru(en)_3)]^{2+}$. Comment on the magnitude of the rate constants.

$$MV^{2+} = CH_3N\underset{}{\bigcirc}-\underset{}{\bigcirc}NCH_3 \quad ^{2+}$$

(deOliveira, L. A. A.; Haim, A. *J. Am. Chem. Soc.* 1982, **104**, 3363–3366.)

5.6 The reduction of $[IrCl_6]^{2-}$ by SCN^- to give $[IrCl_6]^{3-}$, S and CN^- has been studied under pseudo-first-order conditions with an excess of $[SCN^-]$. From the rate data in the table, deduce an empirical rate law for the reaction and hence deduce a chemically reasonable mechanism.

Table. Rate constants at 25°C, 0.10 M ionic strength and pH = 2

10^2 [SCN^-] (M)	10^4 k_{obsd} (s^{-1})	[SCN^-] (M)	10^4 k_{obsd} (s^{-1})
0.558	0.811	1.116	2.12
2.23	6.86	4.46	21.6
8.93	85.6		

5.7 In basic solution, the decomposition of OCl^- is catalyzed by $[Cu(OH)_4]^{2-}$.

$$2\ OCl^- \longrightarrow O_2 + 2\ Cl^-$$

The reaction is complex but the rate law shows that two pathways are important:

$$\text{Rate} = k_1[[Cu(OH)_4]^{2-}][OCl^-][OH^-]^{-1} + k_2[[Cu(OH)_4]^{2-}]^2[OCl^-][OH^-]^{-1}$$

Speculate mechanisms for the two pathways based on these rate laws.

REFERENCES

[1] Wetton, E. A. M.; Higginson, W. C. E. *J. Chem. Soc.* 1965, 5890–5906.
[2] Ashurst, K. G.; Higginson, W. C. E. *J. Chem. Soc.* 1953, 3044–3049.
[3] Cannon, R. D. *Electron Transfer Reactions*, Butterworths, 1980.
[4] Sykes, A. G. *Adv. Inorg. Chem. Radiochem.* 1967, **10**, 153–245.

[5] Bard, A. J.; Parsons, R.; Jordan, J. *Standard Potentials in Aqueous Solution,* Marcel Dekker, 1985.
[6] Baes, C. F.; Mesmer, R. E. *The Hydrolysis of Cations,* Wiley, 1976.
[7] Harkness, A. C.; Halpern, J. *J. Am. Chem. Soc.* 1959, **81**, 3526–3529.
[8] Schwarz, H. A.; Comstock, D.; Yandell, J. K.; Dodson, R. W. *J. Phys. Chem.* 1974, **78**, 488–493.
[9] Daugherty, N. A. *J. Am. Chem. Soc.* 1965, **87**, 5026–5030.
[10] Baker, F. B.; Brewer, W. D.; Newton, T. W. *Inorg. Chem.* 1966, **5**, 1294–1296.
[11] Higginson, W. C. E., Rosseinsky, D. R.; Stead, J. B.; Sykes, A. G. *Discussions of the Faraday Society* 1960, **29**, 49–59.
[12] Ardon, M.; Plane, R. A. *J. Am. Chem. Soc.* 1959, **81**, 3197–3200.
[13] Falcinella, B.; Felgate, P. D.; Laurence, G. S. *J. Chem. Soc., Dalton Trans.* 1974, 1367–1373.
[14] Harbottle, G.; Dodson, R. W. *J. Am. Chem. Soc.* 1951, **73**, 2442–2447.
[15] Dodson, R. W. *J. Am. Chem. Soc.* 1953, **75**, 1795–1797.
[16] Roig, E.; Dodson, R. W. *J. Phys. Chem.* 1961, 65, 2175–2181.
[17] Cercek, B.; Ebert, M.; Swallow, A. J. *J. Chem. Soc. (A)* 1966, 612–615.
[18] Basolo, F.; Morris, M. L., Pearson, R. G. *Discussions of the Faraday Society* 1960, **29**, 80–91.
[19] Mason, W. R.; Johnson, R. C. *Inorg. Chem.* 1965, **4**, 1258–1261.
[20] Rich, R. R.; Taube, H. *J. Am. Chem. Soc.* 1954, **76**, 2608–2611.
[21] Beattie, J. K.; Basolo, F. *Inorg. Chem.* 1971, **10**, 486–491.
[22] Glennon, C. S.; Hand, T. D.; Sykes, A. G. *J. Chem. Soc., Dalton Trans.* 1980, 19–23.
[23] Beattie, J. K.; Basolo, F. *Inorg. Chem.* 1971, **10**, 486–491.
[24] Bakac, A.; Hand, T. D.; Sykes, A. G. *Inorg. Chem.* 1975, 14, 2540–2543.
[25] Peloso, A. *J. Chem. Soc., Dalton Trans.* 1987, 1473–1475.
[26] Halpern, J.; Pribanic, M. *J. Am. Chem. Soc.* 1968, **90**, 5942–5943.
[27] Zhou, M.; Pfennig, B. W.; Steiger, J.; Van Engen, D.; Bocarsley, A. *Inorg. Chem.* 1990, **29**, 2456–2460.
[28] Pfennig, B. W.; Bocarsley, A. B. *J. Phys. Chem.* 1992, **96**, 226–233.
[29] Haight, G. P. *Progr. Inorg. Chem.* 1972, **17**, 93–145.
[30] Birk, J. P. *J. Am. Chem. Soc.* 1969, **91**, 3189–3197.
[31] Espenson, J. H.; King, E. L. *J. Am. Chem. Soc.* 1963, **85**, 3328–3333.
[32] Birk, J. P.; Gasiewski, J. W. *Inorg. Chem.* 1971, **10**, 1586–1589.
[33] Bose, R. N.; Gould, E. S. *Inorg. Chem.* 1985, **24**, 2832–2835.
[34] Bose, R. N.; Neff, V. D.; Gould, E. S. *Inorg. Chem.* 1986, **25**, 165–168.
[35] Ghosh, S. K.; Gould, E. S. *Inorg. Chem.* 1986, **25**, 3357–3359.
[36] Ghosh, M. C.; Gould, E. S. *J. Chem. Soc., Chem. Commun.* 1992, 195–196.
[37] Espenson, J. H. *J. Am. Chem. Soc.* 1964, **86**, 1883–1884.
[38] Espenson, J. H. *J. Am. Chem. Soc.* 1964, **86**, 5101–5107.
[39] Rosseinsky, D. R.; Nicol, M. J. *J. Chem. Soc., A* 1970, 1196–1200.
[40] Gordon, G.; Taube, H. *Inorg. Chem.* 1962, **1**, 69–72.
[41] Espenson, J. H.; Wang, R. T. *Inorg. Chem.* 1972, **11**, 955–959.
[42] Ekstrom, A. *Inorg. Chem.* 1977, **16**, 845–849.

[43] Song, J.-S.; Bullock, R. M.; Creutz, C. *J. Am. Chem. Soc.* 1991, **113**, 9862–9864.
[44] Edidin, R. T.; Sullivan, J. M.; Norton, J. R. *J. Am. Chem. Soc.* 1987, **109**, 3945–3953.
[45] Schwarz, C. L.; Bullock, R. M.; Creutz, C. *J. Am. Chem. Soc.* 1991, **113**, 1225–1236.
[46] Anderson, K. A.; Kirchner, K.; Dodgen, H. W.; Hunt, J. P.; Wherland, S. *Inorg. Chem.* 1992, **31**, 2605–2608.
[47] Eberson, L. *Electron Transfer Reactions in Organic Chemistry*, Springer-Verlag, Berlin, 1987.
[48] Eberson, L. *Adv. Phys. Org. Chem.* 1982, **18**, 79–185.
[49] Fukuzumi, S.; Wong, C. L.; Kochi, J. K. *J. Am. Chem. Soc.* 1980, **102**, 2928–2939.
[50] Kochi, J. K. *Angew. Chem.* 1988, **27**, 1227–1266.
[51] Kochi, J. K. *Acta Chem. Scand.* 1990, **44**, 409–432.
[52] Fukuzumi, S.; Kochi, J. K. *Bull. Chem. Soc. Japan* 1983, **56**, 969–979.
[53] Zahir, K.; Espenson, J. H.; Bakac, A. *J. Am. Chem. Soc.* 1988, **110**, 5059–5063.
[54] Pladziewicz, J. R.; Meyer, T. J.; Broomhead, J. A.; Taube, H. *Inorg. Chem.* 1973, **12**, 639–643.
[55] Stanbury, D. M.; Haas, O.; Taube, H. *Inorg. Chem.* 1980, **19**, 518–524.
[56] Creaser, I. I.; Geue, R.; Harrowfield, J. M.; Herlt, A. J.; Sargeson, A. M.; Snow, M. R.; Springborg, J. *J. Am. Chem. Soc.* 1982, **104**, 6016–6025.
[57] Vu, D. T.; Stanbury, D. M. *Inorg. Chem.* 1987, **26**, 1732–1736.
[58] Goldstein, S.; Czappski, G. *J. Am. Chem. Soc.* 1983, **105**, 7276–7280.
[59] Goldstein, S.; Czappski, G. *Inorg. Chem.* 1985, **24**, 1087–1092.
[60] McDowell, M. S.; Espenson, J. H.; Bakac, A. *Inorg. Chem.* 1984, **23**, 2232–2236.
[61] Bakac, A.; Espenson, J. H.; Creaser, I. I., Sargeson, A. M. *J. Am. Chem. Soc.* 1983, **105**, 7624–7628.
[62] Davies, G.; Sutin, N.; Watkins, K. O. *J. Am. Chem. Soc.* 1970, **92**, 1892–1897.
[63] Kristine, F. J.; Johnson, C. R.; O'Donnell, S.; Shephard, R. E. *Inorg. Chem.* 1980, **19**, 2280–2284.
[64] Sellers, R. M.; Simic, M. G. *J. Am. Chem. Soc.* 1976, **98**, 6145–6150.
[65] Kolaczkowski, R. W.; Plane, R. A. *Inorg. Chem.* 1964, **3**, 322–324.
[66] Anderson, L. B.; Plane, R. A. *Inorg. Chem.* 1964, **3**, 1470–1472.
[67] Samuni, A.; Meisel, D.; Czapski, G. *J. Chem. Soc., Dalton Trans.* 1972, 1273–1277.
[68] Bakac, A.; Espenson, J. H. *Inorg. Chem.* 1983, **22**, 779–783.
[69] Rush, J. D.; Bielski, B. H. *J. Inorg. Chem.* 1985, **24**, 4282–4285.
[71] Davies, G.; Warnqvist, B. *Coord. Chem. Rev.* 1970, **5**, 349–378.
[72] Davies, G. *Coord. Chem. Rev.* 1969, **4**, 199–224.
[73] Nord, G.; Wernberg, O. *J. Chem. Soc., Dalton Trans.* 1972, 866–868.
[74] Nord, G.; Wernberg, O. *J. Chem. Soc., Dalton Trans.* 1975, 845–849.
[75] Ghosh, P. K.; Brunschwig, B. S.; Chou, M.; Creutz, C.; Sutin, N. *J. Am. Chem. Soc.* 1984, **106**, 4772–4783.

[76] Mønsted, O.; Nord, G. *Adv. Inorg. Chem.* 1991, **37**, 381–397.
[77] Lay, P. A.; Sasse, W. H. F. *Inorg. Chem.* 1985, **24**, 4707–4710.
[78] Gilbert, J. A.; Eggleston, D. S.; Murphy, W. R.; Geselowitz, D. A.; Gersten, S. W.; Hodgson, D. J.; Meyer, T. J. *J. Am. Chem. Soc.* 1985, **107**, 3855–3864.
[79] Geselowitz, D.; Meyer, T. J. *Inorg. Chem.* 1990, **29**, 3894–3896.
[80] Davies, G.; Higgins, R.; Loose, D. J. *Inorg. Chem.* 1976, **15**, 700–703.
[81] Mi, L.; Zuberbühler, A. D. *Helv. Chim. Acta* 1991, **74**, 1679–1688.
[82] Wells, C. F.; Fox, D. *J. Chem. Soc., Dalton Trans.* 1977, 1498–1501.
[83] Davies, G.; Watkins, K. O. *J. Phys. Chem.* 1970, **74**, 3388–3392.
[84] Barb, W. G.; Baxendale, J. H.; George, P.; Hargrave, K. R. *Trans. Fard. Soc.* 1951, **47**, 591–601
[85] Bodek, I.; Davies, G. *Inorg. Chem.* 1975, **14**, 2580–2582.
[86] Gilbert, J. A.; Gersten, S. W.; Meyer, T. J. *J. Am. Chem. Soc.* 1982, **104**, 6872–6873.
[87] Gilbert, J.; Roecker, L.; Meyer, T. J. *Inorg. Chem.* 1987, **26**, 1126–1132.
[88] Sykes, A. G.; Weil, J. A. *Prog. Inorg. Chem.* 1970, **13**, 1–106.
[89] Fallab, S.; Mitchell. P. R. *Adv. Inorg. Bioinorg. Mechanisms,* 1984, **3**, 311–377.
[90] Simplicio, J.; Wilkins, R. G. *J. Am. Chem. Soc.* 1969, **91**, 1325–1329.
[91] Miller, F.; Simplicio, J.; Wilkins, R. G. *J. Am. Chem. Soc.* 1969, 91, 1962–1967.
[92] Miller, F.; Wilkins, R. G. *J. Am. Chem. Soc.* 1970, **92**, 2687–2691.
[93] Wong, C.-L.; Switzer, J. A.; Balakrishnan, K. P.; Endicott, J. P. *J. Am. Chem. Soc.* 1980, **102**, 5511–5518.
[94] Bakac, A.; Espenson, J. H. *J. Am. Chem. Soc.* 1990, **112**, 2273–2278.
[95] Marchai, A.; Bakac, A.; Espenson, J. H. *Inorg. Chem.* 1992, **31**, 4146–4168.
[96] Shinohara, N.; Ishii, K.; Hirose, M. *J. Chem. Soc., Chem. Commun.* 1990, 700–701.
[97] McLendon, G.; Mooney, W. F. *Inorg. Chem.* 1980, **19**, 12–15
[98] Kumar, K.; Endicott, J. F. *Inorg. Chem.* 1984, **23**, 2447–2452.
[99] Burgess, J.; Prince, R. H. *J. Chem. Soc. A* 1966, 1772–1775.
[100] Burgess, J.; Prince, R. H. *J. Chem. Soc. A* 1970, 2114–2115.
[101] Fürholz, U.; Haim, A. *Inorg. Chem.* 1987, **26**, 3243–3248.
[102] Pennington, D. E.; Haim, A. *J. Am. Chem. Soc.* 1968, **90**, 3700–3704.
[103] Shakhashiri, B. Z.; Gordon, G. *Inorg. Chem.* 1968, **7**, 2454–2456.
[104] Ondrus, M. G.; Gordon, G. *Inorg. Chem.* 1971, **10**, 474–477.
[105] Ige, J.; Ojo, J. F.; Olubuyide, O. *Canad. J. Chem.* 1979, **57**, 2065–2070.
[106] Cannon, R. D.; Powell, D. B.; Sarawek, K. *Inorg. Chem.* 1981, **20**, 1470–1474.
[107] Woodruff, W. H.; Margerum, D. W.; Milano, M. J.; Pardue, H. L.; Santini, R. E. *Inorg. Chem.* 1973, **12**, 1490–1494.
[108] Woodruff, W. H.; Margerum, D. W. *Inorg. Chem.* 1974, **13**, 2578–2585.
[109] Lappin, A. G.; Osvath, P.; Baral, S. *Inorg. Chem.* 1987, **26**, 3089–3094.
[110] Woodruff, W. H.; Burke, B. A.; Margerum, D. W. *Inorg. Chem.* 1974, **13**, 2573–2577.
[111] Rudgewick-Brown, N.; Cannon, R. D. *J. Chem. Soc., Dalton Trans.* 1984, 479–481.
[112] Gordon, G.; Andrewes, A. *Inorg. Chem.* 1964, **3**, 1733–1737.

[113] Adegite, A.; Ford-Smith, M. H. *J. Chem. Soc., Dalton Trans.* 1973, 134–138.
[114] Adegite, A.; Ford-Smith, M. H. *J. Chem. Soc., Dalton Trans.* 1973, 138–143.
[115] Mason, W. R. *Inorg. Chem.* 1971, **10**, 1914–1917.
[116] Elding, L. I.; Gustafson, L. *Inorg. Chim. Acta* 1976, **19**, 165–171.
[117] Jones, M. M.; Morgan, K. A. *J. Inorg. Nuclear Chem.* 1971, **34**, 259–274.
[118] Morgan, K. A; Jones, M. M. *J. Inorg. Nuclear Chem.* 1971, **34**, 275–296.
[119] Drougge, L.; Elding, L. I. *Inorg. Chem.* 1985, **24**, 2292–2297.
[120] Malin, J. M.; Swinehart, J. H. *Inorg. Chem.* 1969, **8**, 1407–1410.
[121] Conocchioli, T. J.; Hamilton, E. J.; Sutin, N. *J. Am. Chem. Soc.* 1965, **87**, 926–927.
[122] Adegite, A.; Ford-Smith, M. H. *J. Chem. Soc., Dalton Trans.* 1972, 2113–2115.
[123] Adegite, A.; Edeogu, S. *J. Chem. Soc., Dalton Trans.* 1975, 1203–1206.
[124] Carter, P. R.; Davidson, N. *J. Phys. Chem.* 1952, **56**, 877–884.
[125] Nord, G.; Pedersen, B.; Farver, O. *Inorg. Chem.* 1978, **17**, 2233–2238.
[126] Stanbury, D. M.; Wilmarth, W. K.; Khalaf, S.; Po, H. N.; Byrd, J. E. *Inorg. Chem.* 1980, **19**, 2715–2722.
[127] Wilmarth, W. K.; Stanbury, D. M.; Byrd, J. E.; Po, H. N.; Chua, C.-P. *Coord. Chem. Rev.* 1983, **51**, 155–179.
[128] Ferranti, F. *J. Chem. Soc. A* 1970, 134–136.
[129] McAuley, A.; Norman, P. R.; Olubuyide, O. *Inorg. Chem.* 1984, **23**, 1938–1943.
[130] Raycheba, J. M. T.; Margerum, D. W. *Inorg. Chem.* 1981, **20**, 45–51.
[131] Fox, D.; Wells, C. F. *J. Chem. Soc., Dalton Trans* 1989, 151–154.
[132] Sarala, R.; Rabin, S. B.; Stanbury, D. M. *Inorg. Chem.* 1991, **30**, 3999–4007.
[133] Sarala, R.; Stanbury, D. M. *Inorg. Chem.* 1992, **31**, 2771–2777.
[134] Nord, G.; Pedersen, B.; Floryan-Løvborg, E.; Pagsberg, P. *Inorg. Chem.* 1982, **21**, 2327–2330.
[135] Stanbury, D. M. *Inorg. Chem.* 1984, **23**, 2914–2916.
[136] Stanbury, D. M. *Advances Inorg. Chem.* 1989, **33**, 69–138.
[137] Nord, G. *Comments Inorg. Chem.* 1992, **13**, 221–239.
[138] Laurence, G. S.; Ellis, K. J. *J. Chem. Soc., Dalton Trans.* 1972, 2229–2233.
[139] Haines, R. I.; McAuley, A. *Inorg. Chem.* 1986, **25**, 1233–1238.
[140] Fairbank, M. G.; McAuley, A. *Inorg. Chem.* 1987, **26**, 2844–2848.
[141] Raycheba, J. M. T.; Margerum, D. W. *Inorg. Chem.* 1981, **20**, 1441–1446.
[142] Scaife, C. W. J.; Wilkins, R. G. *Inorg. Chem.* 1980, **19**, 3244–3247.
[143] Mehrotra, R. N.; Wilkins, R. G. *Inorg. Chem.* 1980, **19**, 2177–2178.
[144] Pinnell, D.; Jordan, R. B. *Inorg. Chem.* 1979, **18**, 3191–3194.
[145] Balahura, R. J.; Johnson, M. D. *Inorg. Chem.* 1987, **26**, 3860–3863.
[146] Simmons, C. A.; Bakac, A.; Espenson, J. H. *Inorg. Chem.* 1989, **28**, 581–584.
[147] Lednicky, L. A.; Stanbury, D. M. *J. Am. Chem. Soc.* 1983, **105**, 3098–3101.
[148] Stanbury, D. M.; Lednicky, L. A. *J. Am. Chem. Soc.* 1984, **106**, 2847–2853.
[149] Stapp, E. L.; Carlyle, D. W. *Inorg. Chem.* 1974, **13**, 834–837.
[150] Carlyle, D. W. *J. Am. Chem. Soc.* 1972, **94**, 4525–4529.
[151] Creutz, C.; Sutin, N.; Brunschwig, B. S. *J. Am. Chem. Soc.* 1979, **101**, 1297–1298.

[152] Anast, J. M.; Margerum, D. W. *Inorg. Chem.* 1981, **20**, 2319–2326.
[153] Sarala, R.; Islam, M. A.; Rabin, S. B.; Stanbury, D. M. *Inorg. Chem.* 1990, **29**, 1133–1142.
[154] Higginson, W. C. E.; Marshall, J. W. *J. Chem. Soc.* 1957, 447–458.
[155] Veprek-Siska, J.; Wagnerova, D. M.; Eckschlager, K. *Coll. Czech. Chem. Commun.* 1966, **31**, 1248–1255.
[156] Karraker, D. G. *J. Phys. Chem.* 1963, **67**, 871–874.
[157] Carlyle, D. W.; Zeck, O. F. *Inorg. Chem.* 1973, **12**, 2978–2983.
[158] Kraft, J.; van Eldik, R. *Inorg. Chem.* 1989, **28**, 2297–2305.
[159] Kraft, J.; van Eldik, R. *Inorg. Chem.* 1989, **28**, 2306–2312.
[160] Brown, A.; Higginson, W. C. E. *J. Chem. Soc., Chem. Commun.* 1967, 725–726.
[161] Thacker, M. A.; Scott, K. L.; Simpson, M. E.; Murray, R. S.; Higginson, W. C.E. *J. Chem. Soc., Dalton Trans.* 1974, 647–651.
[162] Scott, K. L. *J. Chem. Soc., Dalton Trans.* 1974, 1486–1489.
[163] Joshi, V. K.; van Eldik, R.; Harris, G. M. *Inorg. Chem.* 1986, **25**, 2229–2237.
[164] van Eldi¢k, R.; Harris, G. M. *Inorg. Chem.* 1980, **19**, 880–886.
[165] Gamsjäger, H.; Murmann, R. K. *Adv. Inorg. Bioinorg. Mech.* 1983, **2**, 317–380.
[166] Hoering, T. C.; Ishimori, F. T.; McDonald, H. O. *J. Am. Chem. Soc.* 1956, **78**, 4829–4831.
[167] Kallen, T. W.; Earley, J. E. *Inorg. Chem.* 1971, **10**, 1152–1155.
[168] Endicott, J. F.; Taube, H. *Inorg. Chem.* 1965, **4**, 437–448.
[169] Cope, V. W.; Miller, R. G.; Fraser, R. T. M. *J. Chem. Soc. (A)* 1967, 301–306.
[170] King, W. R.; Garner, C. S. *J. Phys. Chem.* 1954, **58**, 29–33.
[171] Indelli, A.; Ferranti, F.; Secco, F. *J. Phys. Chem.* 1966, **70**, 631–636.
[172] Kasim, A. Y.; Sulfab, Y. *Inorg. Chim. Acta* 1977, **24**, 247–250.
[173] Kassim, A. Y.; Sulfab, Y. *Inorg. Chem.* 1981, **20**, 506–509.
[174] Sulfab, Y. *J. Inorg. Nuclear Chem.* 1976, **38**, 2271–2274.
[175] Thompson, R. C.; Gordon, G. *Inorg. Chem.* 1966, **5**, 557–562.
[176] Thompson, R. C.; Gordon, G. *Inorg. Chem.* 1966, **5**, 562–569.
[177] Ondrus, M. G.; Gordon, G. *Inorg. Chem.* 1972, **11**, 985–989.
[178] Gordon. G.; Tewari, P. H. *J. Phys. Chem.* 1966, **70**, 200–204.
[179] Shakhashiri, B. Z.; Gordon, G. *J. Am. Chem. Soc.* 1969, **91**, 1103–1107.
[180] Birk, J. P.; Kozub, S. G. *Inorg. Chem.* 1973, **12**, 2460–2464.
[181] Birk, J. P.; Kozub, S. G. *Inorg. Chem.* 1978, **17**, 1186–1191.
[182] Thompson, R. C. *Inorg. Chem.* 1971, **10**, 1892–1895.
[183] Birk, J. P. *Inorg. Chem.* 1978, **17**, 504–506.
[184] Bakac, A.; Thornton, A. T.; Sykes, A. G. *Inorg. Chem.* 1976, **15**, 274–278.
[185] Higginson, W. C. E.; McCarthy, D. A. *J. Chem. Soc., Dalton Trans.* 1980, 797–803.
[186] Birk, J. P. *Inorg. Chem.* 1978, **17**, 1372–1374.
[187] Sulfab, Y.; Elfaki, H. A. *Canad. J. Chem.* 1974, **52**, 2001–2004.
[188] Buchacek, R.; Gordon, G. *Inorg. Chem.* 1972, **11**, 2154–2160.
[189] Thompson, R. C. *Adv. Inorg. Bioniorg. Mech.* 1986, **4**, 65–106.
[190] Edwards, J. O. *Inorganic Reaction Mechanisms,* Benjamin, 1964.
[191] Chaffee, E.; Edwards, J. O. *Prog. Inorg. Chem.* 1970, **13**, 205–242.

[192] Edwards, J. O. *Peroxide Reaction Mechanisms,* Wiley, 1962.
[193] McArdle, J. V.; Hoffmann, M. R. *J. Phys. Chem.* 1983, **87,** 5425–5429.
[194] Lister, M. W.; Rosenblum, P. *Can. J. Chem.* 1963, **41,** 3013–3020.
[195] Kumar, K.; Day, R. A., Margerum, D. W. *Inorg. Chem.* 1986, **25,** 4344–4350.
[196] Kumar, K.; Margerum, D. W. *Inorg. Chem.* 1987, **26,** 2706–2711.
[197] Margerum, D. W.; Gray, E. T.; Huffman, R. P. *A.C.S. Symposium Series,* 1978, **82,** 278–291.
[198] Eigen, M.; Kustin, K. *J. Am. Chem. Soc.* 1962, **84,** 1355–1361.
[199] Johnson, D. W.; Margerum, D. W. *Inorg. Chem.* 1991, **30,** 4845–4851.
[200] Yiin, B. S.; Margerum, D. W. *Inorg. Chem.* 1988, **27,** 1670–1672.
[201] Palmer, D. A.; van Eldik, R. *Inorg. Chem.* 1986, **25,** 928–931.
[202] Eigen, M.; Kustin, K. *J. Am. Chem. Soc.* 1962, **84,** 1355–1361.
[203] Wang, Y. L.; Nagy, J. C.; Margerum, D. W. *J. Am. Chem. Soc.* 1989, **111,** 7838–7844.
[204] Troy, R. C.; Kelley, M. D.; Nagy, J. C.; Margerum, D. W. *Inorg. Chem.* 1991, **30,** 4838–4845.
[205] Turner, D. H.; Flynn, G. W.; Sutin, N.; Beitz, J. V. *J. Am. Chem. Soc.* 1972, **94,** 1544–1559.
[206] Margerum, D. W.; Dickson, P. N.; Nagy, J. C.; Kumar, K.; Bowers, C. P.; Fogelman, K. D. *Inorg. Chem.* 1986, **25,** 4900–4904.
[207] Barton, A. F. M.; Wright, G. A. *J. Chem. Soc., A* 1968, 1747–1753.
[208] Young, H. A.; Bray, W. C. *J. Am. Chem. Soc.* 1932, **54,** 4282–4296.
[209] Sigalla, J. *J. Chim. Phys.* 1958, **55,** 758–767.
[210] Williamson, S. F.; King, E. L. *J. Am. Chem. Soc.* 1957, **79,** 5397–5400.
[211] Halperin, J.; Taube, H. *J. Am. Chem. Soc.* 1952, **74,** 375–380.
[212] Appelman E. H.; Kläning, U. K.; Thompson, R. C. *J. Am. Chem. Soc.* 1979, 101, 929–934.
[213] Song, W. M.; Kustin, K.; Epstein, I. R. *J. Phys. Chem.* 1989, **93,** 4698–4700.
[214] Rábai, G.; Beck, M. T. *Inorg. Chem.* 1987, **26,** 1195–1199.
[215] Stanbury, D. M.; Martinez, R.; Tseng, E.; Miller, C. E. *Inorg. Chem.* 1988, **27,** 4277–4280.

Index

acid catalysis
    and proton demand, 101–104, 137, 231, 265–267, 272
    general, 35, 36, 268–270
    specific, 35, 268
    with oxo ion reagents, 228–231, 252, 263–267, 2272
adduct formation, 235
adiabatic electron transfer, 55, 56, 76
Arrhenius equation, 28
atom transfer, 9, 30, 106, 107, 121, 142, 144–147, 232–238, 263–273
    activation parameters, 232
    considered as substitution or redox, 215, 267

base catalysis
    general, 34, 35
    specific, 35, 57, 63, 101–108, 123, 137–139, 142, 151, 218–222
blue copper protein, 200
Boltzmann distribution, 28
bond energies, 235, 271
bridge formation
    aqua and hydroxide, 123, 130, 217, 219, 2369
    chelate, 123, 126
    criteria required for, 125, 128
    double, 123, 127, 132, 134
    halide abnormal order, 129–131
    halide normal order, 123, 128, 143, 147
    and transfer, 121, 1308–131
bridged outer sphere mechanism, 126, 135, 156, 185
bridging
    atom, 121
    group, 123ff, 145
Brønsted relationship, 106

charge transfer, 10
chemical mechanism, 150–152
chromium(II)reductions, 63–66, 93–95, 120–131
chromium(VI)
    properties, 228ff
    oxidation by, 228–232
cobalt(III)
    reduction by iron(II), 134–135
    reduction by chromium(II), 120–128
    reduction by vanadium(II), 131–133
    self-exchange with cobalt(II), 88–90
collision frequency, 56
complementary process, 215
complex formation, 17, 21
compressibility of activation, 30
consecutive electron transfer, 249, 252
copper(II)
    blue protein, 201
    self-exchange with copper(I), 95–101
    self-exchange with copper(III), 95
coupled proton and electron transfer, 106–108
Creutz–Taube ion, 172–173
cytochrome c, 195–199

dead end mechanism, 45, 46, 1208, 139–140, 155, 205
Debye–Hückel, 83
    expression, 31–33, 198
diffusion, 42–43, 51, 87
    rate of, 13, 43

effective frequency, 28, 54, 76, 84–85
    solvent effect on, 83–84
electron transfer
    adiabatic, 55–56, 76
    non-adiabatic, 55, 76, 89–92
    photoinduced, 67–70, 162–163

Index 283

structural changes in, 52, 78–80, 87–89, 95ff, 215, 261
electron transfer energetics, 51, 75–77, 141ff
    inner sphere reorganization, 52, 77–80, 84, 170, 193, 195
    outer sphere reorganization, 52, 77, 81–85, 93–94, 170, 1730, 184, 193, 195
electronic
    coupling, 54, 86–87, 156
    factor, 76, 86–87
    transmission coefficient, 172
electrostatic interaction, 56
enthalpy of activation, 28–30
    in inner sphere electron transfer, 124–126, 132
    in outer sphere electron transfer, 43, 82–83
    in substitution reactions, 14, 30
entropy of activation, 28–30
    in inner sphere electron transfer, 124–126, 132
    in outer sphere electron transfer, 43, 82–83
    in substitution reactions, 14, 30
Eu(III)/(II) self-exchange, 65, 95
exchange of solvent 13–14
excited state
    electron transfer, 67ff, 206–207
    quenching mechanism, 69
    reduction potential, 68, 70

flash photolysis, 21, 22
Frank–Condon, 54–55, 83, 142, 196
    factor, 77ff
free energy of activation, 28–31
    in atom transfer reactions, 232–235
    in inner sphere electron transfer, 141–146
    in outer sphere electron transfer, 51–55
Fuoss equation, 56

group transfer, 123ff, 145

halide ion catalysis, 34–35, 133–134, 141–145, 223–225
halogen
    bridges, 120–122, 129, 143–144, 223–227
    transfer, 120–122, 129–130, 223, 226–227
Hammond postulate, 30–31, 52–53
homolytic bond cleavage, 143
hydrogen atom transfer, 104–108
hydrogen bonds, 93, 94
hydrolysis, 15, 17–19
    effect on rate law, 33–35, 57
    in outer-sphere reactions, 57, 63, 93–95, 101–104
    in inner-sphere reactions, 123, 135–138, 141–142, 218–220

and proton demand, 101–104, 137, 220, 228–232, 263–267
hydroxyl radical, 22, 243

induction of products, 220–221
inner-sphere reaction
    activation parameters, 122–123, 129–132
    acid catalysis, 130, 142, 150
    adjacent attack, 123–125
    advantage over outer-sphere, 144–146
    dead-end mechanism in, 141, 156
    definition of, 120–121
    elementary steps in, 120
    intermediates in, 126, 129, 135, 136, 140–141, 152–153, 156ff
    intramolecular electron transfer, 153, 155ff
    mechanistic criteria for, 131
    non-bridging effects, 126, 134
    organic bridges, 147ff
    precursor formation, 120
    remote attack, 120–123
    self-exchange rates, 128, 141–144
    trans ligand effect in, 121
inner-sphere reorganization, 52, 77–80, 84, 170, 193, 195
intermediate, 10, 20, 21
    in inner-sphere reaction, 129, 132, 135–136, 140–141, 152, 155–158
    in multiple electron transfer reaction, 216, 219–230
    in outer-sphere reaction, 45–48
    role in mechanism determination, 256, 259, 262
intervalence charge transfer, 172, 181, 227
    in ion pairs, 46, 170
    medium effects, 180
    table of, 174–176
intramolecular electron transfer, 170ff
    chemically induced, 180
    comparisons with outer-sphere, 177
    distance dependence, 180–185, 189–191, 209
    driving force dependence, 186, 207–208
    solvent effects, 173
    table of rates and activation parameters, 181–182, 185–186
inverted region, 54, 209–210
ion pair structure by nmr, 50, 74–75, 198–199
ionic strength, 197
    dependence, 10
    effects, 63
iron(II) reductions, 57, 133–135
isotope effects, 10, 36–38
    in atom transfer reactions, 234, 241
    in inner-sphere electron transfer, 122, 150
    in outer-sphere electron transfer, 57–58, 84, 102

284 Index

isotopic labeling, 268, 271

Jahn Teller, 12–13, 15

kinetic product, 72–75, 120, 132–133, 140–141, 180–181, 192, 206

Latimer diagram, 19
linear free energy relationship
    in inner-sphere electron transfer, 66, 132, 142–143, 151, 238
    in outer-sphere electron transfer, 65–67, 132, 149, 225, 236, 241
    Marcus, 55, 56, 64, 71, 121
ligand electrochemical series, 21
ligand field, 10–13, 16, 21
    high spin, 11–12
    low spin, 11, 14
    spin pairing, 10–12
    stabilization, 12, 14, 17
ligand substitution, 10, 13, 14
    inert complexes, 13
    labile complexes, 13

Marcus theory
    deviations at large driving force, 64, 92–93
    deviations for aqua ions, 64, 92–95, 101–102
    normal region, 52, 53
    inverted region, 54, 208–209
McKay relationship, 57
mechanism
    deduction of, 9–10
    dead end complex, 45–46, 141, 156, 205
    elementary steps, 9–10
    inner-sphere, 120–121
    intermediates in, 45, 135–136, 216, 219–230, 256, 259, 262
    outer-sphere, 42
    substitution, 13–15
metalloproteins, 195ff
    cytochrome c, 195–199
    plastocyanin, 200–205
mixed valence complexes, 171
    Robin-Day classification, 171
More-O'Ferrall diagram, 106
multiple electron transfer
    complementary process, 215–216
    consecutive electron transfer, 216–217
    energetics, 232–235
    intermediates in, 216, 219–230
    non-complementary process, 215–216
    structural changes in, 215, 223, 228

nickel(III)/(II) self-exchange reaction, 95
non-adiabatic electron transfer, 55, 76, 89–92, 193
non-complementary process, 216
non-metal reagents
    atom transfer in reactions of, 242, 244, 265–266, 268, 270–273
    isotopic tracer in studies of, 244, 246, 264, 266, 268 273
    reactions with metal complexes, 238f
    reactions with non-metallic species, 267ff

optical electron transfer, 171ff
orbital overlap, 54, 86, 95, 145, 151, 156
orientation effects, 46, 75
    specific binding, 198, 201–204, 209
outer sphere reaction
    activation parameters for, 43–44, 82–84
    bridged, 127, 136, 157, 186
    dead end mechanism in, 45–46
    definition of, 42
    elementary steps in, 42
    intermediates in, 45–48
    precursor formation in, 42, 44, 45–51
outer-sphere reorganization, 52, 77, 81–85, 93–94, 170, 173, 184, 194, 195

parallel inner sphere and outer sphere reactions, 130, 131, 138, 150
peptide
    complexes, 24–25, 95–97
    spacers, 184–193
peroxo complexes, 246–248
photoinduced intramolecular electron transfer, 161–163
platinum(IV)/(II) self-exchange reaction, 223ff
precursor formation
    dynamics, 51, 122, 131–133
    Fuoss equation, 50–51
    thermodynamics, 48–49, 135, 139
pulse radiolysis, 21–22, 186, 205, 222

radical formation
    in reactions of non-metallic species, 235, 273
    by pulse radiolysis, 22
    in organic bridges, 152–155
rate law, 10
    derivation of, 25–27
    limiting first order, 45–48, 129, 135
    pseudo first order, 27
    second order, 42–43, 121
reaction enthalpy, 23–24
reaction entropy, 23–24
reaction volume, 25, 29
reduction potential
    effect of bond length change and charge on, 24
    effect of ligand on, 21
    effect of temperature and pressure on, 23
    measurement by cyclic voltammetry, 15
    Nernst equation, 15

of non-metallic reagents, 239–240, 242, 249–250, 256, 258, 260, 263
of oxo reagents, 228, 263
solvent effects on, 71
table of, 24
resonance transfer, 141, 151
ruthenium(III)
self-exchange with ruthenium(II), 87–88
reduction by chromium(II), 129–130
reduction by vanadium(II), 63–66, 131–133
ruthenium(IV)/(III)/(II) exchange, 104–106

salt effects
in aqueous media, 31–35
in non aqueous media, 70
selectivity, 236
self-exchange rates
calculation of, 64, 65
for non-metallic reagents, 241, 258, 261
gas phase, 72
in non aqueous media, 71, 72, 81, 82
measurement by isotopic labeling, 57, 128, 222, 223
measurement by magnetic resonance, 59, 63, 196
measurement by optical activity, 58
table of, 63
solvent reorganization, 52, 53, 80–82, 93
specific ion effects, 63
stereoselectivity
in outer-sphere reactions, 58, 72–74
in inner-sphere reactions, 141, 142
steric
effects, 223, 235ff
factors, 196
stoichiometry, 25
substitution
in multiple electron transfer, 215
in inner-sphere reactions, 120, 131, 133, 138
in reactions of non-metallic reagents, 263, 273
supporting electrolyte, 31, 34

thallium(III)/(I) exchange, 216ff
titanium(IV)/(III) self-exchange reactions, 103, 104
tunneling, 38, 76, 87, 261

vanadium(II) reductions, 131–133, 225
vanadium(V)/(IV) self-exchange reactions, 101–103
volume of activation
in atom transfer reactions, 232
in inner sphere electron transfer, 134
in outer sphere electron transfer, 43–44, 60–62, 82–84, 150
water
hydrogen bonding with aqua complexes, 93–95
irradiation, 22
oxidation of, 243, 244